ROBUST PROCESS CONTROL

Manfred Morari
Chemical Engineering
California Institute of Technology

Evanghelos Zafiriou
Department of Chemical and Nuclear Engineering
and Systems Research Center
University of Maryland

Prentice Hall
Englewood Cliffs, New Jersey 07632

Library of Congress Cataloging-in-Publication Data

Morari, Manfred.
   Robust process control / Manfred Morari, Evanghelos Zafiriou.
      p.   cm.
   Bibliography: p.
   Includes index.
   ISBN 0-13-782153-0
   I. Chemical process control. I. Zafiriou, Evanghelos.
  II. Title.
   TP155.75.M67 1989   88-22928
   660.2'81—dc19   CIP

*TP
155.75
.M67
1989*

Cover design: Photo Plus Art
Manufacturing buyer: Mary Ann Gloriande

The fonts in this book are Computer Modern Roman, set using Leslie Lamport's LaTeX document preparation facility, with the help of Jan Owen. Some of the figures were prepared by Glenn C. Smith on an Apple Macintosh with the program MacDraw.

TeX is a trademark of the American Mathematical Society.
Macintosh and MacDraw are trademarks of Apple Computer, Inc.

Printed in the United States of America
10 9 8 7 6 5 4 3 2

18326137

ISBN 0-13-782153-0

*3-25-91
Ae*

Prentice-Hall International (UK) Limited, *London*
Prentice-Hall of Australia Pty. Limited, *Sydney*
Prentice-Hall Canada Inc., *Toronto*
Prentice-Hall Hispanoamericana, S.A., *Mexico*
Prentice-Hall of India Private Limited, *New Delhi*
Prentice-Hall of Japan, Inc., *Tokyo*
Simon & Schuster Asia Pte. Ltd., *Singapore*
Editora Prentice-Hall do Brasil, Ltda., *Rio de Janeiro*

In memoriam

CONSTANTIN G. ECONOMOU

1958 – 1986

Contents

PREFACE

The development of the state space approach in the early 1960s made it possible for the first time to solve general linear multivariable control problems with relative ease. The new techniques seemed to hold high promise for practical applications and attracted much interest — at least in the academic community. Fifteen years later several review papers (e.g., Foss, 1973; Kestenbaum et al., 1976) concluded that the impact on the industrial practice had been negligible.

A number of possible reasons for this failure can be identified. For example, there was no smooth transition from the established and proven techniques and tools (PID controllers, Smith Predictor, Relative Gain Array, Decoupler, etc.) to the new ones. This led to great educational difficulties and from today's perspective pointless debates in which the classic frequency domain approach was pitched against the modern state space techniques. More serious than this lack of understanding which persisted for almost two decades was that the new techniques did not address some very fundamental issues which are at the very heart of feedback design. For example, the concept of "nonminimum phase behavior" seemed forgotten. Also, the problem of model uncertainty and model error, which dominates process control, was only addressed via sensitivity analysis, whose validity is limited to infinitesimal model perturbations.

After years of isolated but persistent criticism a new understanding and approach started to emerge in the late 1970s. Some of the key theoretical contributions were the new formulation and parametrization of the optimal control problem by Youla and coworkers (1976), the definition of the H_∞ control problem by Zames (1981), and the introduction of the structured singular value by Doyle (1982). These and other discoveries sparked again much enthusiasm and a flurry of research activity almost comparable to the 1960s. Only time will show the impact on the control practice.

The new approach does address many issues of practical importance like model uncertainty (robustness). Many research challenges remain however. New results appear monthly but the ideas have not matured enough to form a coherent picture. Moreover, the level of abstraction makes the new developments inaccessible to anybody but the researchers in this area.

This book attempts to bridge the gap between a good undergraduate training in process control and the new arriving generation of robust control theory. Though not a textbook, it is intended as a supplement for a graduate level control course. Because the emphasis is on practical control system design methods, sections of the book should also be of interest to the industrial control engineer who wants to learn about the new powerful techniques. Some basic concepts from linear algebra and complex variables are assumed.

The book motivates the new theory with a series of typical process control examples: single-input single-output systems with time delay, multivariable distillation column models, reactor models, etc. However, the applicability of the new techniques is not limited to process control. The emphasis is on the derivation of new analysis and design tools. Preference is given to simple effective techniques over the most recent and most general theoretical discoveries. There is no doubt that the next few years will bring about a number of refinements which will affect some of the details in the book. We strongly believe, however, that the key ideas of robust control as presented here will profoundly impact process control understanding, teaching, and practice in the future. Indeed, some of the concepts have been applied already to industrial systems and some of the algorithms have become integral parts of industrial process control software.

Many colleagues have influenced our thinking on the topic of robust process control. We wish to acknowledge in particular Reuel Shinnar, Coleman Brosilow, and John Doyle. Shinnar made us first aware of the robustness problems. Brosilow convinced us of the advantages of the Internal Model Control structure. Doyle familiarized us with the new powerful theory which makes the practical robustness analysis of multivariable systems possible.

The initial outline for this book was developed with our late colleague Constantin Economou. We missed his high standards, his flair for communicating complex ideas, and his humor. The project was not the same without him.

A number of present and former Caltech graduate students and research fellows have contributed either directly or indirectly to the material. In particular, we would like to mention Sigurd Skogestad, Dan Rivera, Dan Laughlin, Dan Lewin, and Claudio Scali. Brad Holt, Yaman Arkun, and Christos Georgakis have classroom-tested parts of this book and have provided us with valuable criticism. The continuous encouragement by our friends in industry was most important. We wish to acknowledge, in particular, Dave Prett and Carlos Garcia of Shell Development, Dave Smith and Bjorn Tyreus of DuPont, and John Hamer of Kodak.

Much of the research reported in this book was carried out at Caltech. The extraordinary scholarly environment and supportive administration made this endeavor possible. The final sections of the book were completed while the first author enjoyed an appointment as the Gulf Visiting Professor of Chemical Engineering at Carnegie Mellon University. During the past year, the second author held a joint appointment with the Chemical and Nuclear Engineering Department and the Systems Research Center of the University of Maryland. The excellent atmosphere at this institution allowed him to continue his work on the book.

Over the years we received continuous financial support from the National Science Foundation, the Department of Energy, Shell Development, DuPont, and Kodak which enabled us to pursue our research objectives.

Finally we would like to thank Glenn C. Smith for preparing most of the figures, Evangelos Petroutsos for assisting with TeX and LaTeX and Jan Owen who did such an excellent job typing the manuscript after suffering through three different word processing systems.

M. Morari

E. Zafiriou

NOMENCLATURE

Abbreviations

DIC	Decentralized Integral Controllable
IC	Integral Controllable
IM	Interaction Measure
IMC	Internal Model Control
IS	Integral Stabilizable
ISE	Integral Squared Error
LFT	Linear Fractional Transformation
LHP	Left-Half Plane
MIMO	Multi-Input Multi-Output
MP	Minimum Phase
NMP	Nonminimum Phase
NP	Nominal Performance
NS	Nominal Stability
PFE	Partial Fraction Expansion
RGA	Relative Gain Array
RHP	Right-Half Plane
RHS	Right Hand Side
RP	Robust Performance
RS	Robust Stability
SISO	Single-Input Single-Output
SSE	Sum of Squared Errors
SSV	Structured Singular Value
SVD	Singular Value Decomposition
UC	Unit Circle

Symbols

$B(z)$	Factor to preserve system type in $\tilde{q}(z)$ and $\tilde{Q}(z)$
$b_p(s)(b_p^*(z))$	Allpass with the open RHP (outside the UC) poles of $\tilde{p}(s)(\tilde{p}^*(z))$
$b_v(s)(b_v^*(z))$	Allpass with the open RHP (outside the UC) poles of $v(s)(v^*(z))$
$\mathcal{C}$	Field of complex numbers
$\mathcal{C}^{n\times m}$	Field of complex matrices of dimension $n \times m$
$c(s), c(z)(C(s), C(z))$	SISO (MIMO) continuous and discrete classic feedback controllers
$c_f(s)$	Classic feedforward controller
$d(s)$	Effect of $d'(s)$ on $y(s)$
$d'(s)$	Disturbance entering $p_d(s)$
$d^*(z)(d_\gamma^*(z))$	z-transform of $d(s)$ before (after) passing through $\gamma(s)$
$e(s)(e^*(z))$	Continuous (discrete) error signal: Difference between $y(s)(y^*(z))$ and $r(s)(r^*(z))$
$e'(s)$	Weighted error signal $W_2(s)e(s)$
$f(s), f(z)(F(s), F(z))$	SISO (MIMO) continuous and discrete IMC filters
$f_1(s)$	Type 1 one-parameter discrete SISO IMC filter
$G^F(s)$	Transfer function matrix used as argument of μ in the RP SSV criterion for the continuous case
$G_v^F(s)$	The equivalent of $G^F(s)$ for the discrete case; it depends on whether $v = r$ or $v = d$
$G_{11}^F(s)(G_{11}^{*F}(z))$	Transfer function matrix used as argument of μ in the RS SSV condition for the continuous (discrete) case
$h_0(s)(H_0(s))$	SISO (MIMO) zero-order hold
$H_2(H_2^*)$	$H_2^{n\times m}(H_2^{*,n\times m})$ for $n = m = 1$
$H_2^n(H_2^{*,n})$	$H_2^{n\times m}(H_2^{*,n\times m})$ for $m = 1$
$H_2^{n\times m}$	Subspace of $L_2^{n\times m}$, containing the functions with analytic continuations in the RHP (includes all rational, proper, stable transfer function matrices)
$H_2^{*,n\times m}$	Subspace of $L_2^{*,n\times m}$ defined as the orthogonal complement of the $L_2^{*,n\times m}$ subspace which contains functions with analytic continuations inside the UC ($H_2^{*,n\times m}$ contains all rational, strictly proper, stable z-transfer function matrices).
Im	Imaginary part

$\ell_a(s)(L_A(s))$	Additive SISO (MIMO) uncertainty
$\bar{\ell}_a(\omega)(\bar{\ell}_a^*(\omega))$	Bound on the additive uncertainty for the continuous (discrete) case
$L_E(s)(\bar{\ell}_E(\omega))$	(Bound on the) MIMO inverse multiplicative output uncertainty
$L_I(s)(\bar{\ell}_I(\omega))$	(Bound on the) MIMO input multiplicative uncertainty
$\ell_m(s)$	Multiplicative SISO uncertainty
$\bar{\ell}_m(\omega)(\bar{\ell}_m^*(\omega))$	Bound on the multiplicative SISO uncertainty for the continuous (discrete) case
$L_O(s)(\bar{\ell}_O(\omega))$	(Bound on the) MIMO output multiplicative uncertainty
$L_2(L_2^*)$	$L_2^{n \times m}(L_2^{*,n \times m})$ for $m = n = 1$
$L_2^n(L_2^{*,n})$	$L_2^{n \times m}(L_2^{*,n \times m})$ for $m = 1$
$L_2^{n \times m}(L_2^{*,n \times m})$	Space of $n \times m$ matrix valued functions that are square integrable on the imaginary axis (UC)
$p(s)(P(s))$	SISO (MIMO) plant; also denotes the process model in sections where $p = \tilde{p}(P = \tilde{P})$ is assumed
$\tilde{p}(s)(\tilde{P}(s))$	SISO (MIMO) process model
$p^*(z), p_\gamma^*(z)(P^*(z), P_\gamma^*(z))$	Zero-order hold discrete equivalent of $p(s)(P(s))$ without and with $\gamma(s)(\Gamma(s))$ included, respectively
$\tilde{p}^*(z), \tilde{p}_\gamma^*(z)(\tilde{P}^*(z), \tilde{P}_\gamma^*(z))$	Zero-order hold discrete equivalents of $\tilde{p}(s)(\tilde{P}(s))$ without and with $\gamma(s)(\Gamma(s))$ included, respectively
$\tilde{p}_A(s), \tilde{p}_A^*(z)(\tilde{P}_A(s), \tilde{P}_A^*(z))$	Allpass with the NMP elements of $\tilde{p}(s)$ and $\tilde{p}^*(z)(\tilde{P}(s)$ and $\tilde{P}^*(z))$ respectively
$p_d(s), \tilde{p}_d(s)$	True and model of the disturbance effect plant transfer function
$p_m(s), \tilde{p}_m(s)$	True and model of the measurement device transfer function
$\tilde{p}_M(s), \tilde{p}_M^*(z)(\tilde{P}_M(s), \tilde{P}_M^*(z))$	MP factors of $\tilde{p}(s)$ and $\tilde{p}^*(z)$ $(\tilde{P}(s)$ and $\tilde{P}^*(z))$ respectively
$q(s), q(z)(Q(s), Q(z))$	SISO (MIMO) continuous and discrete IMC controllers
$\tilde{q}(s), \tilde{q}(z)(\tilde{Q}(s), \tilde{Q}(z))$	SISO (MIMO) continuous and discrete 1st step IMC controllers
$\tilde{q}_d(s), q_d(s)(\tilde{q}_r(s), q_r(s))$	Disturbance rejection (setpoint tracking) 1st step and overall IMC controllers of the two-degree-of-freedom structure
$q_f(s)$	Feedforward IMC controller

$\tilde{q}_H(z)(\tilde{Q}_H(z))$	SISO (MIMO) H_2^*-optimal IMC controller
$\tilde{q}_-(z)$	Factor included in $\tilde{q}(z)$ and $\tilde{Q}(z)$ to avoid intersample rippling
$\mathcal{R}$	Field of real numbers
Res	Residue
Re	Real part
$r(s)(r^*(z))$	Continuous (discrete) setpoint
T	Sampling time
$u(s)(u(z))$	Continuous (discrete) input to the plant
$v(s)(v'(s))$	External input $r(s)$ or $d(s)$ (normalized to impulse)
$v^*(z)$	z-transform of $v(s)$
$V(s)(V(z))$	Square matrix whose columns consist of external inputs $v(s)(v^*(z))$
$\mathcal{V}(\mathcal{V}')$	Set of inputs $v(s)(v'(s))$ or $v^*(z)$
$v_A(s), v_A^*(z)(V_A(s), V_A(z))$	Allpass with the NMP elements of $v(s), v^*(z)(V(s), V(z))$ respectively
$v_M(s), v_M^*(V_M(s), V_M(z))$	MP factors of $v(s), v^*(z)$ $(V(s), V(z))$ respectively
$w(s)$	Performance weight for the SISO case
$w_d(s)(w_r(s))$	Performance weight for disturbance rejection (setpoint tracking) in the two-degree-of-freedom structure
$W_1(s)(W_2(s))$	Input (Output) weights for the MIMO case
$y(s)(\tilde{y}(s))$	Continuous plant (model) output
$y^*(z), y_\gamma^*(z)(\tilde{y}^*(z), \tilde{y}_\gamma^*(z))$	z-transform of $y(s)(\tilde{y}(s))$ before and after passing through $\gamma(s)$ respectively
$y_m(s)$	Output of measurement device

Greek Characters

$\gamma(s)(\Gamma(s))$	SISO (MIMO) anti-aliasing prefilter
$\Delta(s)(\Delta(z))$	Block diagonal matrix containing the uncertainty for the continuous (discrete) case
$\boldsymbol{\Delta}$	Set of possible Δ's
$\Delta_u(s)(\Delta_u(z))$	Uncertainty block in $\Delta(s)(\Delta(z))$
$\Delta_p(s)$	RP block in $\Delta(s)$
$\epsilon(s), E(s)(\tilde{\epsilon}(s), \tilde{E}(s))$	(Nominal) sensitivity function for the continuous SISO and MIMO cases, respectively
$\bar{\epsilon}(s)$	first step IMC nominal sensitivity function

$\tilde{\epsilon}^*(z)(\tilde{E}^*(z))$	SISO (MIMO) discrete sensitivity function
$\tilde{\epsilon}_r(s), \tilde{E}_r(s)(\tilde{\epsilon}_d(s), \tilde{E}_d(s))$	Setpoint tracking (disturbance rejection) SISO and MIMO (approximate) continuous nominal sensitivity function for sampled-data systems
$E_v(s)(\tilde{E}_v(s))$	(Nominal) MIMO continuous sensitivity function for sampled-data systems; it depends on whether $v = d$ (approximation) or $v = r$
$\zeta(s)(\zeta(z))$	Zero polynomial for the continuous (discrete) case
$\eta(s), H(s)(\tilde{\eta}(s), \tilde{H}(s))$	(Nominal) Complementary sensitivity function for the continuous SISO and MIMO cases respectively
$\bar{\eta}(s)$	first-step IMC complementary sensitivity function
$\eta^*(z)(H^*(z))$	SISO (MIMO) discrete complementary sensitivity function
$\tilde{\eta}_r(s)(\tilde{H}_r(s))$	Setpoint tracking SISO (MIMO) continuous nominal complementary sensitivity function for sampled-data systems
$\kappa(\kappa^*)$	(Minimized) condition number
κ_d	Disturbance condition number
κ_R	Rijnsdorp IM
$\Lambda(A)$	RGA matrix of the matrix A
$\lambda_i(A)$	i^{th} eigenvalue of A
$\lambda_{ij}(A)$	ij-element of $\Lambda(A)$
$\mu(A)$ or $\mu_\Delta(A)$	SSV of A for same specific set $\mathbf{\Delta}$
Π	Set of possible plants
$\pi(s)(\pi(z))$	Pole polynomial for the continuous (discrete) case
$\pi(\omega)(\pi^*(\omega))$	Uncertainty region on the Nyquist plane for the continuous (discrete) case
$\rho(A)$	Spectral radius of A
$\bar{\sigma}(A)$	Maximum singular value of A
$\underline{\sigma}(A)$	Minimum singular value of A
$\sigma_i(A)$	i^{th} singular value of A
$\phi(z)$	Factor included in $f(z)$ to satisfy the requirements on the IMC filter
ω_B	Closed-loop system bandwidth
ω_s	Sampling frequency

Superscripts

H	Complex conjugate transpose of a matrix
T	Transpose of a matrix
$\perp$	Orthogonal complement of a function space

Special Notation

$F(A, \Delta)$	Matrix resulting from the application of a linear fractional transformation on A, Δ with Δ in the feedback path
$\mathcal{L}\{\cdot\}$	Laplace transform of $(\cdot)$
$\mathcal{Z}\{\cdot\}$	Z-transform of $(\cdot)$
$\|\cdot\|$	Norm on $\mathcal{C}^{n \times m}$
$\|\cdot\|_F$	Frobenius norm on $\mathcal{C}^{n \times m}$
$\|\cdot\|_2$	$L_2^{n \times m}$ or $L_2^{*, n \times m}$ norm
$\|\cdot\|_\infty$	∞-norm of a function
$\|\cdot\|_{i2}$	Operator norm induced by $\|\cdot\|_2$
$\{g(s)\}_+(\{g(z)\}_+)$	Projection of an $L_2^n(L_2^{*,n})$ function $g(s)(g(z))$ on $H_2^n(H_2^{*,n})$
$\{g(s)\}_-(\{g(z)\}_-)$	Projection of an $L_2^n(L_2^{*,n})$ function $g(s)(g(z))$ on $(H_2^n)^\perp((H_2^{*,n})^\perp)$
$\forall$	For all
$\ni$	Such that

We also wish to remind the reader that the following statements are equivalent:

$$A \Leftarrow B$$

$$A \text{ if } B$$

$$B \text{ is sufficient for } A$$

$$B \Rightarrow A$$

$$B \text{ only if } A$$

$$A \text{ is necessary for } B$$

$$\text{not } A \quad \Rightarrow \quad \text{not } B$$

Chapter 1

INTRODUCTION

The main ideas covered in this book are motivated and put in historical perspective. A new controller parametrization is outlined and the concept of "robustness" is explained. The scope of the book is defined and hints for its study are provided.

1.1 The Evolution of Control Theory

Feedback control mechanisms have been used for millenia. The first applications of the feedback principle (Mayr, 1970) can be traced back to ancient Greece: the water clock of Ktesibios (ca. 300 B.C.) employed a float regulator. The first automatic feedback controller used in an industrial process was James Watt's flyball governor invented in 1769 for controlling the speed of a steam engine. As these centrifugal governors were refined, they developed some serious problems, then referred to as "hunting" and now known as "instability." This in turn prompted the first mathematical analysis of a feedback system via differential equations by Maxwell (1868). During the same period Vyshnegradskii (1877) formulated a mathematical theory of regulators.

The invention of the negative feedback amplifier at Bell Laboratories by Black prior to World War II marks another milestone in the use of feedback. The observed stability problems ("singing") were explained through the frequency domain analysis techniques by Bode and Nyquist (Nyquist, 1932; Bode, 1964; Black, 1977).

Throughout the 1940s and 1950s feedback design by trial and error was predominant and even today this approach plays a major role. Typically, performance specifications are defined initially and then the controller design is modified iteratively until the best compromise between the frequently conflicting objectives is reached. Every design step is followed by an analysis step and the iterations are guided by experience and rules of thumb. In the late 1950s Newton, Gould, and Kaiser (1957) showed that if the control objectives are formulated in terms of the integral square error (ISE) the "optimal" feedback controller can be found

1

"analytically" — i.e., directly and without trial and error. This approach, which is based on the Wiener-Hopf factorization technique, was subsequently greatly generalized, in particular by Kalman.

The deficiency of these "synthesis" techniques is that typical practical performance objectives are usually much more complex than ISE. Frequently the designer wishes to impose constraints on the closed-loop system response characteristics like overshoot, rise time, and decay ratio. Also a certain "robustness" against changes in the dynamic characteristics of the plant to be controlled is desirable. Often it is possible to achieve these attributes indirectly by frequency-weighting the ISE and by adding to the objective a penalty for excessive movements of the manipulated variable. For that purpose the weights and penalties have to be chosen again iteratively by trial and error. Thus, for a practical design the advantages of such an "optimal control synthesis" procedure over an ad hoc adjustment of controller parameters and a test via simulation can become quite small.

The ISE or, in the more general context, the H_2-optimal control formulation dominated the literature for about 20 years. In the late seventies three major and in some sense related discoveries started a new era of feedback control theory.

- Youla and coworkers (1976) showed that it is possible to parametrize all stabilizing controllers for a particular system in a very effective manner: when searching for a controller with specific properties one can simply and without loss search over the space of *all stable* transfer functions. The parametrization guarantees that the resulting feedback controller *automatically* yields a closed-loop stable system. Thus, the search for a good controller is greatly simplified.

- Zames (1981) postulated that measuring performance in terms of the ∞-norm rather than the traditional 2-norm (ISE) might be much closer to the practical needs. This ushered in the era of H_∞ optimal control.

- Doyle argued that model uncertainty is often described very effectively in terms of norm-bounded perturbations. For these perturbations and the H_∞ performance objective he developed a powerful tool (the structured singular value) for testing "robust stability" (i.e., stability in the presence of model uncertainty) and "robust performance" (i.e., performance in the presence of model uncertainty). This is probably the primary motivation for the modern ∞-norm objective.

In parallel to these developments in mathematics and electrical engineering a new type of algorithm exemplified by Model Algorithmic Control (Richalet et al.,

1978) and Dynamic Matrix Control (Cutler and Ramaker, 1979) was invented in the process industries and applied successfully to complex process control problems. Though these algorithms had a heuristic basis Garcia and Morari (1982) discovered that some of the modern robust control characteristics had been incorporated in them in an ad hoc fashion. Thus, the time was ripe to put these empirical algorithms on a firmer footing and to have them benefit from the rich new theory. The objective of this book is to do that while retaining the features which make these algorithms easy to understand and easy to adjust on-line. At times this required mathematical generality to be sacrificed.

1.2 Controller Parametrization: The IMC Structure

Consider the block diagram shown in Fig. 1.2-1 where the control system is shaded: it includes the two blocks labeled *controller* and *model*. The control system has as its inputs the setpoint and process output (measurement) and as its output the manipulated variable (process input). Let us discuss qualitatively the advantages of such a structure over the classic feedback structure shown in Fig. 1.2-2.

The effect of the parallel path with the *model* is to subtract the effect of the manipulated variables from the process output. If we assume for the moment that the *model* is a *perfect* representation of the *process*, then the feedback signal is equal to the *influence of disturbances* and is not affected by the action of the manipulated variables. Thus, the system is effectively open-loop and the usual stability problems associated with feedback have disappeared. The overall system is stable simply if and only if both the *process* and the *IMC controller* are stable.

Moreover, the IMC controller plays the role of a feedforward controller and can be designed as such. But the IMC controller does not suffer from the disadvantages of feedforward controllers: it can cancel the influence of (unmeasured) disturbances because the feedback signal is equal to this influence and modifies the controller setpoint accordingly.

If the *model* does not mimic the dynamic behavior of the *process* perfectly then the feedback signal expresses both the *influence of (unmeasured) disturbances* and the effect of this model error. The model error gives rise to feedback in the true sense and leads to possible stability problems. This forces the designer to "detune" the ideal feedforward controller for "robustness."

The controller form described here is a special case of the Youla-parametrization mentioned above. It is inherent in all "model predictive" control schemes, in particular Model Algorithmic Control and Dynamic Matrix Control.

1.3   Robustness

Regardless of what design technique is used, controllers are always designed based on (necessarily incomplete) information about the dynamic behavior of the process. This information (i.e., the "model") can have the form of a system of coupled partial differential equations or be simply the process gain and the settling time experienced by the plant operator. The accuracy of this information varies but is never perfect. Moreover, the behavior of the plant itself changes with time (feedstock changes, catalyst activity changes, etc.) and these changes are rarely captured in the models. It is most desirable that the controller be insensitive to this kind of *model uncertainty*, i.e., the controller should be *robust*.

Though the design objective *robustness* seems most practical and reasonable it is essentially absent from the control literature from about 1960 to 1980. This is probably one of the reasons why the design techniques developed in this time period have had negligible effect on the industrial control practice. Since the late 1970s robustness has become a major objective of control research. This book will demonstrate the importance of robustness considerations for process control and will propose design techniques which include robustness as one of the objectives.

1.4   Scope of Book

The book has several goals:

- to alert the reader to the key role model uncertainty and robustness play in the design and successful operation of feedback control systems.

- to outline the basic ideas behind the most recent theoretical advances for addressing robustness systematically in feedback design.

- to describe simple and effective analysis and design techniques for robust feedback controllers.

The different parts of the book aim at different groups of readers:

- the first-year graduate student who wants to broaden his/her horizon, gain some basic insights into feedback control, and get a taste of the direction of the new theory.

- the beginning researcher in the area of robust control who wishes to master the transition from the undergraduate process control course to the monographs by Vidyasagar (1985) and Francis (1987) as well as the current research papers in this area.

- the industrial practitioner who needs easily applicable and proven controller tuning techniques and who is interested in learning about the basic dos and don'ts of multivariable control.

Depending on the reader's objectives and background some chapters might be of more interest than others. Chapters 2 through 4 discuss the basic trade-offs in single-input single-output (SISO) feedback control systems and how they can be addressed in the Internal Model Control framework. These chapters are essential for understanding the rest of the book. Chapter 5 generalizes the results to unstable systems and can be skipped by the application-oriented reader. Chapter 6 proposes some very effective PID tuning rules and discusses more general SISO control structures like Smith predictor, cascade and feedforward control systems.

Chapters 7 through 9 deal with sampled-data systems and mimic Chapters 1 through 6. In the derivations particular emphasis is placed on the behavior of the *continuous* output — e.g., provisions against intersample rippling are built into the controllers.

Chapters 10 and 11 cover basic aspects of robustness of multi-input multi-output (MIMO) systems. These concepts are the key to understanding the unusual phenomena observed in the control of MIMO systems. The MIMO design technique proposed in Chapter 12 is complex and tentative. Undoubtedly, more effective techniques will become available in the future. Chapter 13 discusses the *system properties* which limit the achievable performance of MIMO systems. A knowledge of these properties helps when screening designs according to their operability characteristics. Chapter 14 deals with the design of controllers for MIMO systems with a restricted information structure — e.g., multiloop controllers. Chapter 15 extends the discussion of Chapters 10 through 12 to sampled-data systems. The book concludes with the detailed report on an application to a high-purity distillation column (Chapter 16).

1.5   Some Hints for the Reader

The book assumes that the reader has mastered the material covered in a typical undergraduate control text like the one authored by Stephanopoulos (1984) or Franklin et al. (1986). Furthermore some knowledge of linear algebra (e.g., Strang, 1980) and complex variables (e.g., Churchill & Brown, 1984) is required. Finally, some familiarity with linear operator theory (e.g., Desoer & Vidyasagar, 1975; chapters 1 through 6 of Ramkrishna & Amundson, 1985) is advantageous, but not necessary.

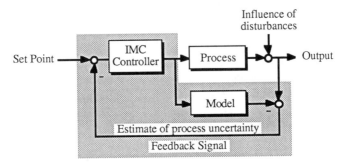

Fig. 1.2-1. Internal Model Control Structure.

Chapters 5, 9 and 13 and in particular the proofs in these chapters are mathematically more demanding and can be skipped at first reading. The chapters on sampled-data systems (7 through 9 and 15) assume that the corresponding material on continuous systems has been studied first. Also the study of the chapters on MIMO systems requires the prior understanding of the SISO concepts. In principle, the SISO material follows as a special case from the MIMO material. For tutorial reasons it is treated separately.

With the exception of this first chapter all references are collected at the chapter end for continuity. Most chapters contain one or several summary sections, usually located immediately before the reference section. This summary section allows the reader to review the main concepts. Often, mastering this summary section is sufficient for the application of the basic techniques covered in the chapter. The equation numbers in the summary section refer to the equation numbers in the main body of the text.

The end of proofs and examples is marked with the symbol □.

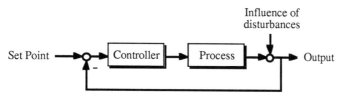

Fig. 1.2-2. Classic Feedback Structure.

Part I

CONTINUOUS SINGLE-INPUT
SINGLE-OUTPUT SYSTEMS

Chapter 2

FUNDAMENTALS OF SISO
FEEDBACK CONTROL

After a review of some basic definitions, the controller design problem will be formulated in general terms. All controller design procedures are based on models of one form or another. These models are necessarily inaccurate. It is of great practical importance that the controller performs well even when the dynamic behavior of the real process differs from that described by its model. In order to accomplish this objective, not only must a process model and the performance specifications be provided for the design, but also some indication of the model accuracy should be made. The attribute "robust" will be used for a property which holds not only for the model but also in the presence of model uncertainty. Mathematical conditions for robust stability and robust performance will be derived.

2.1   Definitions

The block diagram of a typical classic feedback loop is shown in Fig. 2.1-1A. Here c denotes the controller and p the plant transfer function. The transfer function p_d describes the effect of the disturbance d' on the process output y. The measurement device transfer function is symbolized by p_m. Measurement noise n corrupts the measured variable y_m. The controller determines the process input (manipulated variable) u on the basis of the error e. The objective of the feedback loop is to keep the output y close to the reference (setpoint) r.

Commonly we will use the simplified block diagram in Fig. 2.1-1B. Here d denotes the effect of the disturbance on the output. Exact knowledge of the output y is assumed ($p_m = 1, n = 0$).

In general, the transfer functions will be allowed to be rational or irrational — i.e., they may include time delays. In order to be physically realizable, the transfer functions have to be *proper* and *causal*.

11

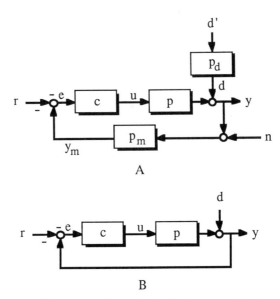

Figure 2.1-1. General (A) and simplified (B) block diagram of feedback control system.

Definition 2.1-1. *A system $g(s)$ is proper if $\lim_{s\to+\infty} g(s)$ is finite. A proper system is strictly proper if $\lim_{s\to+\infty} |g(s)| = 0$ and semi-proper if $\lim_{s\to+\infty} |g(s)| > 0$. All systems which are not proper are improper.*

A system $g(s)$ is improper if the order of the numerator polynomial exceeds the order of the denominator polynomial and proper otherwise. An improper system cannot be realized physically because it contains pure differentiators.

Definition 2.1-2. *A system requiring prediction $(e^{+s\theta})$ is noncausal.*[1] *A system which does not require prediction is causal.*

Some care has to be used in determining if a system is causal or not. A time delay in the denominator of a transfer function does not necessarily imply lack of causality. For example,

$$g(s) = \frac{y(s)}{u(s)} = \frac{e^{s\theta}}{2e^{s\theta} - 1} = \frac{1}{2 - e^{-s\theta}}$$

is causal because the present output y is determined solely by past values of y and the present value of u:

$$y(s) = \frac{1}{2}(e^{-s\theta}y(s) + u(s))$$

---

[1]Generally an improper system is also referred to as noncausal. In this book we will reserve noncausal to denote prediction.

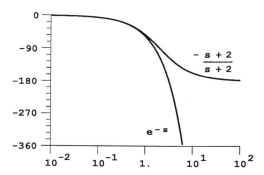

Figure 2.1-2. Phase behavior of two nonminimum phase (NMP) allpass elements.

Definition 2.1-3. *A system $g(s)$ is nonminimum phase (NMP) if its transfer function contains zeros in the Right Half Plane (RHP) or time delays or both. Otherwise a system is minimum phase (MP).*

Systems with an odd number of RHP zeros display *inverse response* characteristics: the initial direction of the step response is opposite to the direction of the final steady state.

The origin of the term "minimum phase" can be best understood in the context of stable systems: if a system $g(s)$ is MP, then there is no other system $g'(s)$ which has the same magnitude ($|g(i\omega)| = |g'(i\omega)| \; \forall \omega$) but a smaller phase lag. In other words, a NMP system $\bar{g}(s)$ does not have the smallest phase lag that is possible for a system with this magnitude ($|\bar{g}(i\omega)|$). For example, the *allpass* systems

$$g_1(s) = e^{-s}, \qquad g_2(s) = \frac{-s+2}{s+2}$$

have

$$|g_1(i\omega)| = |g_2(i\omega)| = 1$$

Their phase behavior is shown in Fig. 2.1-2. Clearly there is another system with the same magnitude (for example, $g'(s) = 1$) which has a smaller (zero) phase lag. Therefore, $g_1(s)$ and $g_2(s)$ are NMP.

2.2 Formulation of Control Problem

For any design procedure to yield a control algorithm which works satisfactorily in a real environment, the following must be specified:

- process model

- model uncertainty bounds

- type of inputs (i.e., setpoints and disturbances)

- performance objectives

Omission of any of these four items invariably leads to controllers which fail in practice. *Every* design or tuning procedure is centered around a *process model* whose complexity can vary. The experience of the operator who "knows" how the plant responds to certain inputs is the simplest kind of model. At the other end of the scale would be, for example, a coupled system of nonlinear partial differential equations. Neglecting *model uncertainty* leads to controllers which are too "tight" and which are likely to become unstable in the real operating environment. It is physically impossible to design controllers which work well for *all* types of setpoint changes and disturbances. The designer has to decide which *inputs* are most important and most frequent, and accept inferior performance if the inputs encountered during the operation are not exactly equal to the inputs assumed for the design. Finally, the designer has to specify what is meant by "good *performance*" for the particular problem at hand. What is good in one case might be entirely unacceptable in another.

2.2.1 Process Model

Chemical processing systems are often most conveniently described by infinite dimensional linear time invariant models (transfer functions with delays). Thus, any book on process control must discuss how to design controllers based on this type of model. However, a rigorous theoretical treatment of time-delay systems is very complex and more of mathematical than of practical interest. The reason is that there are extremely few true time-delay processes encountered in practice. In general, a process modeled with a delay is actually a finite dimensional system of very high order whose dynamic behavior can be *approximated* by a delay. The delay in this case is simply a convenient modeling tool which does not have to be treated rigorously in the design procedure. Also, as we will argue in more detail in the next section, *all* models are inadequate at high frequencies. In the low-frequency range, the delay can be approximated arbitrarily well by a finite dimensional (i.e., delay-free) model. In the high-frequency range the difference between the delay and the finite dimensional approximation can be treated as "model uncertainty" and the controller can be designed accordingly. (One key objective of this book is to show how to account for model uncertainty in the controller design procedure.)

Strictly speaking, most results obtained in this book are only applicable to systems described by rational transfer function models. In a less rigorous manner we will extrapolate the results to systems involving time delays, in particular when this allows us to obtain compact analytical expressions. The mathematically precise reader should interpret the time delays in our equations as a shorthand notation for a very high-order Padé approximation.

2.2.2 Model Uncertainty Description

Linear time invariant models of the type used throughout this book describe actual plant dynamics only approximately. The "model uncertainty" can have several different sources. Most important, real processes are nonlinear. If the process model is obtained via *linearization*, then it is accurate only in the neighborhood of the reference state chosen for the linearization.

In other cases the process might be represented quite accurately by a linear model. However, *different operating conditions* could lead to changes in the parameters in the linear model. For example, increased throughput and flowrates usually result in smaller deadtimes and time constants.

In the two cases above, the sources and the structure of the "uncertainty" are known quite accurately. However, there is always some "true" uncertainty even when the underlying process is essentially linear: the physical parameters are never known exactly and fast dynamic phenomena (e.g., valve dynamics) are usually neglected in the model. Therefore, at high frequencies, even the model order is unknown.

Uncertainty can be described in many different ways: bounds on the parameters of a linear model, bounds on nonlinearities, frequency domain bounds, etc. To account for model uncertainty we will assume in this book that the dynamic behavior of a plant is described not by a single linear time invariant model but by a *family* Π of linear time invariant models. This is a somewhat primitive uncertainty description especially when the effect of nonlinearities is to be captured, but it is the only feasible approach at present.

Throughout this book the family Π of plants will be defined in the frequency domain. We will assume that the transfer function magnitude and phase at a particular frequency ω is not confined to a point but can lie in a *region* $\pi(\omega)$ on the Nyquist plane. In general, the region can have a very complex shape as the following examples illustrate.

Assume, for example, that frequency response experiments allowed the designer to establish an upper and a lower bound on both the magnitude $|p(i\omega)|$ and the phase arg $\{p(i\omega)\}$ of the real plant. At a particular frequency these bounds give rise to the sector shaped region $\pi(\omega)$ in Fig. 2.2-1A.

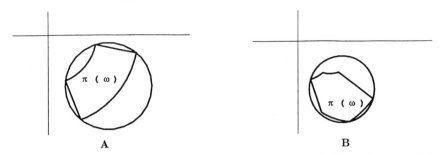

Figure 2.2-1. Uncertainty regions resulting from gain and phase bounds (A) and parametric bounds (B).

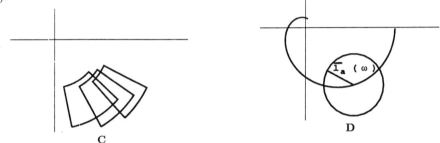

Figure 2.2-1. Uncertainty "band" (C). Uncertainty disk for additive uncertainty (D).

On the other hand, step response experiments might have yielded a first-order model with deadtime ($\tilde{p}(s) = ke^{-s\theta}(\tau s+1)^{-1}$) with upper and lower bounds on the parameters k, θ and τ. The resulting uncertainty region is shown for a particular frequency in Fig. 2.2-1B. The union of all regions $\pi(\omega)$ constitutes the family of possible plants Π and can be viewed as a "fuzzy" Nyquist plot or Nyquist band (Fig. 2.2-1C).

Regions of the general shape seen in these examples have complex mathematical descriptions and are very difficult to deal with in the context of control system design. Therefore, in most of this book we will assume the regions to be disk shaped with radius $\bar{\ell}_a(\omega)$ (Fig. 2.2-1D). Any complex region can be approximated by a disk with more or less conservatism (see Fig. 2.2-1A and B). Algebraically, the family Π of plants described by the disk is defined by:

$$\Pi = \left\{ p : |p(i\omega) - \tilde{p}(i\omega)| \le \bar{\ell}_a(\omega) \right\} \tag{2.2-1}$$

Here $\tilde{p}(i\omega)$ is the *nominal plant* or the *model* defining the center of all the disk-shaped regions. Any member of the family Π satisfies

$$p(i\omega) = \tilde{p}(i\omega) + \ell_a(i\omega) \tag{2.2-2}$$

with

$$|\ell_a(i\omega)| \le \bar{\ell}_a(\omega) \tag{2.2-3}$$

Equation (2.2–2) is referred to as an *additive* uncertainty description and (2.2–3) states a bound on the allowed additive uncertainty.

If we define

$$\ell_m(i\omega) = \frac{\ell_a(i\omega)}{\tilde{p}(i\omega)} \qquad (2.2-4)$$

and

$$\bar{\ell}_m(\omega) = \frac{\bar{\ell}_a(\omega)}{|\tilde{p}(i\omega)|} \qquad (2.2-5)$$

then the family Π (2.2–1) can be represented as

$$\Pi = \left\{ p : \frac{|p(i\omega) - \tilde{p}(i\omega)|}{|\tilde{p}(i\omega)|} \leq \bar{\ell}_m(\omega) \right\} \qquad (2.2-6)$$

Thus, any member of the family Π satisfies

$$p(i\omega) = \tilde{p}(i\omega)(1 + \ell_m(i\omega)) \qquad (2.2-7)$$

with

$$|\ell_m(i\omega)| \leq \bar{\ell}_m(\omega) \qquad (2.2-8)$$

Equation (2.2–7) is referred to as a *multiplicative* uncertainty description and (2.2–8) states a bound on the allowed multiplicative uncertainty. A typical plot of $\bar{\ell}_m(\omega)$ is shown in Fig. 2.2-2. The uncertainty usually increases with frequency and eventually exceeds or becomes equal to unity. The reason is that our models tend to describe well the steady state and low-frequency behavior of processes but become inaccurate for high-frequency inputs.

If a zero is located on the imaginary axis at ω^* for some $p \in \Pi$ then $p(i\omega^*) = 0$ and $|\ell_m(i\omega^*)| = 1$. If $\bar{\ell}_m(\omega) > 1$ for $\omega > \omega^*$, then the set Π allows zeros of $p(s)$ sufficiently far from the the origin "to cross" the imaginary axis. Also, for $\omega > \omega^*$ the disk-shaped regions include the origin which implies that the phase is completely unknown.

On the other hand, if a pole is located on the imaginary axis at ω^* for some $p \in \Pi$ then $|p(i\omega^*)| = \infty$ and therefore $\bar{\ell}_a(\omega^*) = \bar{\ell}_m(\omega^*) = \infty$. Thus the additive and multiplicative uncertainty are unsuitable to describe sets of plants where poles can cross the imaginary axis or equivalently, the number of RHP poles can vary.

Figure 2.2-3 shows a block diagram representation of the two types of uncertainties.

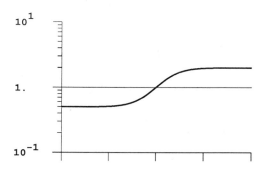

Figure 2.2-2. Typical form of multiplicative uncertainty bound $\bar{\ell}_m$.

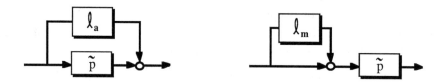

Figure 2.2-3. Additive (ℓ_a) and multiplicative (ℓ_m) uncertainty.

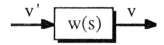

Figure 2.2-4. Weight w transforming normalized input v' to physical input v.

2.2.3   Input Specification

We refer to all external signals entering the feedback loop at some point as inputs. The symbol v will be used to symbolize these inputs which could be setpoint changes or disturbances. For the control system design the inputs have to be specified. The inputs will be *normalized* for notational convenience (Fig. 2.2-4). It will be assumed that an input v entering the control loop is generated by passing the normalized input v' through a transfer function block $w(s)$, sometimes referred to as an *input weight*.

Our design procedures will be able to deal with two types of *normalized* input specifications:

Specific input (impulse):
$$v'(s) = 1 \qquad\qquad (2.2 - 9)$$

Set of bounded inputs (all inputs with 2-norm bounded by unity):
$$\mathcal{V}' = \left\{ v' : \|v'\|_2^2 \triangleq \int_0^\infty v'^2(t)dt \le 1 \right\} \qquad (2.2 - 10)$$

Very often in process control we know quite well what type of inputs to expect. For example, setpoint changes usually have the shape of steps or ramps, disturbances can often be modeled as steps, entering a lag. If the control system design is to be carried out to accommodate a *specific* expected *input* v, such as a step setpoint change, then the designer specifies the input weight w to be an integrator $(1/s)$.

Specific input:
$$v = wv' = w \qquad\qquad (2.2 - 11)$$

However, sometimes the closed-loop performance can be quite sensitive to the choice of input; if the input assumed for the design is not exactly equal to the input encountered in practice the performance can deteriorate significantly. Then it is more meaningful to consider a set of inputs containing the input which is encountered most frequently together with other "similar" inputs. According to (2.2–10) we will consider the

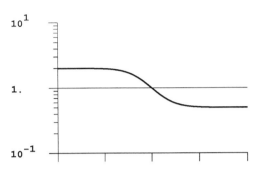

Figure 2.2-5. Typical form of input weight w.

Set of bounded inputs:

$$\mathcal{V} = \left\{ v : \|v'\|_2^2 = \left\| \frac{v}{w} \right\|_2^2 = \frac{1}{2\pi} \int_{-\infty}^{\infty} \left| \frac{v(i\omega)}{w(i\omega)} \right|^2 d\omega \leq 1 \right\} \qquad (2.2 - 12)$$

Note that the frequency integral used in (2.2–12) for the 2-norm is equal to the time integral in (2.2–10) through Parseval's theorem.

The expression (2.2–12) can be interpreted as follows: if the spectrum of the input v is narrow and concentrated near ω^* (i.e., the input looks almost like sin $\omega^* t$) then the input power (amplitude) is limited to $|w(i\omega^*)|$. A typical input weight is shown in Fig. 2.2-5 indicating that high-frequency inputs are expected to have small amplitudes.

Assume, for example, that we expect unit step inputs v. Then we can define a weight w such that a step lies in the set (2.2–12). For example the weight

$$w(s) = \frac{s + \beta}{s\sqrt{2\beta}} \quad \beta > 0 \qquad (2.2 - 13)$$

has the desired characteristics: the input

$$v' = \frac{\sqrt{2\beta}}{s + \beta} \qquad (2.2 - 14)$$

satisfies

$$\|v'\|_2^2 = \frac{1}{2\pi} \int_{-\infty}^{\infty} \frac{2\beta}{\omega^2 + \beta^2} \, d\omega = 1 \qquad (2.2 - 15)$$

and gives rise to a step v when passing through the weight (2.2–13). However, the characterization (2.2–12) with the weight (2.2–13) contains also many other signals v which are "similar" to steps. For example, consider the input

$$v' = \frac{\sqrt{2\alpha}}{s + \alpha} \qquad \alpha > 0, \alpha \neq \beta \qquad (2.2-16)$$

which also satisfies (2.2–10). After passing through the weight the signal becomes

$$v = \sqrt{\frac{\alpha}{\beta} \frac{s + \beta}{(s + \alpha)s}} \qquad (2.2-17)$$

which is a step modified by a lead (for $\alpha > \beta$) or a lag (for $\alpha < \beta$). Thus, if the controller is designed for the set (2.2–12) with weight (2.2–13) it will work well not only for steps but also for "modified steps," for example of the form (2.2–17).

In order for a physically realizable controller to exist, the external signals which are assumed to enter the control loop have to be bounded. Unbounded external signals can only be corrected by unbounded control actions. Therefore, they do not give rise to a meaningful control problem formulation. In general, the effect of the disturbances on the output (d) is bounded and the disturbance transfer function p_d (Fig. 2.1-1A) is stable. If the plant is open loop unstable and a (bounded) disturbance enters at the plant *input*, its effect on the output is usually unbounded and p_d is unstable. In this special case the unstable poles of p_d are a subset of the unstable poles of the plant p. Therefore, a controller with bounded control action can be designed, which provides assymptotic disturbance rejection.

2.2.4 Control Objectives

The ultimate objective of control system design is clearly that the controller works "well" when implemented on the real plant. This goal can be best understood if we decompose it into a series of subobjectives. Because we assume that the model is at least an approximate description of the true plant it is reasonable to require stability when the controller is applied to the plant *model*. Thus, the minimal requirement on the closed-loop system is *nominal* stability. In Sec. 2.3 we will introduce the concept of *internal stability* to derive the conditions for nominal stability which must be satisfied through appropriate controller design.

As a measure of performance we will generally use the Integral Square Error (ISE) because of its mathematical convenience. In Sec. 2.4 we will formulate the optimal control problem for minimizing the ISE.

It is essential that our knowledge about the model uncertainty be incorporated into the controller design procedure. Otherwise the controllers are bound to fail in the real world where the actual plant behavior can be quite different from that

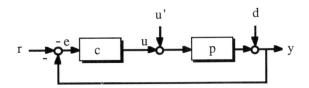

Figure 2.3-1. Block diagram for discussion of internal stability.

of the model. We will require that the controller be designed such that the closed-loop system is stable and meets the performance specifications for *all* members of the family Π of possible plants defined in Sec. 2.2.2. If a closed-loop property holds for a *family* of plants we will refer to it as *robust.* Thus, our controller design objectives are *robust stability* and *robust performance.* In Secs. 2.5 and 2.6 we will derive conditions for robust stability and robust performance, respectively.

2.3 Internal Stability

The signals between the blocks constituting a control system are subject to (possibly very small) errors. In practice it cannot be tolerated that these small errors lead to unbounded signals at some other location in the control system. This motivates the following definition:

Definition 2.3-1. *A control system is internally stable if bounded signals injected at any point of the control system generate bounded responses at any other point.*

Definition 2.3-2. *A linear time invariant control system is internally stable if the transfer functions between any two points of the control system are stable — i.e., have all poles in the open Left Half Plane (LHP).*

In a control system many different points can be selected for signal injection and observation but most of the choices are equivalent for checking internal stability. For example, for the system in Fig. 2.3-1, y and e differ only by a bounded signal (r) and their observation reveals the same information about internal stability.

Also, from the point of view of internal stability, the effect of d and r on u is equivalent. Simple arguments of this type reveal that there are only two "independent" outputs, which can be chosen as y and u, and two "independent" inputs, which can be chosen as r and u'. Thus, the classic feedback system is stable if and only if all elements in the 2×2 transfer matrix in (2.3–1) have all their poles in the open LHP.

$$\begin{pmatrix} y \\ u \end{pmatrix} = \begin{pmatrix} \frac{pc}{1+pc} & \frac{p}{1+pc} \\ \frac{c}{1+pc} & \frac{-pc}{1+pc} \end{pmatrix} \begin{pmatrix} r \\ u' \end{pmatrix} \qquad (2.3-1)$$

It should be noted that the concept of internal stability is more complete than the usual stability concept of undergraduate textbooks where system stability is checked by examining the roots of the characteristic equation $1+pc = 0$. Consider the example

$$c = \frac{-s+1}{s+1} \qquad (2.3-2)$$

$$p = \frac{1}{-s+1} \qquad (2.3-3)$$

for which the characteristic equation

$$1 + pc = 1 + \frac{1}{s+1} = 0 \qquad (2.3-4)$$

has a single root at $s = -2$ which would indicate stability.

On the other hand, the transfer matrix in (2.3–1) evaluated for this example

$$\begin{pmatrix} y \\ u \end{pmatrix} = \begin{pmatrix} \frac{1}{s+2} & \frac{s+1}{(-s+1)(s+2)} \\ \frac{-s+1}{s+2} & \frac{-1}{s+2} \end{pmatrix} \begin{pmatrix} r \\ u' \end{pmatrix} \qquad (2.3-5)$$

shows that any bounded input u' leads to an unbounded output y. Thus, the system is not internally stable.

Note that the RHP zero of the controller (2.3–2) cancels exactly the RHP pole of the plant (2.3–3) when (2.3–4) is formed. The internal stability concept clarifies the fact that even *exact* cancellation is not enough to guarantee the stability of the system.

If both p and c are stable then the unstable poles in (2.3–1) can arise from $(1 + pc)^{-1}$ only. In this special case it is necessary and sufficient for internal stability that all the roots of the characteristic equation $1 + pc = 0$ are in the open LHP.

2.4 Nominal Performance

The most basic objective of a feedback controller is to keep the error between the plant output y and the reference r small when the overall system is affected by external signals r and d. In order to quantify performance, a measure of "smallness" for the error has to be defined. Furthermore, the set of external input signals has to be specified for which the error is to be made small. Control system design techniques differ in the way they measure "magnitude" and how they define the permissible set of external inputs. Two particularly popular approaches are outlined in Secs. 2.4.4 and 2.4.5.

2.4.1   Sensitivity and Complementary Sensitivity Function

If we set $p_d = p_m = 1$, then the most important relationships between the inputs and outputs in Fig. 2.1-1A are

$$\frac{e}{d-r} = \frac{y}{d} = \frac{1}{1 + pc(s)} \triangleq \epsilon(s) \qquad (2.4-1)$$

$$\frac{y}{r-n} = \frac{pc(s)}{1 + pc(s)} \triangleq \eta(s) \qquad (2.4-2)$$

where the error e is defined as $e = y - r$

The *sensitivity function* $\epsilon(s)^2$ relates the external inputs $d - r$ to the error e. It also expresses the effect of the disturbance d on the output y. The sensitivity $\epsilon(s)$ is of primary importance in judging the performance of a feedback controller. It is desirable to make $\epsilon(s)$ as "small" as possible. If pc is strictly proper (which is always the case for physical systems) then

$$\lim_{s \to +\infty} pc = 0 \qquad (2.4-3)$$

This implies that

$$\lim_{\omega \to \infty} |\epsilon(i\omega)| = \lim_{\omega \to \infty} \left| \frac{1}{1 + pc(i\omega)} \right| = 1 \qquad (2.4-4)$$

Thus $|\epsilon|$ can be made small only over a finite frequency range. Physical limitations do not allow "perfect control" ($\epsilon = 0$). A typical plot of $|\epsilon|$ is shown in Fig. 2.4-1.

The frequency ω_B at which $|\epsilon|$ exceeds $1/\sqrt{2}$ will be called the *system bandwidth*.

$$|\epsilon(\omega)| < 1/\sqrt{2} \qquad \forall \omega < \omega_B \qquad (2.4-5)$$

The bandwidth ω_B can serve as a simple closed-loop performance measure.

The *complementary sensitivity function* $\eta(s)$ derives its name from the equality

$$\epsilon(s) + \eta(s) = 1 \qquad (2.4-6)$$

---

[2]The symbol ϵ was chosen because of the Greek word $\epsilon\nu\alpha\iota\sigma\vartheta\eta\sigma\iota\alpha$ for sensitivity. It was then most natural to denote the complementary sensitivity by η.

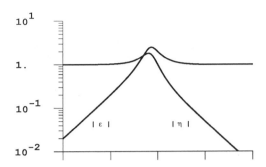

Figure 2.4-1. Typical form of sensitivity ϵ and complementary sensitivity η.

The complementary sensitivity $\eta(s)$ relates the reference r to the output y (2.4–2). From this point of view, $\eta(s)$ should be made as close to unity as possible. However, because of (2.4–3)

$$\lim_{\omega \to \infty} |\eta(i\omega)| = \lim_{\omega \to \infty} \left| \frac{pc(i\omega)}{1 + pc} \right| = 0 \qquad (2.4 - 7)$$

Thus $|\eta|$ can be made equal to unity over a finite frequency range only.

The complementary sensitivity $\eta(s)$ also expresses the effect of measurement noise n on y (2.4–2). From this point of view $\eta(s)$ should be made small. This illustrates one of the basic trade-offs in feedback design: good setpoint tracking and disturbance rejection ($\epsilon \approx 0, \eta \approx 1$) has to be traded off against suppression of measurement noise ($\epsilon \approx 1, \eta \approx 0$). The optimal compromise can be found via stochastic optimal control theory. Experience has shown that this particular trade-off is usually irrelevant in process control. Measurement noise is often small and η resulting from stochastic optimal control considerations is close to unity over a wide frequency range. As we will learn later, model uncertainty also imposes an upper bound on the magnitude of η. With the poor models generally available for process control, this bound is much more restrictive than the measurement noise constraint. Therefore, measurement noise will be largely neglected in this book.

2.4.2 Two-Degree-of-Freedom Controller

From (2.4–1) we see that apart from the sign, r and d have the *same* effect on e.

$$e = \epsilon(d - r) = \frac{1}{1 + pc}(d - r) \qquad (2.4 - 8)$$

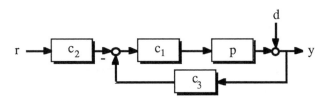

Figure 2.4-2. Block diagram for two-degree-of-freedom controller.

The control system has only "one degree of freedom." If the reference signal r and the disturbance d vary in a similar manner then c can be chosen such that e is satisfactory. If r and d behave differently, for example, if one expects step changes in r and ramp changes in d, then c has to be chosen either for good response to steps in r or good response to ramps in d or else some compromise has to be found. If there are strict requirements on both setpoint tracking and disturbance suppression, an acceptable compromise might not exist. Then additional controller blocks have to be introduced into the system (Fig. 2.4-2) to allow independent controller adjustments for both r and d. The relationship between the inputs and the error, which demonstrates the "two degrees of freedom," is:

$$e = \frac{1}{1 + pc_1c_3}d + \left(\frac{pc_1c_2}{1 + pc_1c_3} - 1\right)r \qquad (2.4-9)$$

Here c_1c_3 can be selected first for good disturbance rejection and then the *pre-filter* c_2 can be chosen independently for good setpoint tracking. As we will show in Chap.5, all three controller blocks are necessary, in general, to minimize the impact of the controller structure on the achievable trajectory tracking and disturbance rejection performance. In some special cases two blocks (c_1 and c_2, or c_1 and c_3) are sufficient.

We see from (2.4–8) that the effects of the inputs d and $-r$ on the error e are the same. Therefore, in the rest of the chapter we will use the symbol v to denote an input, where $v = d$ or $v = -r$. If the control system is to be designed for both trajectory tracking and disturbance rejection, then these two design tasks can be handled sequentially and independently through the two-degree-of-freedom structure.

2.4.3 Asymptotic Properties of the Closed-Loop Response (System Type)

The disturbances encountered in the chemical process industries are frequently slowly varying and can be approximated well by steps or ramps. As a basic closed loop performance specification the error is often required to vanish asymptotically for these types of inputs. Obviously, it is only meaningful to check if this specification is met when the closed loop system is stable. Then the asymptotic properties can be assessed from the Final Value Theorem. For example, for $\epsilon(s)$ not to exhibit any offset to step inputs it is necessary and sufficient that

$$\lim_{s \to 0} \left[s \left(\epsilon(s) \frac{1}{s} \right) \right] = 0 \qquad (2.4-10)$$

It has become customary to define "System Types" to classify the asymptotic behavior.

Definition 2.4-1. *Let $g(s)$ be the open loop transfer function and let m be the largest integer for which*

$$\lim_{s \to 0} s^m g(s) \neq 0$$

Then the system $g(s)$ is said to be of Type m. (In other words, a Type m system has m poles at the origin.)

Theorem 2.4-1. *Let the open loop system $g(s)$ be of Type m and assume that the closed-loop system with unity feedback is stable. Then the sensitivity function $\epsilon(s) = (1 + g(s))^{-1}$ satisfies*

Type m:

$$\lim_{s \to 0} \frac{\epsilon(s)}{s^k} = 0 \qquad 0 \leq k < m \qquad (2.4-11)$$

Also, as $t \to \infty$ the closed loop system tracks perfectly setpoint changes (rejects perfectly disturbances) of the form $\sum_{k=0}^{m} a_k s^{-k}$ where a_k are real constants.

Proof. Follows directly from Final Value Theorem.

Specifically, in order for a closed loop system to track steps without offset the sensitivity function $\epsilon(s)$ has to have a zero at the origin. This is the case if the open loop system $g(s)$ has a pole at the origin. That is, $g(s)$ has to be of Type 1. Similarly offset free tracking of ramps requires $g(s)$ to have two poles at the origin (Type 2).

2.4.4   Linear Quadratic (H_2-) Optimal Control

The idea of optimal control is to express the control objectives in the form of an optimization problem and to determine the controller structure and parameters from its solution. In the formulation practical relevance always has to be sacrificed for mathematical convenience – i.e., an optimization problem which can be solved. Thus the "optimality" of a controller does not imply its usefulness. Whether a controller determined from optimal control principles solves a specific practical problem, depends to a large extent on how well the control system designer has approximated reality to fit the mathematical formalism. For several decades only the so-called *Linear Quadratic (H_2-) Optimal Control* was available. In the seventies the new paradigm of H_∞-*Optimal Control* emerged. Depending on the problem, the assumptions inherent in one or the other optimal control formalism can be more applicable and can allow a better approximation and thus a better solution.

In this section we will concentrate on the H_2-optimal control formulation. The controller c is determined such that the integral square error

$$\|e\|_2^2 = \int_0^\infty e^2(t)dt \qquad (2.4-12)$$

is minimized for a particular input v. Referring back to the introductory discussion of Sec. 2.4. we note that the "measure of smallness" is the 2-norm of the error ($\|e\|_2$) and the "input set" is just one specific input v. Using Parseval's theorem, we can state the optimization problem in the frequency domain

$$\min_c \|e\|_2^2 = \min_c \frac{1}{2\pi} \int_{-\infty}^\infty |e(i\omega)|^2 d\omega \qquad (2.4-13)$$

or upon substitution for e from (2.4–8) and for v from (2.2–11)

$$\min_c \|\epsilon w\|_2^2 = \min_c \frac{1}{2\pi} \int_{-\infty}^\infty |\epsilon(i\omega)w(i\omega)|^2 d\omega \qquad (2.4-14)$$

For a comparison with $H_\infty-$ optimal control and for the extension to the multivariable case an alternate interpretation of (2.4–14) is useful. The integrand is the magnitude of the sensitivity function *weighted* by w, where w depends on the input v. The integral operation generates the *average* of this weighted magnitude. Thus, the H_2-optimal controller minimizes the average magnitude or, in mathematical terms, *minimizes the 2-norm of the sensitivity function ϵ weighted by w.*

Throughout the book we will adopt this latter more mathematical interpretation of (2.4–14) because it is more practical. If w is chosen to represent a particular input according to (2.2–11) then it is rare that the controller which solves (2.4–14) satisfies *all* the practical design requirements. For example, the minimum ISE response might be too oscillatory or the overshoot too large. Therefore, it is common practice for the designer to use w as a "tuning parameter." The weight w is then varied until the closed-loop response characteristics are satisfactory. Naturally, the input type is a very effective guide for the weight selection. The advantage of using w for tuning rather than setting up an alternate optimization problem which addresses the practical requirements more directly, is that very effective methods are available for solving (2.2–14).

For the optimization (2.4–14) clearly only the magnitude of the weight w matters but not its phase characteristics — i.e., whether it is MP or NMP, stable or unstable. However, if w reflects the effect of a specific unstable input v, then the optimal ϵ has to satisfy certain internal stability requirements to guarantee that the error e vanishes as $t \to \infty$ and the objective (2.4–12) is bounded (see Chap. 5).

Usually the problem (2.4–13) leads to controllers which are too aggressive because emphasis is placed on the error only and not on the manipulated variable. More general objective functions than (2.4–12), including, for example, a term u^2, to penalize excessive input variations, are well known and are not discussed here. Entirely equivalent stochastic formulations are also available.

2.4.5 H_∞-Optimal Control

Here, the inputs v are assumed to belong to the set $\mathcal{V}$ of norm-bounded functions with a frequency dependent weight w as discussed in Sec. 2.2.3.

$$\mathcal{V} = \left\{ v : \|v'\|_2^2 = \left\| \frac{v}{w} \right\|_2^2 = \frac{1}{2\pi} \int_{-\infty}^{\infty} \left| \frac{v(i\omega)}{w(i\omega)} \right|^2 d\omega \le 1 \right\} \qquad (2.4-15)$$

The input class defined by (2.4–15) contains a large variety of spectra including possibly steps, pulses and narrow-band signals of various frequencies. This input characterization is sometimes more meaningful than what is required for the H_2-optimal control problem. Recall that there we assumed a single fixed input.

Each input $v \in \mathcal{V}$ gives rise to an error e. The H_∞-optimal controller is designed to minimize the *worst* error which can result from any input $v \in \mathcal{V}$:

$$\min_c \sup_{v \in \mathcal{V}} \|e\|_2 = \min_c \sup_{v' \in \mathcal{V}'} \|\epsilon w v'\|_2 \qquad (2.4-16)$$

Here we performed the same substitution for e as in (2.4–14). The worst error can be bounded as follows

$$\sup_{v' \in \mathcal{V}'} \|\epsilon w v'\|_2^2 = \sup_{v' \in \mathcal{V}'} \frac{1}{2\pi} \int_{-\infty}^{\infty} |\epsilon w v'|^2 d\omega \le \sup_{\omega} |\epsilon w|^2 \sup_{v' \in \mathcal{V}'} \frac{1}{2\pi} \int_{-\infty}^{\infty} |v'|^2 d\omega$$

Then, using the definition of the set $\mathcal{V}'$ (2.2–10) we find

$$\sup_{v' \in \mathcal{V}'} \|\epsilon w v'\|_2^2 \leq \sup_\omega |\epsilon w|^2 \qquad (2.4-17)$$

It can be shown that there exists a sequence of inputs $v_k^* \in \mathcal{V}'$ for which $\lim_{k \to \infty} \|\epsilon w v_k^*\|_2 = \sup_\omega |\epsilon w|$. Thus (2.4–17) is actually an equality. The characteristics of v_k^* can be understood from the following argument. Let us assume that $|\epsilon w(i\omega)|$ reaches its maximum for $\omega = \omega^*$. Then, the spectrum of the input v_k^* which makes (2.4–17) an equality as $k \to \infty$ has to be increasingly narrow-band and concentrated near ω^* as k increases.

We define the ∞-norm of the weighted sensitivity function

$$\|\epsilon w\|_\infty \overset{\Delta}{=} \sup_\omega |\epsilon w| \qquad (2.4-18)$$

and from (2.4–16)-(2.4–18) the H_∞-optimal control problem is

$$\min_c \|\epsilon w\|_\infty = \min_c \sup_\omega |\epsilon w(i\omega)| \qquad (2.4-19)$$

Thus the H_∞-optimal controller minimizes the maximum magnitude or, in mathematical terms, *minimizes the ∞-norm of the sensitivity function ϵ weighted by w*.

According to this frequency domain interpretation the H_2-optimal controller minimizes the *average* value and the H_∞-optimal controller the *peak value* of the magnitude of the weighted sensitivity function. For the optimization (2.4–19) clearly only the magnitude of the weight w matters, but not its phase characteristics — i.e., whether it is MP or NMP, stable or unstable.

As in the case of H_2-optimal control it is more practical to think of w as a tuning parameter affecting the solution of (2.4–19) rather than to adhere to the formal problem definition (2.4–16). Let us assume that for a particular w the optimal value of the objective function (2.4–18) is k. Then (2.4–19) implies that $|\epsilon| \leq k|w|^{-1}, \forall \omega$. This suggests that the designer can impose a bound on the sensitivity function

$$|\epsilon(i\omega)| < |w(i\omega)|^{-1} \qquad \forall \omega \qquad (2.4-20)$$

which the H_∞-optimal controller aims to satisfy. If a controller is found such that

$$\|\epsilon w\|_\infty < 1 \qquad (2.4-21)$$

then the bound (2.4–20) is met. Therefore, the H_∞-performance requirement is usually written in the form (2.4–21).

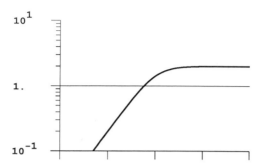

Figure 2.4-3. Typical bound (w^{-1}) imposed on sensitivity function.

Naturally, the types of inputs which are expected will guide the selection of the weight w. Figure 2.4-3 shows a typical weight. It implies that the expected inputs are mostly slowly varying and that high frequency signals have smaller amplitudes. The bound w^{-1} forces the sensitivity function to be small in the frequency range where large inputs are anticipated. The weight w in Fig. 2.2-5 would indicate much milder performance requirements. Expressing the control objective as a bound on the sensitivity function ϵ has much practical appeal. Often the designer wishes to specify a minimum bandwidth and to limit the maximum peak of the sensitivity function to avoid excessive disturbance amplification. In the H_∞-optimal control formulation this can be done explicitly. Most important, however, this new formalism allows one to address the robust controller design problem in a consistent framework, as will be discussed in the next section.

Recall that for high frequencies $|pc|$ is small and therefore

$$|\epsilon| = \left|\frac{1}{1+pc}\right| \cong 1 \quad \omega \quad \text{large} \qquad (2.4-22)$$

Thus tight performance specifications are only meaningful in the low-frequency range, where $|\tilde{p}c|$ is large and $|\epsilon| \cong |pc|^{-1}$. Then the performance specification (2.4–21) reduces to

$$|pc| > |w| \quad \omega \quad \text{small} \qquad (2.4-23)$$

That is, the loop gain $|pc|$ has to be shaped to fall above the performance weight $|w|$.

2.5 Robust Stability

We wish to derive conditions for the robust stability of the family Π of plants defined by (2.2–6). For this purpose we will employ the Nyquist stability criterion.

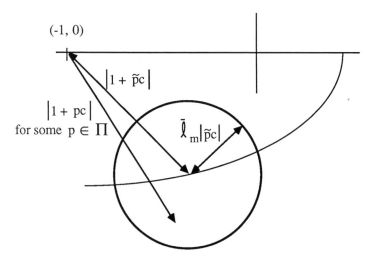

Figure 2.5-1. Diagram for graphical derivation of robust stability and robust performance conditions.

Let us denote by $N(k, g(s))$ the net number of clockwise encirclements of the point $(k, 0)$ by the image of the Nyquist D contour under $g(s)$. We will assume that all $p \epsilon \Pi$ have the same number (n) of RHP poles. Then we have robust stability for a specific controller c if and only if

$$N(-1, pc) = -n \qquad \forall p \in \Pi \qquad (2.5-1)$$

In particular it is necessary that the nominal plant $\tilde{p}$ be closed loop stable:

$$N(-1, \tilde{p}c) = -n \qquad (2.5-2)$$

Assuming that (2.5–2) holds, (2.5–1) will hold if and only if the Nyquist band which comprises all $p \in \Pi$ does not include the point (-1,0). From the simple geometric argument in Fig. 2.5-1 it is clear that this will be the case if and only if the distance of $\tilde{p}c$ from the point (-1,0) — i.e., $|1 + \tilde{p}c|$, exceeds the disk radius $|\tilde{p}c(i\omega)|\bar{\ell}_m(\omega)$:

$$|1 + \tilde{p}c(i\omega)| > |\tilde{p}c(i\omega)|\bar{\ell}_m(\omega) \qquad \forall \omega$$

or

$$\left| \frac{\tilde{p}c(i\omega)}{1 + \tilde{p}c(i\omega)} \right| \bar{\ell}_m(\omega) = |\tilde{\eta}(i\omega)|\bar{\ell}_m(\omega) < 1 \qquad \forall \omega \qquad (2.5-3)$$

where $\tilde{\eta} = \tilde{p}c(1 + \tilde{p}c)^{-1}$. We have proven the following theorem.

Theorem 2.5-1 (Robust Stability). *Assume that all plants p in the family* Π

$$\Pi = \left\{ p : \left| \frac{p - \tilde{p}(i\omega)}{\tilde{p}} \right| \leq \bar{\ell}_m(\omega) \right\} \qquad (2.2-6)$$

have the same number of RHP poles and that a particular controller c stabilizes the nominal plant $\tilde{p}$. Then the system is robustly stable with the controller c if and only if the complementary sensitivity function $\tilde{\eta}(s)$ for the nominal plant $\tilde{p}$ satisfies the following bound

$$\|\tilde{\eta}\bar{\ell}_m\|_\infty \triangleq \sup_\omega |\tilde{\eta}\bar{\ell}_m(\omega)| < 1 \qquad (2.5-4)$$

Let us clarify what it means that Thm. 2.5-1 is not only sufficient for robust stability but also necessary. If condition (2.5-4) is violated then in the set Π defined by (2.2-6) there exists a plant p for which the closed-loop system with the controller c is unstable. However, if the set Π was obtained by approximating the true uncertainty regions with disks (see Sec. 2.2.2) and therefore contains plants not present in the original uncertainty set then condition (2.5-4) is generally only sufficient for the original uncertainty set. Nevertheless, the necessity implies that at least for the set Π defined by (2.2-6), this is the tightest robust stability condition that can be derived.

Theorem 2.5-1 is most instructive for several reasons. First note that robust stability imposes a bound on the ∞-*norm of the complementary sensitivity function $\tilde{\eta}$ weighted by $\bar{\ell}_m$.* Recall, on the other hand, that for the H_2-optimal control formulation the performance specifications were expressed in terms of the *2-norm of the sensitivity function $\tilde{\epsilon}$ weighted by w.* The different norms are an indication that performance, robust stability and robust performance cannot be treated "jointly" when the ISE is employed as a performance objective. One of the main motivations for introducing H_∞-optimal control is that the robust stability condition and the performance specification are expressed in terms of the same norm.

Furthermore recall that measurement noise also tends to impose a bound on the magnitude of η (see 2.4–2). However, in process control the constraint imposed by model uncertainty (2.5-4) tends to dominate.

Also note that for high frequencies $|\tilde{p}c|$ is small and therefore

$$|\tilde{\eta}| = \left| \frac{\tilde{p}c}{1 + \tilde{p}c} \right| \cong |\tilde{p}c| \quad \omega \quad \text{large} \qquad (2.5-5)$$

Thus, for large ω (2.5-4) reduces to

$$|\tilde{p}c| < \frac{1}{|\bar{\ell}_m|} \quad \omega \quad \text{large} \qquad (2.5-6)$$

The design implication is that the controller gain for high frequencies is limited by uncertainty. The loop gain $|\tilde{p}c|$ has to be shaped to fall below the uncertainty bound $|\bar{\ell}_m|^{-1}$.

Finally note that in the H_∞-optimal control framework we want to minimize $\|\tilde{\epsilon}w\|_\infty$ for performance (see (2.4–19)) and $\|\tilde{\eta}\bar{\ell}_m\|_\infty$ for robustness. A trade-off between performance and robustness arises from the fact that $\tilde{\epsilon}$ and $\tilde{\eta}$ are not independent (see (2.4–6)) and making one small will automatically make the other one large. This problem is inherent in feedback control and cannot be removed by clever controller design. The objective of a design method for robust control systems is to reach the best compromise between the conflicting objectives of performance and robustness.

2.6   Robust Performance

Robust stability is the minimum requirement a control system has to satisfy to be useful in a practical environment where model uncertainty is an important issue. However, robust stability alone is not enough. If the bound (2.5–4) is satisfied for a family Π, then there exists a particular plant $p \in \Pi$ for which the closed-loop system is on the verge of instability and for which the performance is arbitrarily poor. Thus we also have to make sure that some performance specifications are met for *all* plants in the family Π. We will derive conditions for robust performance when the performance is measured in terms of both the 2-norm and the ∞-norm.

2.6.1 H_2 Performance Objective

Expanding the scope of Sec. 2.4.4 dealing with a single plant p, we are now looking at the family of plants Π defined by (2.2–6). We set as our objective to design the controller such that the error resulting from a specific input d is minimized for the "worst" plant in the family. The worst plant is the one giving rise to the largest error.

$$\min_c \max_{p \in \Pi} \int_0^\infty e^2 dt = \min_c \max_{p \in \Pi} \frac{1}{2\pi} \int_{-\infty}^\infty \left| \frac{1}{1+pc} \right|^2 |v|^2 d\omega \qquad (2.6-1)$$

The integral is maximized by maximizing the integrand at every frequency. From geometric arguments (Fig. 2.5-1) we find

$$|1+pc| \geq |1+\tilde{p}c| - |\tilde{p}c|\bar{\ell}_m, \qquad \forall p \in \Pi \qquad (2.6-2)$$

or

$$|\epsilon| = \left| \frac{1}{1+pc} \right| \leq \frac{|\tilde{\epsilon}|}{1 - |\tilde{\eta}|\bar{\ell}_m}, \qquad \forall p \in \Pi \qquad (2.6-3)$$

where $\tilde{\epsilon} = (1 + \tilde{p}c)^{-1}$. Note that when the robust stability condition (2.5–4) holds the denominator of the RHS of (2.6–3) is positive. Thus (2.6–1) can be rewritten as

$$\min_c \frac{1}{2\pi} \int_{-\infty}^{\infty} \left| \frac{1}{1 - |\tilde{\eta}||\bar{\ell}_m|} \right|^2 |\tilde{\epsilon} v|^2 d\omega \qquad (2.6 - 4)$$

Because the controller c appears in a very complex manner in (2.6–4), there is little hope of finding a simple solution to this problem. However, for a specific controller c, we can use (2.6–3) to determine the worst error which can result from a specific input v for a set of plants Π.

$$\max_{p \in \Pi} \|e\|_2^2 = \frac{1}{2\pi} \int_{-\infty}^{\infty} \left| \frac{1}{1 - |\tilde{\eta}||\bar{\ell}_m|} \right|^2 |\tilde{\epsilon} v|^2 d\omega \qquad (2.6 - 5)$$

The first factor in the integral accounts for the effect of model error. Because of the robust stability condition (2.5–4) the integrand is always bounded. As expected, the observed error increases relative to the nominal error as the stability boundary is approached.

Equation (2.6–5) is a handy analysis tool. After a controller has been designed, a performance bound in terms of ISE can be established. The error which is observed on the unknown "real" system as a result of a specific input v is always less than the established bound – provided the uncertainty description is correct.

2.6.2 H_∞ Performance Objective

If the performance specifications are stated in the H_∞ framework, then we will require that (2.4–21) be met for the "worst" plant

$$\max_{p \in \Pi} \|\epsilon w\|_\infty = \max_{p \in \Pi} \sup_\omega |\epsilon w(i\omega)| < 1 \qquad (2.6 - 6)$$

or

$$\|\epsilon w\|_\infty = \sup_\omega |\epsilon w(i\omega)| < 1 \qquad \forall p \in \Pi \qquad (2.6 - 7)$$

Using (2.6–3), (2.6–7) can be rewritten as

$$\frac{|\tilde{\epsilon} w|}{1 - |\tilde{\eta}||\bar{\ell}_m|} < 1 \qquad \forall \omega \qquad (2.6 - 8)$$

or

$$|\tilde{\eta}\bar{\ell}_m| + |\tilde{\epsilon} w| < 1 \qquad \forall \omega \qquad (2.6 - 9)$$

Theorem 2.6-1 (Robust Performance). *Assume that all plants in the family*

$$\Pi = \left\{ p : \left| \frac{p(i\omega) - \tilde{p}(i\omega)}{\tilde{p}(i\omega)} \right| \leq \bar{\ell}_m(\omega) \right\} \qquad (2.2 - 6)$$

have the same number of RHP poles. Then the closed-loop system will meet the performance specification

$$\|\epsilon w\|_\infty = \sup_\omega |\epsilon w| < 1 \qquad \forall p \in \Pi$$

if and only if the nominal system is closed-loop stable and the sensitivity function $\tilde{\epsilon}$ and the complementary sensitivity function $\tilde{\eta}$ satisfy

$$|\tilde{\eta}\bar{\ell}_m| + |\tilde{\epsilon}w| < 1 \qquad \forall\omega \qquad\qquad (2.6-10)$$

Trivially, robust performance (2.6–10) implies robust stability (2.5–4) and nominal performance (2.4–20). The interdependence of $\tilde{\epsilon}$ and $\tilde{\eta}$ makes it a challenge to meet (2.6–10). Improving the nominal performance (decreasing $|\tilde{\epsilon}w|$) worsens the robustness (increases $|\tilde{\eta}\bar{\ell}_m|$) and pushes the system closer to the point of instability for some plant $p \in \Pi$.

If $\bar{\ell}_m(0) < 1$ then the bound (2.6–10) can be satisfied for low frequencies by making $|\tilde{p}c|$ large; see (2.4–23). If $w < 1$ for high frequencies the bound (2.6–10) is met by choosing $|\tilde{p}c|$ small; see (2.5–6). By shaping the loop gain $|\tilde{p}c|$ to be large for low frequencies and small for high frequencies the robust performance specifications can be met assuming that they are not too tight (w large) in a frequency range where the uncertainty $\bar{\ell}_m$ is large (design by "loop shaping").

A controller which optimizes robust performance solves

$$\min_c \sup_\omega (|\tilde{\eta}\bar{\ell}_m| + |\tilde{\epsilon}w|) \qquad\qquad (2.6-11)$$

Compare problem (2.6–11) for optimizing robust performance with problem (2.4–19) for optimizing nominal performance. For *nominal performance* the weighted ∞-norm of the sensitivity function is to be minimized. For *robust performance* the ∞-norm ceases to be a suitable measure. As we will learn in Chap. 11 the appropriate measure for robust performance is the Structured Singular Value (SSV) which will be discussed in the context of multivariable systems. The LHS of (2.6–10) turns out to be the SSV of a particular matrix.

We see from (2.6–10) that we can attain robust performance simply by satisfying both the robust stability test (2.5–4) and the nominal performance test (2.4–20) with some margin: If $|\tilde{\eta}\bar{\ell}_m| < \alpha(\omega)$ and $|\tilde{\epsilon}w| < 1-\alpha(\omega)$ where $\alpha \leq 1, \forall\omega$, then robust performance is automatically guaranteed. Thus, for SISO systems, robust performance is not an objective which is particularly difficult to deal with. This will not be true for MIMO systems.

2.7 Summary

For the design of a feedback controller the following essentials have to be specified:

- process model

- model uncertainty bounds

- type of inputs

- performance objectives

All design techniques aim to make the error resulting from certain external inputs small. This is equivalent to making the sensitivity function

$$\epsilon = \frac{e}{d-r} = \frac{1}{1+pc} \qquad (2.4-1)$$

small. The H_2-optimal controller minimizes the ISE for a particular input or equivalently it minimizes the 2-norm ("average") of ϵ weighted by the input.

H_2:

$$\min_c \int_0^\infty e^2 dt = \min_c \|\epsilon w\|_2^2 \qquad (2.4-14)$$

The H_∞-optimal controller minimizes the worst ISE which can result from a set of inputs or equivalently it minimizes the ∞-norm ("peak") of ϵ weighted by w.

H_∞:

$$\min_c \|\epsilon w\|_\infty = \min_c \sup_\omega |\epsilon w| \qquad (2.4-19)$$

For the suppression of measurement noise and in particular for robust stability the complementary sensitivity $\tilde{\eta}$ should be small in the frequency range where the model uncertainty measured by $\bar{\ell}_m$ is large.

Robust Stability:

$$\|\tilde{\eta}\bar{\ell}_m\|_\infty = \sup_\omega |\tilde{\eta}\bar{\ell}_m| < 1 \qquad (2.5-4)$$

Because

$$\tilde{\epsilon} + \tilde{\eta} = 1 \qquad (2.4-6)$$

a basic tradeoff between nominal performance and robust stability arises.

Robust Performance:

$$|\tilde{\eta}\bar{\ell}_m| + |\tilde{\epsilon}w| < 1 \qquad \forall\omega \qquad (2.6-10)$$

Robust performance is achieved simply by satisfying both nominal performance ($|\tilde{\epsilon}w| < 1$) and robust stability ($|\tilde{\eta}\bar{\ell}_m| < 1$) with some margin.

2.8   References

2.1. The inverse response behavior of NMP systems is proven, for example, by Holt & Morari (1985a).

2.2.2. The uncertainty description via norm-bounded perturbations ("disks") was popularized by Doyle & Stein (1981).

2.3.   Desoer & Lin (1985) investigated the stability of a large variety of feedback structures in the same manner as in this section.

2.4.2. Horowitz (1963) is generally credited with having first proposed and studied the two-degree-of-freedom controller.

2.4.4. Newton first presented an analytic feedback design technique to minimize an H_2 objective. The procedure is described in the book by Newton, Gould & Kaiser (1957).

2.4.5. The H_∞-optimal control problem was first formulated by Zames (1981). Closely related ideas proposed by other authors are discussed in his paper. A possible limiting response which proves that (2.4–17) is an equality is proposed by Desoer & Vidyasagar (1975).

2.5. The robust stability condition (at least for the MIMO case) is generally attributed to Doyle & Stein (1981).

2.6.2. The robust performance condition follows as a special case from the MIMO Structured Singular Value Analysis proposed by Doyle (1984, 1985).

Chapter 3

THE SISO IMC STRUCTURE FOR STABLE SYSTEMS

The Internal Model Control (IMC) structure will be introduced as an alternative to the classic feedback structure. Its main advantage is that closed-loop stability is assured simply by choosing a stable IMC controller. Also, closed-loop performance characteristics (like settling time) are related *directly* to controller parameters, which makes on-line tuning of the IMC controller very convenient. A two-step design procedure will be proposed. In the first step the controller is designed for optimal setpoint tracking (disturbance rejection) without regards for input saturation or model uncertainty. In the second step the controller is detuned for robust performance. While this procedure has no inherent optimality characteristics, it constitutes a simple and transparent approach for finding controllers which satisfy all typical practical requirements.

3.1 IMC Structure

The block diagram of the IMC loop is shown in Fig. 3.1-1A. Here p denotes the plant and p_m the measurement device transfer functions. In general neither p nor p_m are known exactly but only their nominal models $\tilde{p}$ and $\tilde{p}_m$ are available. The transfer function p_d describes the effect of the disturbance d' on the process output y. The measurement of y is corrupted by measurement noise n. The controller q determines the value of the input (manipulated variable) u. The control objective is to keep y close to the reference (setpoint) r.

Commonly we will use the simplified block diagram in Fig. 3.1-1B. Here d denotes the effect of the disturbance on the output. Exact knowledge of the output y is assumed ($p_m = 1, n = 0$).

Note that the complete control system to be implemented through computer software or analog hardware is contained in the shaded box in Fig. 3.1-1B. Because in addition to the controller q it includes the plant model $\tilde{p}$ explicitly we

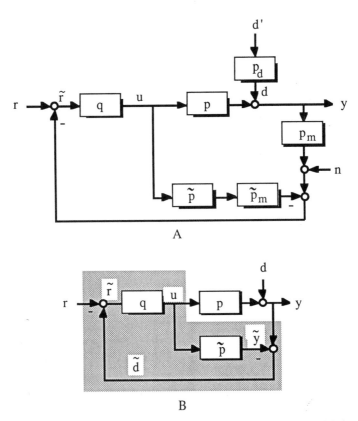

Figure 3.1-1. General (A) and simplified (B) block diagram of Internal Model Control system.

refer to this feedback configuration as *Internal Model Control* (IMC).

The feedback signal is

$$\tilde{d} = (p - \tilde{p})u + d \qquad (3.1-1)$$

If the model is exact $(p = \tilde{p})$ and there are no disturbances $(d = 0)$, then the model output $\tilde{y}$ and the process output y are the same and the feedback signal $\tilde{d}$ is zero. Thus, the control system is open-loop when there is no uncertainty — i.e., no model uncertainty and no unknown inputs d. This demonstrates very instructively that for open-loop stable processes feedback is *only* needed because of uncertainty. If a process and all its inputs are known perfectly, there is no need for feedback control. The feedback signal $\tilde{d}$ expresses the uncertainty about the process.

3.2 Stability Conditions for IMC

3.2.1 Internal Stability

In order to test for internal stability we examine the transfer functions between all possible system inputs and outputs. From the block diagram in Fig. 3.2-1 we find that there are three independent system inputs and three independent outputs. As shown in Fig. 3.1-1 we choose the independent inputs to be r, u_1, and u_2 and the independent outputs y, u, and $\tilde{y}$. If there is no model error $(p = \tilde{p})$, then the inputs and outputs are related through the following transfer matrix.

$$\begin{pmatrix} y \\ u \\ \tilde{y} \end{pmatrix} = \begin{pmatrix} pq & (1-pq)p & p \\ q & -pq & 0 \\ pq & -p^2q & p \end{pmatrix} \begin{pmatrix} r \\ u_1 \\ u_2 \end{pmatrix} \qquad (3.2-1)$$

Theorem 3.2-1 follows trivially by inspection.

Theorem 3.2-1. *Assume the model is perfect $(p = \tilde{p})$. Then the IMC system in Fig. 3.1-1B is internally stable if and only if both the plant p and the controller q are stable.*

This result is not unexpected. Recall that we pointed out in the last section that the IMC system is effectively open-loop when there is no uncertainty. Because the stabilization of open-loop unstable systems requires feedback the IMC structure cannot be applied in this case.

It can be argued that the lack of model uncertainty is an artificial assumption. Uncertainty gives rise to feedback and thus it could be possible to stabilize an unstable system with IMC. However in any practical situation it is unacceptable to rely on model uncertainty for stability.

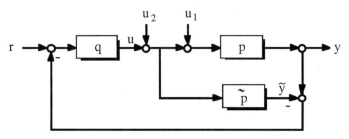

Figure 3.2-1. Block diagram for derivation of internal stability conditions.

3.2.2   Relationship with "Classic Feedback"

The manipulations necessary to transform the block diagram in Fig. 3.2-2A into the one in Fig. 3.2-2B leave the signals u and y unaffected. If we combine the two blocks q and $\tilde{p}$ in Fig. 3.2-2B, which are both part of the control system, into one block c then we obtain a classic feedback control system with

$$c = \frac{q}{1 - \tilde{p}q} \qquad (3.2-2)$$

On the other hand if we add the two blocks $\tilde{p}$ to the classic feedback system in Fig. 3.2-2C the signals u and y also remain unaffected. The IMC structure follows with

$$q = \frac{c}{1 + \tilde{p}c} \qquad (3.2-3)$$

Thus, in the way the outputs u and y react to inputs r and d the classic feedback structure and the IMC structure are *entirely equivalent* and the controllers c and q are related through (3.2–2) and (3.2–3) respectively.

Consider specifically the case that p is stable and that $p = \tilde{p}$.

i. Assume that the IMC structure (Fig. 3.2-2A) is internally stable — i.e., that q is stable. Then the equivalent classic feedback structure (Fig. 3.2-2B) is stable because the internal signals u and y are unaffected by the transformation.

ii. Assume that the classic feedback structure (Fig. 3.2-2B) is internally stable. Then (2.3–1) is stable and therefore q defined by (3.2–3) is stable. Thus the equivalent IMC structure is internally stable.

We have proven the following theorem.

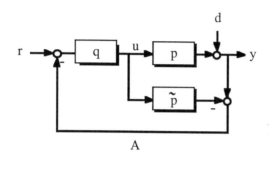

A

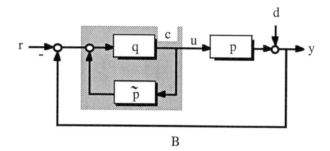

B

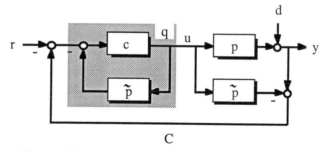

C

Figure 3.2-2. Alternate representations of the IMC structure.

Theorem 3.2-2. *Assume that p is stable and that $p = \tilde{p}$. Then the classic feedback system with controller*

$$c = \frac{q}{1 - \tilde{p}q} \qquad\qquad (3.2 - 2)$$

is internally stable if and only if q is stable.

This theorem has very profound implications for control system design. Usually a design procedure involves the search for a controller c such that the closed-loop system has certain desired properties. This search is greatly complicated by the fact that only those controllers c are allowed for which the closed-loop system is stable. Theorem 3.2-2 provides a simple parametrization of *all* stabilizing controllers c for the plant p in terms of the stable transfer function q. Thus, instead of searching for c it is possible to search for q without any loss of generality. This search is much simpler because stability of q automatically guarantees the stability of the closed loop system with c determined from (3.2–2).

3.2.3 Implementation: Classic Feedback Versus IMC

In the previous section we showed that IMC provides a convenient parametrization of the classic feedback controller. However, there are significant advantages to not only *designing* the controller via IMC but also *implementing* it in the IMC structure as a model block $\tilde{p}$ and a controller block q.

In any practical situation u is constrained by some upper and lower bounds (e.g., valve saturation). It is well known that these constraints can cause problems for the classic feedback controller. If the controller includes integral action they lead to "reset windup." Even if there is no integral component, a system which is stable without constraints can be destabilized when constraints become active. In the classic feedback configuration these issues have to be dealt with in an ad-hoc fashion ("anti-reset windup").

If the controller is implemented in the IMC configuration, input constraints do not cause any problems provided the *actual* (constrained) plant input is sent to the model $\tilde{p}$ rather than the input *computed* by the controller q (Fig. 3.2-3). Then, under the assumption that $p = \tilde{p}$, the IMC structure remains effectively open-loop and stability is guaranteed by the stability of p and q, as is the case in the absence of constraints. Omitting the saturation block in front of the model block would cause the model to behave differently from the plant when the inputs reach constraints. This model/plant mismatch can lead to instability as we will show later.

In the structure in Fig. 3.2-3 there is no information feedback: the feedback signal is always $\tilde{d}$ and the controller q is entirely unaware of the effect of its

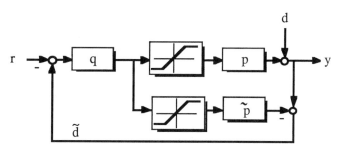

Figure 3.2-3. IMC implementation in the presence of actuator constraints.

actions. In particular, it does not know if and when the manipulated variable saturates. It is reasonable to assume that the closed-loop performance could be improved if the controller adjusted its actions when the manipulated variable has saturated. Thus, while the structure in Fig. 3.2-3 guarantees global stability (for $p = \tilde{p}$), performance might suffer somewhat compared to other anti-windup schemes.

It should also be noted that even if q, p, and $\tilde{p}$ are general *nonlinear* operators the IMC system would still be effectively open-loop (as long as $p = \tilde{p}$) and the stability of p and q would imply closed-loop stability. Thus the IMC structure offers the opportunity of implementing complex nonlinear control algorithms without generating complex stability issues.

3.3   Performance of IMC

3.3.1   Sensitivity and Complementary Sensitivity Function

For IMC (Fig. 3.1-1) we easily find the following transfer functions relating inputs and outputs:

$$y = \frac{pq}{1 + q(p - \tilde{p})}r + \frac{1 - \tilde{p}q}{1 + q(p - \tilde{p})}d \qquad (3.3 - 1)$$

The sensitivity function $\epsilon(s)$ relates the external inputs r and d to the error $e = y - r$

$$\frac{e}{d - r} = \frac{y}{d} = \frac{1 - \tilde{p}q}{1 + q(p - \tilde{p})} \triangleq \epsilon(s) \qquad (3.3 - 2)$$

We find the complementary sensitivity function $\eta(s)$ by subtracting $\epsilon(s)$ from unity:

$$\frac{y}{r} = \frac{pq}{1 + q(p - \tilde{p})} \triangleq \eta(s) \qquad (3.3 - 3)$$

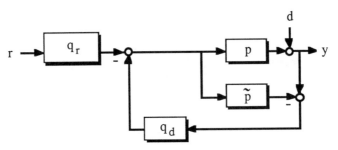

Figure 3.3-1. Two-degree-of-freedom IMC.

When the model is exact $(p = \tilde{p})$, (3.3–2) and (3.3–3) reduce to

$$\tilde{\epsilon}(s) = 1 - \tilde{p}q \tag{3.3-4}$$

$$\tilde{\eta}(s) = \tilde{p}q \tag{3.3-5}$$

We concluded in Chap.2 that $\epsilon(s)$ determines performance and $\eta(s)$ robustness. Through the IMC parametrization the controller q is related to $\epsilon(s)$ and $\eta(s)$ in a very simple (linear) manner which will make the design of q easy. The effect of the classic controller c on $\epsilon(s)$ and $\eta(s)$ (2.4–1 and 2.4–2) is much more complex.

3.3.2   Two-Degree-of-Freedom Controller

If both good tracking of r and good disturbance rejection are important and if the dynamic characteristics of the two inputs r and d are substantially different it is advantageous to introduce another controller block (Fig. 3.3-1). The effects of r and d on the error are described by

$$e = \frac{1 - \tilde{p}q_d}{1 + q_d(p - \tilde{p})}d - \left[1 - \frac{pq_r}{1 + q_d(p - \tilde{p})} \right] r \tag{3.3-6}$$

For $p = \tilde{p}$ (3.3–6) becomes

$$e = (1 - pq_d)d - (1 - pq_r)r \tag{3.3-7}$$

Here q_d is designed for disturbance rejection and q_r for setpoint tracking.

3.3.3   Asymptotic Properties of Closed-Loop Response (System Type)

"System types" were defined in Sec. 2.4.3 to classify the asymptotic closed-loop behavior.

Type m:

$$\lim_{s \to 0} \frac{\epsilon(s)}{s^k} = 0 \qquad 0 \leq k < m \qquad (3.3-8)$$

Using (3.3–2) this definition becomes

Type m:

$$\lim_{s \to 0} \frac{1 - \tilde{p}q}{1 + q(p - \tilde{p})} \frac{1}{s^k} = 0 \qquad 0 \leq k < m \qquad (3.3-9)$$

Condition (3.3–9) is satisfied if and only if $(1 - \tilde{p}q)$ has m zeros at the origin which is the case if and only if

Type m:

$$\lim_{s \to 0} \frac{d^k}{ds^k}(1 - \tilde{p}q) = 0 \qquad 0 \leq k < m \qquad (3.3-10)$$

Specifically we obtain

Type 1:

$$\lim_{s \to 0} \tilde{p}q = 1 \qquad (3.3-11)$$

Thus, in order to track asymptotically constant inputs error free the controller steady state gain has to be the inverse of the model steady-state gain.

Type 2:

$$\lim_{s \to 0} \tilde{p}q = 1, \quad \lim_{s \to 0} \frac{d}{ds}(\tilde{p}q) = 0 \qquad (3.3-12)$$

Note that (3.3–11) and (3.3–12) are necessary and sufficient for off set free tracking of steps and ramps respectively even when model error is present.

3.3.4 The Concept of "Perfect Control"

In Chap.2 we identified the sensitivity function ϵ as a direct indicator of performance. The feedback controller is designed to make some measure (e.g., a weighted norm) of the sensitivity function small. For IMC

$$\epsilon = \frac{e}{d - r} = \frac{1 - \tilde{p}q}{1 + q(p - \tilde{p})} \qquad (3.3-2)$$

Algebraically ϵ can be made identically equal to zero by choosing the model inverse as the controller

$$q = \frac{1}{\tilde{p}} \qquad (3.3-13)$$

This would imply "perfect control"

$$e(t) = 0 \qquad \forall \quad r, d, t \qquad (3.3-14)$$

which we know from practical experience to be impossible. A reason for the infeasibility of perfect control is evident from the classic feedback controller c generated by (3.3–13).

$$c = \frac{q}{1 - \tilde{p}q} = \infty \quad \text{for } q = \tilde{p}^{-1} \qquad (3.3-15)$$

Though "perfect control" cannot be achieved, it is of great theoretical and practical interest to determine how closely this ideal can be approached. Obviously the closeness depends on the controller q (or c). If c is a PI controller, then better performance (a "smaller" ϵ) is expected than when c is just a P controller. In the age of digital control when very complex controllers can be implemented with relative ease, it does not seem to be meaningful to investigate what performance is achievable with particular simplistic controllers. What is much more important is how *plant characteristics* limit the achievable performance *independent* of controller complexity.

This information would be useful in several different ways: If the potential performance improvement through an advanced control strategy is to be assessed, the current performance (for example, with a PI controller) could be compared against the *best* achievable performance with *any* controller. If this comparison does not reveal any significant differences, *no* advanced control algorithm will be able to generate significant benefits. This judgment can be made without actually considering or designing any particular advanced controllers. The analysis would also imply that real performance improvements can come only from a design change of the plant itself rather than from modifications of the control system.

Our earlier discussions of $q = \tilde{p}^{-1}$ indicate that achievable performance is related to plant invertibility. If inversion were feasible then perfect control would be possible. Limits on invertibility are limits on achievable performance. There are three reasons why the use of the model inverse as controller is generally infeasible and therefore perfect control impossible.

i. *Nonminimum-phase (NMP) Characteristics of the Plant.* If p contains a time delay, p^{-1} is noncausal and cannot be used as a controller. If p contains RHP zeros then p^{-1} is unstable. However, for the stability of IMC q has to be stable and therefore $q = p^{-1}$ is not viable. Thus, when NMP elements are present in the plant, the sensitivity function cannot be made zero — not even in an approximate sense.

ii. *Constraints on the Inputs.* If p is strictly proper then q is improper — i.e., it involves pure differentiators which cannot be realized. This does not appear to be a severe handicap because poles can always be added to the

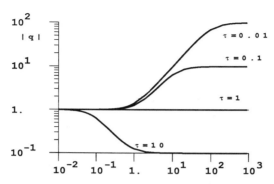

Figure 3.3-2. q' approaches the improper q as $\tau \to 0$.

denominator of q such that q approximates p^{-1} arbitrarily closely without being improper. For example if $p = (s+1)^{-1}$, $q = (s+1)$ is improper but

$$q' = \frac{s+1}{\tau s + 1}$$

is proper and "approaches" q as τ approaches zero. The Bode plot in Fig. 3.3-2 indicates that the excursions of the input u are excessive for high frequency inputs when τ is small. Thus, the constraints on the manipulated variable impose a limit on how closely q' can approximate q.

iii. *Model Uncertainty.* The controller is designed based on a process model $\tilde{p}$ which is an approximate description of the real plant p. We saw in Chap. 2 that model uncertainty constrains the choice of the complementary sensitivity function $\tilde{\eta}(s)$. In particular, for robust stability $\tilde{\eta}$ has to satisfy

$$|\tilde{\eta}(i\omega)\bar{\ell}_m(\omega)| < 1 \qquad \forall \omega \tag{3.3 - 16}$$

when the multiplicative uncertainty is bounded by $\bar{\ell}_m(\omega)$:

$$\left|\frac{p(i\omega) - \tilde{p}(i\omega)}{\tilde{p}(i\omega)}\right| \leq \bar{\ell}_m(\omega) \tag{3.3 - 17}$$

When we rewrite (3.3–16) in terms of $\tilde{\epsilon}$

$$|(1 - \tilde{\epsilon})\bar{\ell}_m| < 1 \qquad \forall \omega \tag{3.3 - 18}$$

we see that this bound is satisfied for $\tilde{\epsilon} = 0$ ("perfect control") as long as $\bar{\ell}_m < 1$. However, for large frequencies $\bar{\ell}_m$ always exceeds unity because

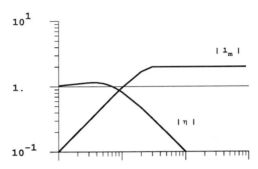

Figure 3.3-3. Complementary sensitivity $\tilde{\eta}$ satisfying robustness bound $\bar{\ell}_m$.

there is complete phase uncertainty, and therefore $\tilde{\epsilon}$ cannot be zero. Figure 3.3-3 shows a typical plot of how $|\tilde{\eta}|$ has to decrease as $\bar{\ell}_m$ increases. The frequency range over which good control ($|\tilde{\eta}| \approx 1$) is possible, is limited by model uncertainty.

The general conclusion is that perfect control is only feasible in the low frequency range. For high frequencies the perfect controller has to be detuned.

One of the attractive features of IMC is that it reveals the fundamental limitations on performance with clarity and simplicity. This insight is difficult to obtain from a study of the classic feedback structure.

3.4   Outline of the IMC Design Procedure

3.4.1   Basic Design Philosophy

In Chap.2 we discussed the objectives of control system design.

Nominal Performance:
  Specification:
$$\|\tilde{\epsilon}w\|_\alpha < 1 \qquad \alpha = 2, \infty \tag{3.4 - 1}$$

  Optimal Design Problem:

$$\min_q \|\tilde{\epsilon}w\|_\alpha = \min_q \|(1 - \tilde{p}q)w\|_\alpha \quad \alpha = 2, \infty \tag{3.4 - 2}$$

Robust Stability:
  Specification:

$$\|\tilde{\eta}\bar{\ell}_m\|_\infty = \|\tilde{p}q\bar{\ell}_m\|_\infty < 1 \tag{3.4 - 3}$$

Robust Performance:
  Specification:

$$\|\epsilon w\|_\infty < 1 \qquad \forall p \in \Pi \tag{3.4 - 4}$$

or

$$|\tilde\epsilon w| + |\tilde\eta \bar\ell_m| = |(1 - \tilde p q)w| + |\tilde p q \bar\ell_m| < 1 \quad \forall \omega \tag{3.4 - 5}$$

  Optimal Design Problem:

$$\min_{\tilde q} \sup_\omega (|\tilde\epsilon w| + |\tilde\eta \bar\ell_m|) = \min_{\tilde q} \sup_\omega (|(1 - \tilde p q)w| + |\tilde p q \bar\ell_m|) \tag{3.4 - 6}$$

For good nominal performance a weighted norm of the sensitivity function $\tilde\epsilon$ should be made small (3.4–1). In the popular optimal control formulations either the 2-norm or the ∞-norm is minimized (3.4–2). For robust stability the complementary sensitivity function $\tilde\eta$ is constrained by modeling error (3.4–3). For robust performance a more complex constraint (3.4–5) involving both $\tilde\epsilon$ and $\tilde\eta$ has to be met.

Clearly what we need in practice is robust performance. However, there are several hurdles, which prevent us at present from designing controllers by solving the optimal robust performance problem (3.4–6). First, the efficient solution of (3.4–6) is still an active area of research. For the general multivariable case no reliable solution procedures are available yet. Furthermore, the specification of a frequency-dependent weight w, which results in desirable time domain response characteristics, when the ∞-norm is minimized, involves much trial and error. Finally, in an application environment it is desirable to make the "degree of robustness" adjustable on-line because the plant-model mismatch tends to increase with time after the controller has been installed. Solving (3.4–6) repeatedly with different functions w and $\bar\ell_m$ is not practical under these circumstances.

Therefore, for IMC design we use a two-step approach which has no inherent optimality characteristics but should provide a good engineering approximation to the optimal solution. It guarantees robustness but the performance is generally not optimal in any sense.

STEP 1: Nominal Performance
  The controller $\tilde q$ is selected to yield a "good" system response for the input(s) of interest, without regard for constraints and model uncertainty. Generally we will choose $\tilde q$ such that it is Integral-Square-Error (ISE) or H_2-optimal

$$\min_{\tilde q} \|(1 - \tilde p \tilde q)w\|_2 \tag{3.4 - 7}$$

The optimal sensitivity function becomes

$$\bar{\epsilon} \triangleq 1 - \tilde{p}\tilde{q} \tag{3.4 - 8}$$

and the optimal complementary sensitivity function

$$\bar{\eta} \triangleq \tilde{p}\tilde{q} \tag{3.4 - 9}$$

STEP 2: Robust Stability and Performance
  At high frequencies, when $\bar{\ell}_m$ exceeds unity $\bar{\eta}$ has to be rolled off. Therefore, $\tilde{q}$ is augmented by a low-pass filter f

$$q \triangleq \tilde{q}f \tag{3.4 - 10}$$

The order of f is such that q is proper and its roll-off frequency low enough to satisfy the robust stability constraint (3.4–3). With the filter, $\tilde{\eta}$ and $\tilde{\epsilon}$ become

$$\tilde{\eta} = \tilde{p}q = \tilde{p}\tilde{q}f \tag{3.4 - 11}$$

$$\tilde{\epsilon} = 1 - \tilde{p}q = 1 - \tilde{p}\tilde{q}f \tag{3.4 - 12}$$

  The function of f is to detune the controller, to sacrifice performance (increase $|\tilde{\epsilon}|$) for robustness (decrease $|\tilde{\eta}|$). Clearly, an open-loop stable system can be made arbitrarily robust, simply by turning off the controller. Thus (3.4–3) can always be satisfied. However, because of the associated loss of performance (3.4–5) will generally not be met. To design the filter for robust performance (3.4–5) is difficult. Indeed, there might not exist *any* filter to satisfy (3.4–5) for the particular choices of weights w and $\bar{\ell}_m$. In our design procedures we will tend to relax performance (decrease w) in exchange for simple search techniques for the appropriate filter. This will yield robust stability and reasonable, albeit not optimal, performance.

3.4.2   Two-Degree-of-Freedom Design

For the two-controller structure in Fig. 3.3-1 we derived the relationship

$$e = \frac{1 - \tilde{p}q_d}{1 + q_d(p - \tilde{p})}d - \left[1 - \frac{pq_r}{1 + q_d(p - \tilde{p})}\right]r \tag{3.3 - 6}$$

which becomes for $p = \tilde{p}$

$$e = (1 - pq_d)d - (1 - pq_r)r \tag{3.3 - 7}$$

The design problems for nominal performance are independent. If the H_∞ performance objective is chosen then q_r and q_d are designed to satisfy

$$\|(1 - \tilde{p}q_r)w_r\|_\infty < 1 \qquad\qquad (3.4 - 13)$$

$$\|(1 - \tilde{p}q_d)w_d\|_\infty < 1 \qquad\qquad (3.4 - 14)$$

Robust stability depends only on q_d

$$\|\tilde{p}q_d\bar{\ell}_m\|_\infty < 1 \qquad\qquad (3.4 - 15)$$

Because the "prefilter" q_r is not in the feedback path, it has no effect on stability.

Consider now the design for robust performance. Let us first design q_d for robust performance just as described in the last section. The robust performance specification for disturbance rejection is

$$\|\epsilon w_d\|_\infty < 1 \qquad \forall p \in \Pi \qquad\qquad (3.4 - 16)$$

which is equivalent to

$$|\tilde{\epsilon}w_d| + |\tilde{\eta}\bar{\ell}_m| < 1 \qquad \forall\omega \qquad\qquad (3.4 - 17)$$

Next let us design q_r for robust tracking performance. The performance specification is

$$\left|\left[1 - \frac{\tilde{p}q_r(1 + \ell_m)}{1 + \tilde{p}q_d\ell_m}\right]w_r\right| < 1 \qquad \forall \ell_m \in \Lambda_m, \quad \forall\omega \qquad\qquad (3.4 - 18)$$

where the set $\Lambda_m(i\omega)$ is defined by

$$\Lambda_m(i\omega) = \{\ell_m(i\omega): \quad |\ell_m(i\omega)| < \bar{\ell}_m(\omega)\} \qquad\qquad (3.4 - 19)$$

From (3.4–18) we derive the condition

$$\Leftrightarrow |[1 + \tilde{p}q_d\ell_m - \tilde{p}q_r(1 + \ell_m)] \cdot w_r| < |1 + \tilde{p}q_d\ell_m| \qquad \forall \ell_m \in \Lambda_m, \forall\omega$$

$$\Leftarrow |[1 + \tilde{p}q_d\ell_m - \tilde{p}q_r(1 + \ell_m)] \cdot w_r| + |\tilde{p}q_d\ell_m| < 1 \qquad \forall \ell_m \in \Lambda_m, \forall\omega$$

$$\Leftarrow |(1 - \tilde{p}q_r)w_r| + |\tilde{p}q_d\bar{\ell}_m|\left[1 + \left|1 - \frac{q_r}{q_d}\right|w_r\right] < 1 \qquad \forall\omega \qquad (3.4 - 20)$$

Note that because of the last two steps in the derivation, (3.4–20) is only *sufficient* for robust setpoint tracking performance. A necessary and sufficient condition can be derived via the Structured Singular Value, which will be introduced in the discussion of multivariable systems.

When $q_r = q_d$, (3.4–20) reduces to the necessary and sufficient condition for robust performance derived in Sec. 2.6. The first term in (3.4–20) expresses nominal performance. The second term is proportional to the multiplicative uncertainty $\bar{\ell}_m$ and to the disturbance controller magnitude — i.e., the *disturbance* controller q_d affects the robust *setpoint tracking* performance. Thus the designs of q_d and q_r are *not* independent when the objective is robust performance. For the proper tradeoff they have to be designed jointly rather than sequentially as we did here.

3.4.3   Design in the Presence of Measurement Device Dynamics

If the measurement device dynamics can be identified independent from the process dynamics (for example a time delay caused by a chromatograph employed for composition analysis) and if the real process output rather than the measured one is to be controlled then the IMC structure should be modified as shown in Fig. 3.1-1A. Through the addition of the measuring device the closed-loop relationship becomes

$$y = \frac{1 - \tilde{p}'q}{1 + q(p' - \tilde{p}')}d + \frac{pq}{1 + q(p' - \tilde{p}')}r \qquad (3.4-21)$$

where the augmented plant and model are defined as

$$p' = pp_m \qquad (3.4-22)$$

$$\tilde{p}' = \tilde{p}\tilde{p}_m \qquad (3.4-23)$$

For $p = \tilde{p}, p_m = \tilde{p}_m$ we find from (3.4–21)

$$e = (1 - p'q)d - (1 - pq)r \qquad (3.4-24)$$

The design of q for nominal tracking performance is unaffected by the measuring device. This should be intuitively obvious from the "open-loop nature" of IMC. For nominal disturbance-rejection performance q must be designed for the augmented plant p'.

Note that even when d and r are identical signals (e.g., steps) the optimal q's for setpoint tracking and disturbance rejection are generally different because of the measurement device dynamics. Thus a two-degree-of-freedom controller is needed for optimal performance.

The robust stability and robust disturbance rejection problem are in the standard form studied in Sec.3.4.1. Using similar arguments as in Sec.3.4.2 a sufficient condition for robust tracking can be derived. A necessary and sufficient condition can be expressed in terms of the Structured Singular Value which we will discuss later.

3.5 Summary

In the absence of model uncertainty, the IMC structure (Fig. 3.1-1) is effectively open-loop. Therefore, it can only be implemented on open-loop stable processes. For a stable process and a perfect model, closed-loop stability is assured by a stable IMC controller. The performance achievable with IMC is identical to that achievable with any classic feedback controller selected such that

$$c = \frac{q}{1 - \tilde{p}q} \qquad (3.2 - 2)$$

and vice versa.

The discussion of the IMC structure revealed the following factors which prevent perfect control: NMP plant characteristics, input constraints and model uncertainty.

A two-step design procedure is proposed. In the first step the controller $\tilde{q}$ is selected for good response without regard for constraints and model uncertainty. Often $\tilde{q}$ will be chosen to be H_2-optimal:

$$\min_{\tilde{q}} \|(1 - \tilde{p}\tilde{q})w\|_2 \qquad (3.4 - 7)$$

In the second step the controller $\tilde{q}$ is augmented by a low-pass filter f

$$q = \tilde{q}f \qquad (3.4 - 10)$$

to provide the rolloff necessary for robustness and milder action of the manipulated variables. The filter can be adjusted on-line if necessary.

3.6 References

An excellent historical review of the origins of the IMC structure which has as its special characteristic a model in parallel with the plant is provided by Frank (1974). It appears to have been discovered by several people simultaneously in the late fifties. Newton, Gould and Kaiser (1957) use the structure to transform the closed loop system into an open loop one so that the results of Wiener can be applied to find the H_2-optimal controller q. The Smith Predictor (Smith, 1957) also contains a process model in parallel with the plant (see Chap.6). Independently Zirwas (1958) and Giloi (1959) suggested the predictor structure for the control of systems with time delay.

Frank (1984) first realized the general power of this structure, fully exploits it and extends the work by Newton et al. (1957) to handle persistent disturbances and setpoints. Youla et al. (1976) extend the convenient "q-parametrization" of the controller c to handle unstable plants. In 1981 Zames ushers in the era of H_∞-control utilizing for his developments the q-parametrization. At present it is used in all robust controller design methodologies.

Unaware of all these developments the process industries both in France (Richalet et al., 1978) and in the United States (Cutler and Ramaker, 1979; Prett and Gillette, 1979) exploit the advantages of the parallel model/plant arrangement, Brosilow (1979) utilizes the Smith predictor parametrization to develop a robust design procedure and Garcia and Morari (1982, 1985 a,b) unify all these concepts and refer to the structure in Fig. 3.1-1 as Internal Model Control (IMC) because the process *model* is explicitly an *internal* part of the controller.

3.2.3. Alternate ways to deal with the windup problem are discussed by Doyle, Smith & Enns (1987). A general anti-windup scheme was proposed recently by Hanus, Kinnaert & Henrotte (1987).

Chapter 4

SISO IMC DESIGN FOR STABLE SYSTEMS

The IMC design procedure consists of two steps

STEP 1: Nominal Performance

$\tilde{q}$ is selected to yield a "good" system response for the input(s) of interest, without regard for constraints and model uncertainty.

STEP 2: Robust Stability and Performance

$\tilde{q}$ is augmented by a low-pass filter f $(q = \tilde{q}f)$ to achieve robust stability and robust performance.

We will proceed with a detailed discussion of these two steps.

4.1 Nominal Performance

For SISO systems we will generally choose $\tilde{q}$ such that it is H_2 optimal for a particular input v where $v = d$ or $v = -r$. Thus $\tilde{q}$ has to solve

$$\min_{\tilde{q}} \|e\|_2 = \min_{\tilde{q}} \|(1 - \tilde{p}\tilde{q})v\|_2 \qquad (4.1-1)$$

subject to the constraint that $\tilde{q}$ is stable and causal. Problem (4.1–1) reaches its absolute minimum (zero) for

$$\tilde{q} = \frac{1}{\tilde{p}} \qquad (4.1-2)$$

However, the model inverse is an acceptable solution only for MP systems. For NMP systems the exact inverse (4.1–2) is unstable and/or noncausal. The objective function cannot be made zero and an "approximate inverse" of $\tilde{p}$ has to be found such that the weighted 2-norm of the sensitivity function is minimized. Note that for NMP systems the optimal solution depends on the weight (the input v), while for MP systems (4.1–2) is optimal *independent* of the weight.

4.1.1 H_2-Optimal Control

In the next chapter we will derive a fairly simple general analytic procedure to solve the problem (4.1–1). In this chapter we will just state the solution and demonstrate the results for some important special cases.

Theorem 4.1-1. *Assume that $\tilde{p}$ is stable. Factor $\tilde{p}$ into an allpass portion $\tilde{p}_A$ and a MP portion $\tilde{p}_M$.*

$$\tilde{p} = \tilde{p}_A \tilde{p}_M \qquad (4.1-3)$$

so that $\tilde{p}_A$ includes all the RHP zeros and delays of $\tilde{p}$ and

$$|\tilde{p}_A(i\omega)| = 1 \qquad \forall \omega \qquad (4.1-4)$$

In general $\tilde{p}_A$ has the form

$$\tilde{p}_A(s) = e^{-s\theta} \prod_i \frac{-s + \zeta_i}{s + \zeta_i^H} \qquad \mathrm{Re}(\zeta_i),\ \theta > 0 \qquad (4.1-5)$$

where the superscript H denotes complex conjugate.

  Factor the input v similarly

$$v = v_A v_M \qquad (4.1-6)$$

The controller $\tilde{q}$ which solves (4.1–1) is given by

$$\tilde{q} = (\tilde{p}_M v_M)^{-1} \left\{ \tilde{p}_A^{-1} v_M \right\}_* \qquad (4.1-7)$$

where the operator $\{\cdot\}_*$ denotes that after a partial fraction expansion of the operand all terms involving the poles of $\tilde{p}_A^{-1}$ are omitted.

In general the optimal controller $\tilde{q}$ is not proper. This is of no concern at this point because it is to be augmented by a low-pass filter in Step 2 of the IMC design procedure.

Note that the objective function (4.1–1) and the optimal controller (4.1–7) are unaffected by NMP elements in v. For example, for the design of $\tilde{q}$ and the optimal closed-loop performance it makes no difference if a step disturbance enters directly at the plant output or after passing through a time delay. We will learn in Chap.6 that this is exactly the situation when feedforward control can be used effectively.

The analytic expression (4.1–7) allows us to make some general observations regarding the pole-zero configuration of the closed-loop system with the H_2-optimal controller $\tilde{q}$.

Theorem 4.1-2. *The H_2-optimal complementary sensitivity function*

$$\bar{\eta} = \tilde{p}\tilde{q} = \tilde{p}_A v_M^{-1} \left\{ \tilde{p}_A^{-1} v_M \right\}_* \qquad (4.1-8)$$

has the following properties:

1. *The poles of $\bar{\eta}$ are at the mirror images of the plant RHP zeros and at the zeros of v_M.*

2. *$\bar{\eta}$ has RHP zeros at the plant RHP zeros. If $\bar{\eta}$ has any additional RHP zeros they are at mirror images of zeros of v_M.*

3. *If v_M has no finite zeros, then $\bar{\eta}$ has no RHP zeros except those of the plant. This implies that $\tilde{q}$ is MP.*

Proof.

1. The potential poles of $\bar{\eta}$ are the poles of $\tilde{p}_A$, the zeros of v_M and the poles of $\{\tilde{p}_A^{-1}v_M\}_*$, which are the poles of v_M. However, the poles of v_M are cancelled by zeros of v_M^{-1} premultiplying $\{\cdot\}_*$.

2. Assume that $\tilde{q}_1$ is stable and minimizes

$$\|(1 - \tilde{p}\tilde{q}_1)v\|_2 \qquad (4.1-9)$$

The RHP zeros of $\tilde{p}$ appear in $\bar{\eta} = \tilde{p}\tilde{q}_1$ in the form of an allpass. Define a new plant $\hat{p} = \tilde{p}\tilde{q}_1$. The stable $\hat{q}_2$ which minimizes

$$\|(1 - \hat{p}\hat{q}_2)v\|_2 \qquad (4.1-10)$$

is unity (if it were not, $\tilde{q}_1$ could not be optimal). The RHP zeros of $\hat{p}$ have to appear in $\hat{\eta} = \hat{p}\hat{q}_2 = \hat{p} = \bar{\eta}$ in the form of an allpass. Because of 1. all RHP zeros in addition to the plant RHP zeros have to be located at the mirror images of the zeros of v_M.

3. Follows directly from 2. □

RHP zeros are generally considered as "bad" for performance. However, 2. implies that depending on the input v_M the optimal controller $\tilde{q}$ can sometimes *add* RHP zeros.

Example 4.1-1. $\tilde{p} = \tilde{p}_A = \frac{-s+2}{s+2}, v_M = \frac{s+1}{s}$

$$\bar{q} = \frac{s}{s+1}\left\{\frac{s+2}{-s+2}\frac{s+1}{s}\right\}_* = \frac{s}{s+1}\left\{-1 + \frac{1}{s} + \frac{6}{-s+2}\right\}_* = \frac{-s+1}{s+1}$$

$$\tilde{\eta} = \frac{(-s+2)(-s+1)}{(s+2)(s+1)} \qquad (4.1-11)$$

The controller $\tilde{q}$ added a RHP zero at $(+1,0)$. As predicted by Thm. 4.1-2 it is at the mirror image of the disturbance zero at $(-1,0)$. □

Table 4.1-1. H_2–optimal controller $\tilde{q}$ for some typical input forms ($p = \tilde{p}$).

Input v_M	Controller $\tilde{q}$	Compl. Sensitivity $\bar{\eta}$
$\frac{1}{s}$	p_M^{-1}	p_A
$\frac{1}{\tau s+1}$	$p_M^{-1} \cdot p_A^{-1}\big|_{s=-\frac{1}{\tau}}$	$p_A \cdot p_A^{-1}\big|_{s=-\frac{1}{\tau}}$
$\frac{1}{s(\tau s+1)}$	$p_M^{-1}\left[1 + (1 - p_A^{-1}\big|_{s=-\frac{1}{\tau}})\tau s\right]$	$p_A\left[1 + (1 - p_A^{-1}\big|_{s=-\frac{1}{\tau}})\tau s\right]$
$\frac{1}{s^2}$	$p_M^{-1}(1 - s \cdot \frac{dp_A}{ds}\big|_{s=0})$	$p_A(1 - s \cdot \frac{dp_A}{ds}\big|_{s=0})$

4.1.2 Design for Specific Input Forms

The evaluation of (4.1–7) for specific inputs v_M yields the results shown in Table 4.1-1.

Consider, for example, the special case of a step input $v_M = s^{-1}$

$$\tilde{q} = p_M^{-1}s\left\{p_A^{-1}s^{-1}\right\}_* = p_M^{-1}sp_A^{-1}(0)s^{-1} = p_M^{-1} \qquad (4.1 - 12)$$

which agrees with the corresponding entry in Table 4.1-1. Consider next the input $v_M = (\tau s + 1)^{-1}$

$$\tilde{q} = p_M^{-1}(\tau s + 1)\left\{p_A^{-1}(\tau s + 1)^{-1}\right\}_* = p_M^{-1}(\tau s + 1) \cdot p_A^{-1}\big|_{s=-\frac{1}{\tau}} \cdot (\tau s + 1)^{-1}$$

$$= \left[p_M \cdot p_A\big|_{s=-\frac{1}{\tau}}\right]^{-1} \qquad (4.1 - 13)$$

This result also agrees with Table 4.1-1. The derivation of $\tilde{q}$ for the other inputs is left as an exercise.

Tables 4.1-2 and 4.1-3 illustrate the application of the general formulas to the specific cases of a system with delay and a system with a single RHP zero. The following observations can be made:

1. For the ISE to be bounded, the error has to vanish as $t \to \infty$. This implies that the optimal controllers for A and C have to generate a Type 1 system. The optimal controller for D has to lead to a Type 2 system. Indeed we find for A and C.

Type 1:

$$\lim_{s \to 0} p\tilde{q} = 1$$

and for D

Type 2:

$$\lim_{s \to 0} p\tilde{q} = 1 \quad \text{and} \quad \lim_{s \to 0} \frac{d}{ds}(p\tilde{q}) = 0$$

2. From the plots of $|\bar{\epsilon}|$ it is evident that the performance of the optimal controllers can be quite bad for inputs other than the one they are designed for. In particular, high frequency inputs tend to be amplified. For cases A and B the amplification is less or equal to 2, for cases C and D it increases with frequency.

3. For "sluggish" inputs (cases C and D) the high performance is paid for with poor robustness as the large values of $|\bar{\eta}|$ demonstrate.

4.1.3 Minimum Error Norm for Step Inputs

We learned that for MP systems the H_2-norm of the error can be made arbitrarily small. For stable NMP systems subjected to steps, a simple formula can be derived to compute the minimum error norm. First, we will show that the contributions of different NMP elements to the ISE are additive. Then, we will compute the ISE arising from a single RHP zero.

Theorem 4.1-3. *Let the plant be given by $p = p_1 p_2 p_M$ where both p_1 and p_2 are allpass. Then for step inputs*

$$\min_{\tilde{q}} \|e\|_2^2 = \|(1 - p_1 p_2)s^{-1}\|_2^2 = \|(1 - p_1)s^{-1}\|_2^2 + \|(1 - p_2)s^{-1}\|_2^2 \qquad (4.1 - 14)$$

The proof of this theorem can be found in Sec. 5.2.3.

If $p_A = e^{-s\theta}$ then the effect of any control action is not felt before $t = \theta$. The optimal $\tilde{q}$ reduces the error to zero for $t > \theta$. Thus, for a unit step input

$$\min_{\tilde{q}} \|e\|_2^2 = \theta \qquad (4.1 - 15)$$

For RHP zeros the minimum error norm can be found via the calculus of residues.

Theorem 4.1-4. *Let $p_A = \frac{-s+\zeta}{s+\zeta^H}$ then for a unit step input*

$$\min_{\tilde{q}} \|e\|_2^2 = \frac{2\text{Re}(\zeta)}{|\zeta|^2} \qquad (4.1 - 16)$$

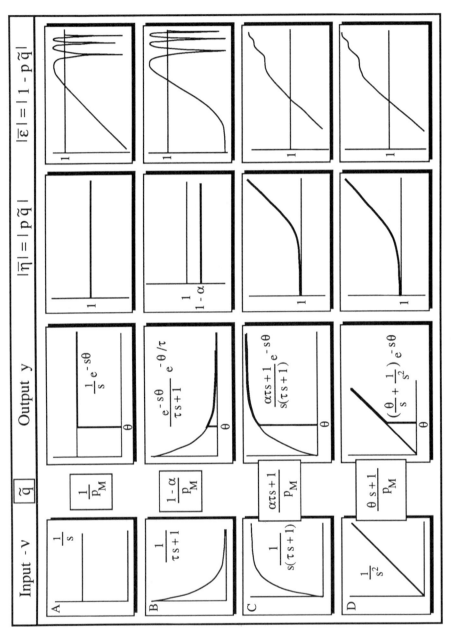

Table 4.1-2. Optimal controller $\tilde{q}$, output y, sensitivity $\bar{\epsilon}$ and complementary sensitivity $\bar{\eta}$ for different inputs v. The system is $p = \tilde{p} = p_M e^{-s\theta}$. $\alpha = 1 - e^{-\theta/\tau}$.

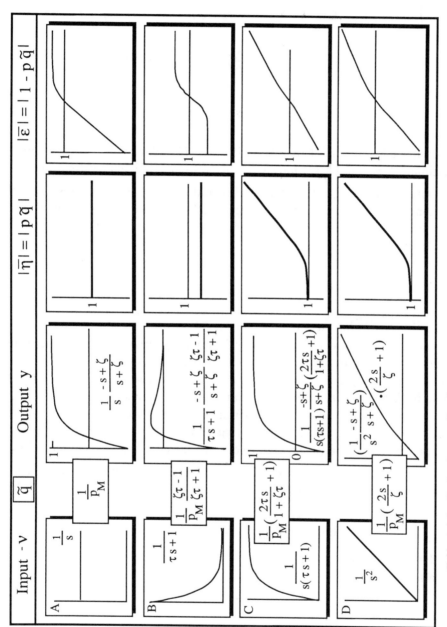

Table 4.1-3. Optimal controller $\tilde{q}$, output y, sensitivity $\bar{\epsilon}$ and complementary sensitivity $\bar{\eta}$ for different inputs v. The system is $\tilde{p} = p = p_M \frac{-s+\zeta}{s+\zeta}$.

Proof.

$$\|e\|_2^2 = \frac{1}{2\pi i} \int_{-i\infty}^{+i\infty} |(1-p_A)s^{-1}|^2 \, ds = \frac{1}{2\pi i} \int_{-i\infty}^{+i\infty} \left| 1 - \frac{-s+\zeta}{s+\zeta^H} \right|^2 |s^{-1}|^2 ds$$

$$= \frac{1}{2\pi i} \int_{-i\infty}^{+i\infty} \left[2 - \frac{-s+\zeta}{s+\zeta^H} - \frac{s+\zeta^H}{-s+\zeta} \right] |s^{-1}|^2 ds$$

$$= \text{Res} \left[\frac{s+\zeta^H}{-s+\zeta} \frac{1}{s^2} \right] \Big|_{s=\zeta} = \frac{2\text{Re}(\zeta)}{|\zeta|^2}$$

$\square$

Theorems 4.1-3 and 4.1-4 together with (4.1–15) allow one to easily compute the minimum ISE ($\min_{\tilde{q}} \|e\|_2^2$) for any stable system subjected to step inputs.

It is interesting to note that the detrimental effect of RHP zeros is inversely proportional to their distance from the origin. Zeros close to the origin are very bad for closed-loop performance, zeros which are far away from the origin have a negligible effect.

4.2   The IMC Filter

4.2.1   Filter Form

In general, the complementary sensitivity functions $\bar{\eta}$ obtained for the H_2-optimal controller $\tilde{q}$ exhibit undesirable high frequency behavior (see Tables 4.1-2 and 4.1-3). For robustness $\tilde{q}$ has to be augmented by a low-pass filter f ($q = \tilde{q}f$). In principle both the structure and the parameters of f should be determined such that an optimal compromise between performance and robustness is reached. To simplify the design task we fix the filter structure and search over a small number of filter parameters (usually just one) to obtain desired robustness characteristics.

It is logical to choose f such that the closed-loop system retains its asymptotic tracking properties (Type m). For systems of Type m, f has to satisfy (3.3–10).

$$\lim_{s \to 0} \frac{d^k}{ds^k}(1 - \tilde{p}\tilde{q}f) = 0 \quad 0 \le k < m \qquad (4.2-1)$$

If $\tilde{q}$ is designed to satisfy (4.2–1) for $f = 1$ then the conditions on f for (4.2–1) to be satisfied are

$$\lim_{s \to 0} \frac{d^k}{ds^k}(1 - f) = 0 \quad 0 \le k < m \qquad (4.2-2)$$

Thus

Type 1:

$$f(0) = 1 \qquad (4.2-3)$$

Type 2:

$$f(0) = 1 \quad \text{and} \quad \lim_{s \to 0} \frac{df}{ds} = 0 \qquad (4.2-4)$$

Typically we will use one-parameter filters with unity steady-state gain of the form

$$f = (\beta_{m-1}s^{m-1} + \ldots \beta_1 s + 1)\frac{1}{(\lambda s + 1)^n} \qquad (4.2-5)$$

where λ is the adjustable filter parameter, n is selected large enough to make q proper and β_i is chosen to satisfy (4.2-2). The simplest filters of the form (4.2-5) which satisfy (4.2-2) are

Type 1:

$$f(s) = \frac{1}{(\lambda s + 1)^n} \qquad (4.2-6)$$

Type 2:

$$f(s) = \frac{n\lambda s + 1}{(\lambda s + 1)^n} \qquad (4.2-7)$$

4.2.2   A Qualitative Interpretation of the Function of the Filter

With the filter f and the H_2-optimal $\tilde{q}$ the complementary sensitivity function becomes

$$\tilde{\eta} = \frac{y}{r} = \tilde{p}\tilde{q}f \qquad (4.2-8)$$

Examples of $\tilde{\eta} = \tilde{p}\tilde{q}$ for typical inputs were shown without the appropriate filter in Table 4.1-1. Note that in *all* cases the poles of $\tilde{\eta}$, which are the closed-loop system poles, are the poles of the system allpass p_A and the poles of the filter at $-\lambda^{-1}$. Adjusting λ is equivalent to adjusting the speed of the closed-loop response. When $\lambda = 0$ the system response is H_2 optimal. When $\lambda \ll \zeta_i^{-1}$ then the poles $-\zeta_i$ of the allpass will dominate the response and the filter will have little effect. When $\lambda \gg \zeta_i^{-1}$ the filter dominates the response and the closed-loop time constant becomes equal to λ. This is illustrated in Fig. 4.2-1.

Next we will describe techniques to adjust λ on the basis of model uncertainty information. However, because of its directness and intuitive appeal it is conceivable that λ is left for on-line adjustment by the operating personnel. With a good model a high speed of response can be demanded from the system. When the knowledge about the system — i.e., the model, is poor then the speed of response has to be decreased.

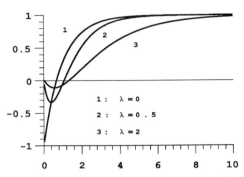

Figure 4.2-1. Step response of $\tilde{p}\tilde{q}f = (-s+1)(s+1)^{-1}(\lambda s+1)^{-1}$.

4.3 Robust Stability

4.3.1 Norm-Bounded Uncertainty Regions

Theorem 2.5-1 states that the closed-loop system is robustly stable if and only if

$$\|\tilde{\eta}\bar{\ell}_m\|_\infty = \sup_\omega |\tilde{\eta}\bar{\ell}_m(\omega)| < 1 \qquad (4.3-1)$$

For IMC $\tilde{\eta} = \tilde{p}q = \tilde{p}\tilde{q}f$ and (4.3–1) becomes

$$|\tilde{p}\tilde{q}f\bar{\ell}_m(\omega)| < 1 \qquad \forall\omega \qquad (4.3-2)$$

Corollary 4.3-1. *Assume that the family of stable plants* Π *is described by*

$$\Pi = \left\{ p : \left| \frac{p(i\omega) - \tilde{p}(i\omega)}{\tilde{p}(i\omega)} \right| < \bar{\ell}_m(\omega) \right\} \qquad (4.3-3)$$

and that $\tilde{q}$ *is stable. Then the closed-loop system is robustly stable if and only if the IMC filter satisfies*

$$|f| < \frac{1}{|\tilde{p}\tilde{q}\bar{\ell}_m|} \qquad \forall\omega \qquad (4.3-4)$$

An f can always be found such that (4.3–4) is satisfied. However $|f|$ small implies $|\tilde{\eta}|$ small and thus poor performance. If performance requirements have to be met, the uncertainty expressed through $\bar{\ell}_m$ has to be limited for robust stability to be possible. As a simple performance specification let us require the closed-loop system to be of Type 1 — i.e., $\tilde{p}(0)\tilde{q}(0) = f(0) = 1$. Then we find directly from (4.3–4) the following corollary.

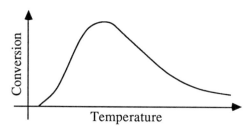

Figure 4.3-1. Conversion vs. Temperature for CSTR with exothermic reversible reaction.

Corollary 4.3-2. *Assume $\bar{\ell}_m(\omega)$ is continuous. There exists a filter f such that the closed-loop system is Type 1 and robustly stable for the family Π described by (4.3-3) if and only if $\bar{\ell}_m(0) < 1$.*

Corollary 4.3-2 implies that it is possible to design a feedback controller which guarantees a zero tracking error for steps if and only if the process steady-state gain uncertainty satisfies

$$\left| \frac{p(0) - \tilde{p}(0)}{\tilde{p}(0)} \right| < 1 \qquad (4.3-5)$$

i.e., if the gain error does not exceed 100% or equivalently if the gain does not change sign. When the gain changes sign a positive feedback loop results and the system becomes unstable.

Example 4.3-1. Consider an exothermic continuous flow stirred tank reactor (CSTR) where conversion is related to reactor temperature as shown in Fig. 4.3-1. Assume that conversion is controlled by adjusting the reactor temperature. The steady-state gain of the linearized reactor model is equal to the slope of the curve in Fig. 4.3-1 and depends on the operating point. It is positive for low temperatures and negative for high temperatures. According to Cor. 4.3-2 it is not possible to design a controller for Type 1 performance which is stable over the whole operating range.   □

The uncertainty bound $\bar{\ell}_m$ generally increases with frequency. How small $|f|$ has to be made to satisfy (4.3-4) for robust stability depends on the H_2-optimal complementary sensitivity function for the nominal system $|\tilde{p}\tilde{q}|$ and the uncertainty bound $\bar{\ell}_m$. Tables 4.1-2 and 4.1-3 show plots of $|\tilde{p}\tilde{q}|$ with the optimal controllers $\tilde{q}$ designed for different inputs. For the same uncertainty, cases C and D require more drastic filter action than A and B because $|\tilde{\eta}|$ increases with frequency.

From another point of view we can say that designing controllers specifically

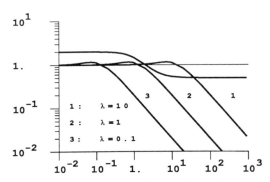

Figure 4.3-2. Robust stability bound (4.3–4) and Type 2 filter f for different values of the filter time constant λ.

for slowly rising inputs (C and D) might not bring any performance advantages over designing controllers for steps (A) unless the model uncertainty is sufficiently small such that large values of $|\tilde{\eta}|$ can be tolerated at high frequencies.

The filter designer chooses first the appropriate form of the filter for Type 1 (4.2–6) or Type 2 (4.2–7) behavior and then increases the filter time constant until (4.3–4) is satisfied (Fig. 4.3-2). In some cases it can be prudent to adjust the filter order n to meet the constraint (4.3–4) with the least conservatism.

4.3.2   General Uncertainty Regions

The controller c is to be designed to guarantee the stability of a family Π of plants surrounding a nominal plant $\tilde{p}$. All members of the family Π are assumed to be open-loop stable. At each frequency ω the magnitude and phase of the plant lies in a region $\pi(\omega)$ on the Nyquist plane. The union of the regions forms a *Nyquist band* and constitutes the family Π.

At each frequency ω the controller $c(i\omega)$ is a complex number. The multiplication of a region with a number ($\pi(\omega)c(i\omega)$) stretches and rotates the region (Fig. 4.3-3). These operations can be carried out easily numerically and the resulting band $\pi(\omega)c(i\omega)$ can be displayed on a graphics screen.

Theorem 4.3-1. *The closed-loop system is robustly stable for the family Π if and only if the Nyquist band $\pi(\omega)c(i\omega)$ does not encircle or cover (-1,0).*

For the controller c the IMC parametrization is employed

$$c = \frac{\tilde{q}f}{1 - \tilde{p}\tilde{q}f} \qquad (4.3-6)$$

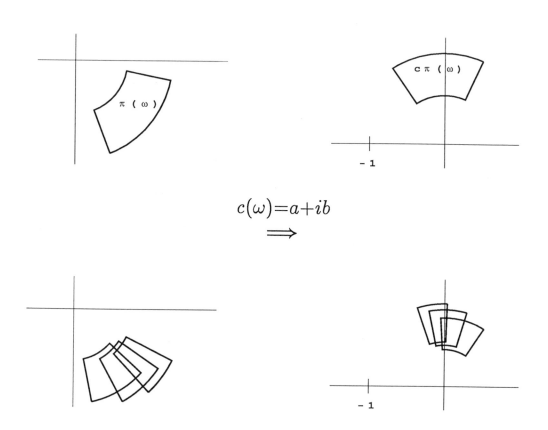

Figure 4.3-3. Transformation of region $\pi(\omega)$ by multiplication with complex number $c(i\omega)$.

It is intuitively obvious and it also follows from our discussions in the previous section that there always exists some filter (e.g., the trivial case of $f = 0$) for which robust stability can be achieved. In order for the closed-loop system to be robustly stable and of Type 1 the family of steady-state gains $\pi(0)$ has to be restricted. The following corollary follows directly from Cor. 4.3-2 and the subsequent discussion.

Corollary 4.3-3. *A controller c exists such that the closed loop system is Type 1 and robustly stable for the family Π if and only if all the plant gains in $\pi(0)$ have the same sign.*

Because the nominal plant $\tilde{p}$ appears in the controller c it has a great effect on the closed stability and performance of the family Π. In Sec. 4.3.1. $\tilde{p}$ was naturally defined as the center of the disk shaped uncertainty regions. For the general uncertainty region considered in this section there is no natural center and $\tilde{p}$ has to be selected judiciously. For example, $\tilde{p}(0)$ should be chosen as the center of the segment $\pi(0)$ of the real line. If the conditions of Cor. 4.3-3 are satisfied and if the IMC parametrization (4.3–6) is used for this $\tilde{p}$ then the existence of a Type 1 filter (4.2–6) is guaranteed for which the system will be robustly stable. This follows directly from Cor. 4.3-2.

The advantage of using general uncertainty regions is that tighter stability bounds can be established than when the regions are approximated by disks. The disadvantage is the need for a graphical rather than an analytic/numerical procedure.

4.4   Robust Performance

4.4.1   Norm-Bounded Uncertainty Regions

Theorem 2.6-1 states the following necessary and sufficient condition for robust performance

$$|\tilde{\eta}\bar{\ell}_m| + |\tilde{\epsilon}w| < 1 \quad \forall \omega \tag{4.4-1}$$

For IMC $\tilde{\eta} = \tilde{p}\tilde{q}f$ and the following corollary results.

Corollary 4.4-1. *Assume that the family Π of stable plants is described by*

$$\Pi = \left\{ p : \left| \frac{p(i\omega) - \tilde{p}(i\omega)}{\tilde{p}(i\omega)} \right| < \bar{\ell}_m(\omega) \right\} \tag{4.4-2}$$

Then the closed loop system will meet the performance specification

$$\|\epsilon w\|_\infty = \sup_\omega |\epsilon w| < 1 \qquad \forall p \in \Pi \tag{4.4-3}$$

if and only if

$$|\tilde{p}\tilde{q}f\bar{\ell}_m| + |(1 - \tilde{p}\tilde{q}f)w| < 1 \qquad \forall \omega \qquad (4.4-4)$$

Increasing the filter time constant λ will tend to decrease the first term and increase the second term in (4.4-4). Thus, depending on $\bar{\ell}_m$ and w there might not be any λ for which (4.4-4) is satisfied. On the other hand, assume that the controller/filter is selected such that the system is of Type 1 or higher ($\tilde{p}\tilde{q}f(0) = 1$) and that $\bar{\ell}_m(0) < 1$. Then "robust performance at $\omega = 0$" can be achieved for any w.

Corollary 4.4-2. *There exists a filter f such that the closed-loop system exhibits "robust performance at $\omega = 0$" for any weight w if and only if $\bar{\ell}_m(0) < 1$.*

Corollary 4.4-2 simply implies that if the system is robustly stable with a controller c with integral action (Type 1) then the steady-state performance is perfect even when there is modelling error.

For IMC the shapes of $\tilde{\epsilon}$ and $\tilde{\eta}$ are essentially fixed by the H_2-optimal controller $\tilde{q}$ and the filter f which has only a single adjustable parameter. With only one degree of freedom available in the controller it is not always meaningful to define a very complex performance weight which has to be satisfied. Thus, we often select w to be a constant. This implies that the filter parameter has to be selected such that the maximum peak of the sensitivity function does not exceed w^{-1} for all plants in the uncertainty set. Because this objective is generally satisfied for the trivial choice $\lambda = \infty$ — i.e., open loop — the following simple filter design procedure is proposed.

1. Check that (4.4-4) can be satisfied at $\omega = 0$ for $f(0) = 1$ and $w = 0$. This guarantees the existence of a filter for Type 1 performance and robust stability.

2. Select a constant performance weight w and adjust the filter time constant so that (4.4-4) is met as an equality. (In the "worst case" the maximum peak of ϵ is exactly equal to w^{-1}.) Some typical iterations in the design procedure are illustrated in Fig. 4.4-1.

For general nonconstant weights w it is possible that (4.4-4) cannot be satisfied for any λ. Then either the filter structure is too simplistic to meet the performance specifications or the specifications are too tight for the uncertainty which is present. After the filter parameter λ has been determined a bound on the error norm can be established (see Sec. 2.6) for all plants in the family Π subjected to a specific input v.

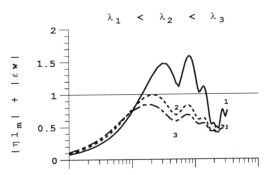

Figure 4.4-1. Robust performance bound (4.4–4) for different values of the filter time constant λ.

Though in the context of process control it is rarely justified to use anything more complicated than a one-parameter filter (4.2–6 or 4.2–7) our design procedure is in no way limited to this case. More complex filters have the potential of satisfying the robust stability constraint with a smaller loss of performance but at the expense of a more complicated search procedure for the appropriate filter parameters.

Also, in general, w can be selected in a manner which reflects the performance specifications more accurately than is possible with a constant w. It is often meaningful to base w on the H_2-optimal $\bar{e}$ obtained for the typical system input:

$$w = \frac{\alpha}{|\bar{e}|} \qquad \alpha > 0 \qquad\qquad\qquad (4.4 - 5)$$

The parameter α is selected by the designer and can be interpreted as follows. If the robust performance objective (4.4–3) with the weight (4.4–5) is met then

$$|\epsilon| < \frac{|\bar{e}|}{\alpha} \qquad \forall \omega, \quad \forall p \in \Pi \qquad\qquad (4.4 - 6)$$

and for a specific input v

$$\|e\|_2^2 = \frac{1}{2\pi} \int_{-\infty}^{\infty} |\epsilon|^2 |v|^2 d\omega$$

$$\leq \frac{1}{\alpha^2} \frac{1}{2\pi} \int_{-\infty}^{\infty} |\bar{e}|^2 |v|^2 d\omega \qquad \forall p \in \Pi \qquad (4.4 - 7)$$

Equation (4.4–7) implies that for all plants $p \in \Pi$ the error norm exceeds the minimum error norm by not more than a factor α^{-1}.

The choice (4.4–5) is not always practical. For example, for a system with a delay, w would be unbounded at nonzero frequencies (see Table 4.1-2) and the

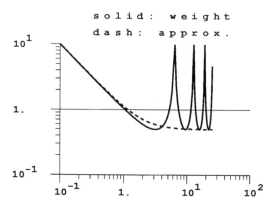

Figure 4.4-2. Weight and approximate weight for a system with time delay subject to a step input.

robust performance specifications could not be met by any controller. In this case w can be chosen as a reasonable approximation of (4.4–5). Let us assume that we are interested in a system with time delay subjected to step inputs — i.e., $\bar{\epsilon} = 1 - e^{-\theta s}$. Then a meaningful weight can be derived from (4.4–5) when a Padé approximation is used for the time delay in $\bar{\epsilon}$.

$$w = \alpha \left| 1 - \frac{-\theta s + 2}{\theta s + 2} \right|^{-1} = \alpha \left| \frac{\theta s + 2}{2\theta s} \right| \qquad (4.4-8)$$

The effect of the approximation is illustrated in Fig. 4.4-2.

4.4.2 General Uncertainty Regions

The controller c is to be designed to guarantee robust stability and robust performance for the family Π described by the Nyquist band discussed in Sec. 4.3.2. For robust performance it is required that

$$|\epsilon w| < 1 \qquad \forall \omega, \quad \forall p \in \Pi \qquad (4.4-9)$$

or

$$|1 + \pi c| > w \qquad \forall \omega \qquad (4.4-10)$$

where $\pi c(i\omega)$ denotes the *Nyquist region* at frequency ω which was constructed in Sec. 4.3.2 by multiplying the region $\pi(\omega)$ with the complex number $c(i\omega)$. For robust stability the Nyquist band πc must not encircle or cover (-1,0), for robust performance the distance of (all points in) πc from the point (-1,0), i.e., $|1 + \pi c|$, has to exceed the specified minimum w.

Theorem 4.4-1. *The closed-loop system satisfies the robust performance condition*

$$|\epsilon w| < 1 \qquad \forall \omega, \quad \forall p \in \Pi \qquad (4.4-9)$$

if and only if it is robustly stable for the family Π *and the Nyquist band* $\pi(\omega)c(\omega)$
retains a distance of at least $w(\omega)$ *from the point (-1,0).*

The controller c is parametrized by the IMC filter (4.3–6). As a design procedure w is generally selected to be constant and the regions $\pi(\omega)c(\omega)$ are displayed
on a graphics screen for different values of the filter time constant λ. Then λ is
increased just enough to make the minimum distance from (-1,0) equal to w.

4.5   Summary of IMC Design Procedure

Required Information:

1. Process model $\tilde{p}$

2. Type of inputs (setpoints and disturbances) v affecting the process output
   — e.g., steps, ramps, steps entering through first-order lag, and so on.

3. Performance specifications:

   (i) Closed-loop system Type (1 or 2)

   (ii) Frequency dependent performance weight or in the case of a one-
        parameter filter simply the maximum allowed peak height (w^{-1}) of the
        sensitivity function ϵ (typically $0.3 < w < 0.9$).

4. Uncertainty information $\bar{\ell}_m(\omega)$.

   Family of plants considered for robust design

$$\Pi = \left\{ p : \left| \frac{p(i\omega) - \tilde{p}(i\omega)}{\tilde{p}(i\omega)} \right| \le \bar{\ell}_m(\omega) \right\}$$

Design Procedure:

STEP 1: Nominal Performance
   The H_2-optimal controller $\tilde{q}$ is determined which minimizes

$$\|(1 - \tilde{p}\tilde{q})v\|_2$$

for the specified input v. If the input v is of a standard type, $\tilde{q}$ can be obtained
directly from Table 4.1-1, 4.1-2, or 4.1-3. Otherwise Thm. 4.1-1 can be used.

$$\tilde{q} = (\tilde{p}_M d_M)^{-1} \left\{ \tilde{p}_A^{-1} d_M \right\}_* \qquad (4.1-7)$$

STEP 2: Robust Stability and Robust Performance

The controller $\tilde{q}$ is augmented by the IMC filter f

$$q = \tilde{q}f$$

where for asymptotically constant inputs

Type 1:

$$f = \frac{1}{(\lambda s + 1)^n} \qquad (4.2-6)$$

and for asymptotically ramp-like inputs

Type 2:

$$f = \frac{n\lambda s + 1}{(\lambda s + 1)^n} \qquad (4.2-7)$$

Here n is selected large enough for q to be proper. For MP systems and in the absence of modeling error λ is the closed-loop time constant. For NMP systems λ becomes the dominant closed-loop time constant when it is made large enough. In general, increasing λ slows down the system and makes it more robust.

Robust Stability: Check if

$$|\tilde{p}\tilde{q}f\bar{\ell}_m| < 1 \qquad \text{for } \omega = 0 \qquad (4.5-1)$$

This condition is necessary and sufficient for a filter time constant $\lambda > 0$ to exist for which the system is robustly stable.

Robust Performance: Increase λ just enough to meet the condition

$$|\tilde{p}\tilde{q}f\bar{\ell}_m| + |(1 - \tilde{p}\tilde{q}f)w| < 1 \qquad \forall \omega \qquad (4.4-4)$$

i.e., choose λ to make (4.4-4) an equality for some specific value(s) of ω.

4.6 Application: IMC Design for a First-Order Deadtime System

This particular example system was chosen because of its relevance for process control. We will demonstrate the control system design for both the exact uncertainty regions and for norm-bounded uncertainties — i.e., disk-shaped approximations of the exact uncertainty regions.

4.6.1 Deadtime Uncertainty

We will follow exactly the summary of the IMC Design Procedure in Sec. 4.5.

Information:

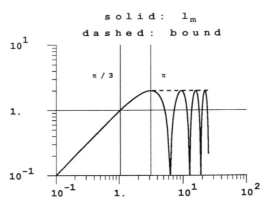

Figure 4.6-1. Multiplicative uncertainty ℓ_m for $\delta = \bar{\delta}$ and bound $\bar{\ell}_m$ versus $\omega\bar{\delta}$.

1. Process model $\tilde{p} = \frac{ke^{-s\theta}}{\tau s+1}$

2. Type of input: step

3. Performance specification

   (i) Type 1 (no offset for steps)
   (ii) Maximum peak of ϵ: $w^{-1} = 2.5$

4. Uncertainty information: Only deadtime uncertainty

$$\theta = \tilde{\theta} + \delta \quad \text{where} \quad |\delta| \leq \bar{\delta} \tag{4.6-1}$$

$$\frac{p - \tilde{p}}{\tilde{p}} = e^{-s\delta} - 1 \tag{4.6-2}$$

$$\ell_m(i\omega) = e^{-i\omega\delta} - 1 \qquad |\delta| \leq \bar{\delta} \tag{4.6-3}$$

Figure 4.6-1 shows the plot of $|\ell_m(i\omega)|$ for $\delta = \bar{\delta}$. It is easily seen that

$$\bar{\ell}_m(\omega) = |e^{-i\omega\bar{\delta}} - 1| \qquad \omega\bar{\delta} \leq \pi \tag{4.6-4a}$$

$$\bar{\ell}_m(\omega) = 2 \qquad \omega\bar{\delta} \geq \pi \tag{4.6-4b}$$

A plot of the bound $\bar{\ell}_m(\omega)$ is also shown in Fig. 4.6-1. Note the characteristic increase with frequency.

Design Procedure

STEP 1: Nominal Performance

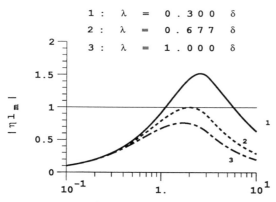

Figure 4.6-2. Robust stability bound $|\tilde{\eta}\bar{\ell}_m|$ for different values of the filter parameter λ versus $\omega\bar{\delta}$.

For step inputs the H_2-optimal controller is provided by Thm. 4.1-1 or case A in Tab. 4.1-2.

$$\tilde{p}_A = e^{-s\tilde{\theta}} \qquad (4.6-5)$$

$$\tilde{q} = \tilde{p}_M^{-1} = k^{-1}(\tau s + 1) \qquad (4.6-6)$$

$$\tilde{p}\tilde{q} = e^{-s\theta} \qquad (4.6-7)$$

STEP 2: Robust Stability and Robust Performance

Type 1 filter:

$$f = \frac{1}{\lambda s + 1} \qquad (4.6-8)$$

A first-order filter is sufficient to make q proper.

Robust Stability:

$$|\tilde{p}\tilde{q}f\bar{\ell}_m| = |f\bar{\ell}_m| < 1 \qquad (4.6-9)$$

Figure 4.6-1 immediately suggests a filter with a corner frequency $\omega = \pi/3\bar{\delta} \cong 1/\bar{\delta}$ or $\lambda = \bar{\delta}$. A more careful search reveals $\lambda > 0.67\bar{\delta}$ for robust stability (Fig. 4.6-2)

Robust Performance:

The constraint

$$|\tilde{p}\tilde{q}f\bar{\ell}_m| + |(1 - \tilde{p}\tilde{q}f)w| < 1$$

or

$$|f\bar{\ell}_m| + |(1 - e^{-i\omega\tilde{\theta}}f)w| < 1 \qquad (4.6-10)$$

is shown in Fig. 4.6-3A and B for a 15% and 30% maximum deadtime error respectively ($\tilde{\theta} = 1$, $\bar{\delta} = 0.15$ and 0.3).

Note from (4.6–9) and (4.6–4) that λ for robust stability depends only on the deadtime error $\bar{\delta}$ but not on the absolute value of the deadtime $\tilde{\theta}$. For robust

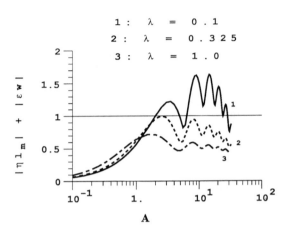

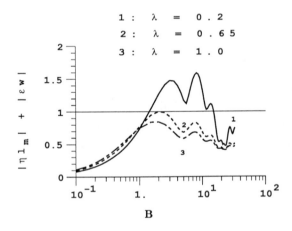

Figure 4.6-3. Robust performance test for 15% (A) and 30% (B) deadtime uncertainty for different filter parameters.

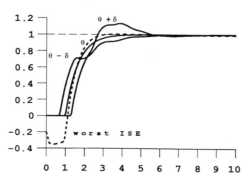

Figure 4.6-4. System response for $\lambda = 0.65$ and various deadtimes. "Worst-ISE" response is also shown.

performance (4.6–10) both $\bar{\delta}$ and $\tilde{\theta}$ have to be taken into consideration. The parameters which are known exactly, k and τ, do not affect the filter design. Typical closed-loop responses with $\lambda = 0.65$ chosen for robust performance are shown in Fig. 4.6-4. Also shown is the response of the plant contained in the set bounded by (4.6–4) which gives rise to the largest error norm. According to the arguments in Sec. 2.6 the frequency response of this plant is determined by locating at each frequency the point in the uncertainty disk around $\tilde{p}c$ which is closest to (-1,0) — i.e., the point p which satisfies

$$|1 + pc| = |1 + \tilde{p}c| - |\tilde{p}\bar{\ell}_m| \qquad (4.6 - 11)$$

It is intuitively clear that with pure deadtime uncertainty the worst error should occur for the plant with the maximum deadtime. The "worst-ISE" response shown in Fig. 4.6-4 illustrates the conservativeness arising from the norm-bound approximation of the deadtime uncertainty.

Next we will compute the filter parameters for robust stability and robust performance when we account for the deadtime uncertainty exactly. Figure 4.6-5A shows the Nyquist band where each region $\pi(\omega)$ is just a line for the case of pure deadtime uncertainty. Figures 4.6-5B and C show the transformed band $\pi(\omega)c(i\omega)$ with

$$c = \frac{\tilde{q}f}{1 - \tilde{p}\tilde{q}f} = \frac{s + 1}{\lambda s + 1 - e^{-s}} \qquad (4.6 - 12)$$

Figure 4.6-5B shows the Nyquist band for the minimum λ required for robust stability. In Fig. 4.6-5C λ is chosen for robust performance. In Fig. 4.6-6 some simulations for different values of the uncertain parameter θ with $\lambda = 0.5$ are shown. Table 4.6-1 compares the values of λ for robust stability and performance found by the two procedures.

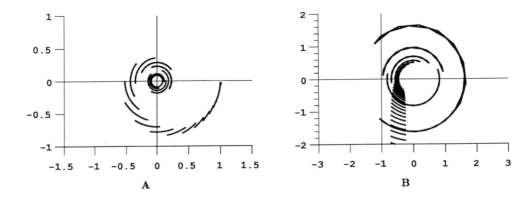

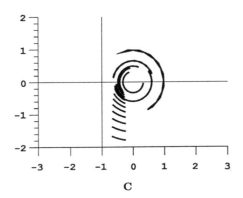

Figure 4.6-5. Nyquist bands for system (A), system with controller tuned for robust stability (B) and system with controller tuned for robust performance (C).

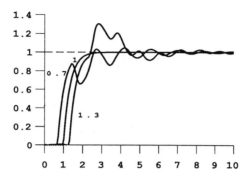

Figure 4.6-6. Response of closed loop system tuned for robust performance for various deadtimes.

Table 4.6-1. Filter parameter λ for robust stability and robust performance when the deadtime uncertainty ($\bar{\delta} = 0.3$) is treated exactly and when it is approximated by a norm-bound.

	Stability	Performance $w^{-1} = 2.5$	Performance $w^{-1} = 2.0$
Exact	0.20	0.34	0.50
Approximate	0.20	0.65	0.99

From this and other examples it can be concluded that the choice $\lambda = \bar{\delta}$ is a good rule of thumb which leads to good performance in practice.

4.6.2 Three-Parameter Uncertainty

The problem is identical with the one treated in the last section except that $w^{-1} = 2.0$ and all three parameters in the model

$$\tilde{p} = \frac{\tilde{k} e^{-s\tilde{\theta}}}{\tilde{\tau} s + 1}$$

are uncertain

$$0.8 \leq k \leq 1.2$$

$$0.7 \leq \tau \leq 1.3 \tag{4.6 - 13}$$

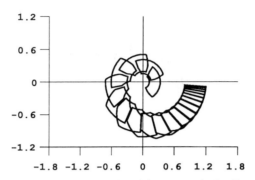

Figure 4.6-7. Nyquist band for system (4.6–13).

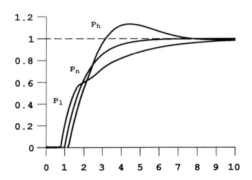

Figure 4.6-8. Response of closed-loop system tuned for robust performance for various system parameters.

$$0.8 \leq \theta \leq 1.2$$

The Nyquist band can be seen in Fig. 4.6-7. With the controller (4.6–12) the filter time constants for robust stability and performance are 0.21 and 1.03, respectively. Simulations with representative sample parameters from the set (4.6–13) are shown in Fig. 4.6-8.

Alternatively we can approximate the regions in Fig. 4.6-7 by disks centered at the nominal plant $\tilde{p}$ defined by the mean parameter values $\tilde{k}, \tilde{\tau}$ and $\tilde{\theta}$. The disk radii translate into the multiplicative uncertainty bound $\bar{\ell}_m(\omega)$ in Fig. 4.6-9. Going through a procedure equivalent to that demonstrated in the previous section, we find that for robust stability λ has to exceed 0.23 and for robust performance λ is 1.51. These values can be seen to be slightly more conservative than those obtained with the exact uncertainty description (Tab. 4.6-2).

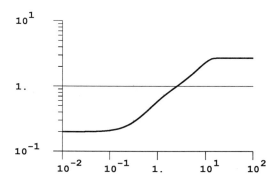

Figure 4.6-9. Multiplicative uncertainty bound $\bar{\ell}_m$ generated from Nyquist band in Fig. 4.6-7.

Table 4.6-2. Filter parameter λ for robust stability and robust performance when the uncertainty is treated exactly and when it is approximated by a norm bound.

	Stability	Performance
Exact	0.21	1.03
Approximate	0.23	1.51

4.7   References

4.1.1. The expression (4.1–7) for the H_2-optimal controller was derived by Frank (1974) who follows the work by Newton, Gould & Kaiser (1957).

4.1.3. Theorem 4.1-2 was obtained from Frank (1974).

4.3.2.   Several researchers have proposed controller-design techniques using arbitrarily shaped regions $\pi(\omega)$ to represent model uncertainty. The procedure described here was developed by Laughlin et al. (1986) and is very similar to that outlined by Chen (1984). Laughlin et al. review other techniques.

All example calculations in this chapter were carried out by D. Laughlin and C. Scali.

Chapter 5

SISO IMC DESIGN FOR UNSTABLE SYSTEMS

There are two main challenges in extending IMC to open-loop unstable systems. First, the parametrization of all controllers c, for which the closed-loop system is stable, in terms of the IMC controller q is not straightforward: stability of q does not guarantee closed-loop stability when p is unstable. Second, designing the controller for robust stability is significantly more difficult. When p is stable we can simply detune the controller (even to the degree of almost turning it off) to make it arbitrarily robust. When p is unstable there is a limit to how much we can detune the controller; some minimal controller gain is needed just to stabilize the system.

5.1 Parametrization of All Stabilizing Controllers

5.1.1 Conditions for Internal Stability

For stable systems the IMC parametrization offers two advantages. The search for the controller which yields good closed-loop behavior is greatly simplified: if the search is carried out over all possible stable and unstable feedback controllers c, closed-loop stability has to be checked for each choice of c. On the other hand, any stable IMC controller q automatically yields a feedback controller c for which the closed-loop system is stable. Furthermore, if the IMC structure is also used for the controller implementation, closed-loop stability is preserved in the presence of constraints and controller nonlinearities, contrary to what happens for the classic feedback structure.

As we move on to unstable systems we have to abandon the IMC structure for the control system *implementation*. This is apparent from Thm. 3.2-1 where we established that stability of both p and q is required for internal stability. It is also obvious from Fig. 3.2-1. If $p = \tilde{p}$, then the input signal u_2 does not generate any feedback signal and, therefore, no control action u. The signals $\tilde{y}$ and $y = pu_2$

grow without bound for an unstable plant.

On the other hand even for unstable systems we can use the IMC parametrization for the *design* of the feedback controller c. Let us establish the conditions q must satisfy for the closed-loop system with

$$c = \frac{q}{1 - \tilde{p}q} \qquad (5.1-1)$$

to be internally stable. Substituting (5.1–1) into (2.3–1) we find for $p = \tilde{p}$

$$\begin{pmatrix} y \\ u \end{pmatrix} = \begin{pmatrix} pq & (1 - pq)p \\ q & -pq \end{pmatrix} \begin{pmatrix} r \\ u' \end{pmatrix} \qquad (5.1-2)$$

For internal stability all four individual transfer functions in (5.1–2) have to be stable. This implies:

(i) q stable.

(ii) pq stable $\Rightarrow$ q has to cancel all RHP poles of p or equivalently: q has to have zeros wherever p has unstable poles.

(iii) $(1 - pq)p$ stable $\Rightarrow$ $(1 - pq)$ has to cancel all RHP poles of p or equivalently: $(1 - pq)$ has to have zeros wherever p has unstable poles.

Now assume that (iii) holds but (ii) does not. Then pq is unbounded at the unstable poles of p. If pq is unbounded, $(1 - pq)$ is unbounded at the unstable poles. This contradicts the assumption that $(1 - pq) = 0$ at the unstable poles of p (iii). Thus (iii) implies (ii). These arguments are summarized in the following theorem.

Theorem 5.1-1. *Assume that the model is perfect $(p = \tilde{p})$ and that p has k unstable poles at $\pi_1, \ldots, \pi_k$. The feedback system with the controller $c = q(1 - pq)^{-1}$ is internally stable if and only if*

(i) q is stable

(ii) $(1 - pq)$ has k RHP zeros at $\pi_i, \ldots, \pi_k$.

Note that Thm. 5.1-1 reduces to Thm. 3.2-1 when p is stable. Condition (ii) is quite instructive: recall that $\tilde{\eta} = \tilde{p}q$ is the complementary sensitivity function of the system which should be small for robustness. For unstable systems the sensitivity function cannot be chosen arbitrarily, but it has to satisfy (ii) – i.e., it must have a value of unity at all RHP plant poles; the achievable robustness is *constrained* by the RHP plant poles. We will elaborate further on this issue later in this chapter.

5.1.2   Controller parametrization

IMC controllers q which satisfy condition (ii) of Thm. 5.1-1 are not particularly easy to find. Thus, at this point the IMC parametrization does not offer any advantages. In this section we will derive a convenient expression for *all q's* satisfying (ii).

Define the allpass

$$b_p(s) = \prod_{i=1}^{k} \frac{-s + \pi_i}{s + \pi_i^H} \qquad (5.1-3)$$

where π_i are the poles of $\tilde{p}$ in the open RHP.

Theorem 5.1-2. *Assume that $p = \tilde{p}$ has k unstable poles at $\pi_1, \ldots, \pi_k$ in the open RHP and ℓ poles at the origin. Assume furthermore that there exists a q_0 such that $c = q_0(1 - pq_0)^{-1}$ stabilizes p. Then all controllers which stabilize p are parametrized by*

$$c = q(1 - pq)^{-1} \qquad (5.1-1)$$

$$q = q_0 + b_p^2 s^{2\ell} q_1 \qquad (5.1-4)$$

where q_1 is any arbitrary stable transfer function.

Proof.$\Rightarrow$ Assume that q and q_0 stabilize p. Then Thm. 5.1-1 implies that q, $q_0, (1 - pq)b_p^{-1}s^{-\ell}$, and $(1 - pq_0)b_p^{-1}s^{-\ell}$ are stable. Thus the differences $q - q_0$ and $p(q - q_0)b_p^{-1}s^{-\ell}$ are stable. Define

$$pb_p s^{\ell}(q - q_0)b_p^{-2}s^{-2\ell} \stackrel{\Delta}{=} pb_p s^{\ell} q_1 \qquad (5.1-5)$$

By construction (5.1–5) is stable. Therefore any RHP poles of q_1 are a subset of the RHP zeros of p. Because $q - q_0$ is stable, any RHP poles of $q_1 = (q - q_0)b_p^{-2}s^{-2\ell}$ must be a subset of the RHP poles of p. Because p cannot have RHP poles and zeros at the same location, q_1 is stable.

$\Leftarrow$ Assume q_0 stabilizes p and q_1 is stable. Then by Thm. 5.1-1, $q_0, q_1, (1 - pq_0)b_p^{-1}s^{-\ell}$ are stable. Also

$$q = q_0 + b_p^2 s^{2\ell} q_1$$

is stable. Then

$$(1 - pq)b_p^{-1}s^{-\ell} = (1 - pq_0)b_p^{-1}s^{-\ell} - pb_p^2 s^{2\ell} q_1 b_p^{-1}s^{-\ell} = (1 - pq_0)b_p^{-1}s^{-\ell} - pb_p s^{\ell} q_1$$

is stable. Therefore q is stabilizing.   □

Note that Thm. 5.1-2 assumes the existence of a stabilizing $q_0(s)$. The construction of the H_2-optimal controller in Sec. 5.2.1 will serve as a proof that such a controller always exists.

Any q constructed from (5.1–4) automatically satisfies the conditions of Thm. 5.1-1. After an initial stabilizing controller q_0 has been found the search for the appropriate controller q which yields desirable closed-loop performance characteristics is again very simple. All stable transfer functions q_1 yield controllers for which the closed-loop system is stable. There are also no disadvantages to using (5.1–4): there is *no* controller c which stabilizes p and which is not parametrized by a stable q_1.

In the special case of an open-loop stable system $b_p = 1$ and $\ell = 0$. The controller $q_0 = 0$ is stabilizing and (5.1–4) yields $q = q_1$, the IMC-parametrization for stable systems, as expected.

Note that q_1 in (5.1–4) is not limited to be proper and q can be improper in general. Because in Step 2 of the IMC design procedure a filter is always added for properness and robustness we are not concerned with properness at this point and hence improper controllers are permitted.

5.2   Nominal Performance

The controller design philosophy for unstable systems is the same as for stable ones. First a controller $\tilde{q}$ is selected to minimize the ISE or equivalently the 2-norm of the sensitivity function weighted by the input of interest. Then $\tilde{q}$ is augmented by a low-pass filter to achieve robust stability and performance. In this section we will derive the formulas for the H_2-optimal $\tilde{q}$.

5.2.1 H_2-Optimal Controller

Theorem 5.2-1. *Let $p = \tilde{p}$ have k poles at $\pi_1, \ldots, \pi_k$ in the open RHP and a pole of multiplicity ℓ at the origin. Define*

$$b_p = \prod_{i=1}^{k} \frac{-s + \pi_i}{s + \pi_i^H} \qquad (5.2-1)$$

and factor the plant into an allpass portion p_A and a MP portion p_M such that

$$p = p_A p_M \qquad (5.2-2)$$

Factor the input v similarly

$$v = v_A v_M \qquad (5.2-3)$$

Assume without loss of generality that the open RHP poles of the input v are the first k' poles π_i of the plant in the open RHP[1] and define accordingly

$$b_v = \prod_{i=1}^{k'} \frac{-s + \pi_i}{s + \pi_i^H} \qquad (5.2-4)$$

Assume further that v has at least ℓ poles at the origin.[2] The controller $\tilde{q}$ which minimizes the objective

$$\|e\|_2^2 = \int_0^\infty e^2 dt = \frac{1}{2\pi} \int_{-\infty}^\infty |e|^2 d\omega = \|(1 - p\tilde{q})v\|_2^2 \qquad (5.2-5)$$

is given by

$$\tilde{q} = b_p (p_M b_v v_M)^{-1} \left\{ (b_p p_A)^{-1} b_v v_M \right\}_* \qquad (5.2-6)$$

where the operator $\{\cdot\}_*$ denotes that after a partial fraction expansion of the operand all terms involving the poles of p_A^{-1} are omitted.

The following proof of (5.2–6) is somewhat involved and can be skipped by result-oriented readers without loss of continuity.

Proof. Before proceeding with the proof we have to establish some preliminary definitions and results. Let $L_2(i\mathcal{R})$ be the Hilbert space of scalar-valued functions on $i\mathcal{R}$ with inner product

$$< f, g > = \frac{1}{2\pi} \int_{-\infty}^\infty f(i\omega)^H g(i\omega) d\omega \qquad (5.2-7)$$

for which $< f, f >$ is finite.

Let $H_2(i\mathcal{R})$ be the subspace of functions $f(s)$ analytic in $\mathrm{Re}(s) > 0$. The orthogonal complement of $H_2(i\mathcal{R})$, denoted by $H_2^\perp(i\mathcal{R})$, is the subspace of functions analytic for $\mathrm{Re}(s) < 0$. Thus, if $f_1 \in H_2(i\mathcal{R})$ and $f_2 \in H_2^\perp(i\mathcal{R})$ then

$$< f_1, f_2 > = \frac{1}{2\pi} \int_{-\infty}^\infty f_1(i\omega)^H f_2(i\omega) d\omega = 0 \qquad (5.2-8)$$

(Equation (5.2–8) can be easily established via the calculus of residues).

Any $f(s) \in L_2$ is analytic on the imaginary axis. Thus any $f(s)$ can be decomposed uniquely into two orthogonal functions $\{f\}_+ \in H_2$ and $\{f\}_- \in H_2^\perp$

$$f(s) = \{f\}_+ + \{f\}_- \qquad (5.2-9)$$

---

[1] If this assumption were not made the problem would not be meaningful: Unbounded controller action would be necessary for the error to vanish as $t \to \infty$. See also the discussion at the end of Sec. 2.2.3.

[2] This assumption is made so that disturbances occurring at the plant *input* can be rejected with vanishing error as $t \to \infty$.

where $\{\cdot\}_+$($\{\cdot\}_-$) denotes that after a partial fraction expansion only the terms with poles in the LHP (RHP) are retained. The 2-norm of f induced by the inner product (5.2-7) can be computed as

$$\|f\|_2^2 = <f,f> = <\{f\}_- + \{f\}_+, \{f\}_- + \{f\}_+> = <\{f\}_-, \{f\}_->$$

$$+ <\{f\}_+, \{f\}_+> = \|\{f\}_-\|_2^2 + \|\{f\}_+\|_2^2 \qquad (5.2-10)$$

where the cross terms $<\{f\}_+, \{f\}_-> = <\{f\}_-, \{f\}_+> = 0$ because of orthogonality.

According to Thm. 5.1-2 all stabilizing controllers of p are parametrized by

$$\tilde{q} = q_0 + b_p^2 s^{2\ell} q_1 \qquad (5.2-11)$$

where q_1 is an arbitrary stable transfer function and q_0 is a stabilizing controller — i.e., it is chosen to be stable and such that $(1 - pq_0)$ has zeros at the unstable poles of the plant p. We will also assume that q_0 is such that $(1 - pq_0)v$ is stable. Not every stabilizing q_0 may have this property, since v may have more poles at the origin than p.

These properties of q_0 can be summarized by assuming that q_0 is chosen stable and such that $(1 - pq_0)$ has zeros at the unstable poles of $b_p^{-1}b_v v_M$ (here b_v has been included to avoid duplication of the k' unstable poles which are present in both p and v). The final construction of the H_2-optimal q serves as proof for the existence of a q_0 with such properties.

By substituting (5.2–11) into (5.2–5) the objective function which is to be minimized by the appropriate choice of a stable q_1 becomes

$$\|(1 - p\tilde{q})v\|_2^2 = \frac{1}{2\pi} \int_{-\infty}^{\infty} |1 - p(q_0 + b_p^2 s^{2\ell} q_1)|^2 |v|^2 d\omega$$

The integrand can be multiplied by the allpass $|b_p p_A b_v^{-1} v_A|^{-2} = 1$ without changing the value of the integral:

$$\frac{1}{2\pi} \int_{-\infty}^{\infty} |b_p^{-1} p_A^{-1}(1 - pq_0) b_v v_M - p_M b_p s^{2\ell} q_1 b_v v_M|^2 d\omega$$

$$\triangleq \frac{1}{2\pi} \int_{-\infty}^{\infty} |f_1 - f_2 q_1|^2 d\omega = \|f_1 - f_2 q_1\|_2^2 \qquad (5.2-12)$$

The optimal q_1 should yield a finite integral (5.2–12), which means that the integrand, $f_1 - f_2 q_1$, must be strictly proper and not have any poles at the origin. By inspection, if $f_2 q_1$ has any unstable poles, they are at the origin. Since all unstable poles of f_1 (the poles of p_A^{-1}) are in the *open* RHP, it cannot cancel any poles at the origin of $f_2 q_1$. Thus $f_2 q_1$ must be stable in order for q_1 to be optimal.

We will assume that q_1 has this property and verify later that our solution has indeed the property.

Let us now consider the requirement that the optimal q_1 must make $f_1 - f_2 q_1$ strictly proper. One can easily see that a constant term may be present in the partial fraction expansion (PFE) of f_1, while both constant and improper terms may be present in the PFE of $f_2 q_1$. The constant term in the PFE of f_1 is equal to $\lim_{s \to \infty} f_1(s) \triangleq f_1(\infty)$. Then in order for q_1 to make $f_1 - f_2 q_1$ strictly proper, q_1 must make $f_2 q_1$ proper and satisfy

$$\lim_{s \to \infty} (f_2 q_1) \triangleq (f_2 q_1)(\infty) = f_1(\infty) \qquad (5.2-13)$$

We can now write (5.2-12) as

$$\|(1 - p\tilde{q})v\|_2^2 = \|\hat{f}_1 - (f_2 q_1 - f_1(\infty))\|_2^2 \qquad (5.2-14)$$

where $\hat{f}_1(s) \triangleq f_1(s) - f_1(\infty)$ is an L_2 function, which can be decomposed according to (5.2-9)

$$\hat{f}_1 = \{\hat{f}_1\}_+ + \{\hat{f}_1\}_-$$

Since the optimal q_1 makes $f_2 q_1$ proper and satisfies (5.2-13), $(f_2 q_1 - f_1(\infty))$ is in H_2. Use of this fact and the decomposition of $\hat{f}_1$ yields after application of (5.2-10) to (5.2-14):

$$\|(1 - p\tilde{q}v\|_2^2 = \|\{\hat{f}_1\}_-\|_2^2 + \|\{\hat{f}_1\}_+ - (f_2 q_1 - f_1(\infty))\|_2^2$$

Since $\{\hat{f}_1\}_-$ is independent of q_1, the obvious solution q_1 which minimizes $\|(1 - p\tilde{q})v\|_2^2$ can be obtained from the condition

$$\|\{\hat{f}_1\}_+ - (f_2 q_1 - f_1(\infty))\|_2 = 0$$

or

$$f_2 q_1 = \{\hat{f}_1\}_+ + f_1(\infty) \triangleq \{f_1\}_{\infty+}$$

or

$$q_1 = (p_M b_p s^{2\ell} b_v v_M)^{-1} \{b_p^{-1} p_A^{-1} (1 - p q_0) b_v v_M\}_{\infty+}$$

where the notation $\{\cdot\}_{\infty+}$ is used to indicate that after a PFE is taken, all strictly proper stable terms as well as the constant term are kept.

Now we have to check that this obvious solution is permissible. Because, by assumption, v_M has at least ℓ poles at the origin, q_1 is stable as required. Also, simple substitution shows that $f_2 q_1$ is proper and stable and that q_1 satisfies (5.2-13). Thus the above obvious solution is indeed the optimal solution. Substituting the expression for q_1 into (5.2-11) we find

$$\tilde{q} = q_0 + b_p(p_M b_v v_M)^{-1}\left\{b_p^{-1}p_A^{-1}(1 - pq_0)b_v v_M\right\}_{\infty+}$$

$$= b_p(p_M b_v v_M)^{-1}[b_p^{-1}p_A^{-1}pq_0 b_v v_M - \left\{b_p^{-1}p_A^{-1}pq_0 b_v v_M\right\}_{\infty+} + \left\{b_p^{-1}p_A^{-1}b_v v_M\right\}_{\infty+}]$$

$$= b_p(p_M b_v v_M)^{-1}[\left\{b_p^{-1}p_A^{-1}pq_0 b_v v_M\right\}_{0-} + \left\{b_p^{-1}p_A^{-1}b_v v_M\right\}_{\infty+}] \qquad (5.2-15)$$

where $\{\cdot\}_{0-}$ indicates that in the partial fraction expansion all poles in the closed RHP are retained. For (5.2–15) these poles are the poles of $b_p^{-1}b_v v_M$ in the closed RHP; $p_A^{-1}pq_0 = p_M q_0$ is stable because q_0 is a stabilizing controller. Indeed, q_0 is chosen such that $pq_0 = 1$ at all unstable poles of $b_p^{-1}b_v v_M$. Thus (5.2–15) simplifies to

$$\tilde{q} = b_p(p_M b_v v_M)^{-1}\{b_p^{-1}p_A^{-1}b_v v_M\}_* \qquad (5.2-6)$$

$\square$

Consider the optimal complementary sensitivity function

$$\bar{\eta} = p\tilde{q} = p_A b_p(b_v v_M)^{-1}\{b_p^{-1}p_A^{-1}b_v v_M\}_* \qquad (5.2-16)$$

We note that the optimal closed-loop pole configuration depends in general on both the plant and the input characteristics. The poles of $\{b_p^{-1}p_A^{-1}b_v v_M\}_*$ are the poles of $b_p^{-1}b_v v_M$. These poles are cancelled by the zeros of $b_p(b_v v_M)^{-1}$ premultiplying $\{\cdot\}_*$ in (5.2–16). Therefore, unless the system is MP, the finite optimal closed-loop poles are the poles of $p_A b_p(b_v v_M)^{-1}$. Thus, some of the optimal closed-loop poles (the poles of $p_A b_p$) are at all the mirror images of the system RHP zeros and poles. The location of the other finite closed-loop poles depends on the zeros of the inputs.

These observations illustrate that without some fundamental insight of the kind attempted here, it is very difficult to state a priori the closed-loop pole locations which will lead to good nominal performance characteristics. We will see in the next section that, independent of RHP poles and input type, the optimal controller pushes all closed-loop poles of MP systems toward infinity. On the other hand, for nontrivial systems with several RHP zeros and poles a trial and error search for reasonable closed-loop pole locations can be a futile task.

5.2.2   Design for Common Input Forms

In this section we will interpret the general optimal solution (5.2–6) for specific systems and inputs.

i) *MP System:* $p_A = 1, p_M = p$

$$\left\{ b_p^{-1} p_A^{-1} b_v v_M \right\}_* = b_p^{-1} b_v v_M$$

$$\tilde{q} = p^{-1} \tag{5.2-17}$$

For MP systems the optimal controller $\tilde{q}$ is the system inverse regardless of whether the open-loop system is stable or not. Perfect control ($\epsilon = 0$) can be achieved for any MP system in the limit as improper controllers are used. The RHP poles impose no performance limitations for MP systems.

ii) *Stable System:* $b_p = b_v = 1$

$$\tilde{q} = (p_M v_M)^{-1} \left\{ p_A^{-1} v_M \right\}_* \tag{5.2-18}$$

This formula was stated in Thm. 4.1-1.

iii) *Integrator:* $p_M = s^{-1}$

$$\tilde{q} = (p_M v_M)^{-1} \left\{ p_A^{-1} v_M \right\}_* \tag{5.2-19}$$

This formula is the same as that for stable systems (5.2–18). Thus Tables 4.1-1, 2 and 3 can be used as well for plants which have one pole at the origin. However, some care is required in defining the appropriate input v_M. Recall that according to our convention v ($v = d$ or $v = -r$) acts at the process output (see Fig. 2.1-1B). If the external input acts at the process input (for example, u' in Fig. 2.3-1) then the equivalent disturbance acting at the process output has to include the plant pole at the origin. If there is a step disturbance (s^{-1}) at the process input and the process is a pure integrator, then $v_M = s^{-2}$, a ramp. If the process model is $p = (\tau s + 1)^{-1} s^{-1}$, then $v_M = (\tau s + 1)^{-1} s^{-2}$. Let us compute the H_2-optimal $\tilde{q}$ for this input.

$$\left\{ \frac{p_A^{-1}}{s^2(\tau s + 1)} \right\}_* = \frac{p_A^{-1}|_{s=0}}{s^2} + \frac{\tau^2 \cdot p_A^{-1}|_{s=-\frac{1}{\tau}}}{\tau s + 1} - \left[\frac{dp_A}{ds} \Big|_{s=0} + \tau \right] \cdot \frac{1}{s}$$

$$\tilde{q} = p_M^{-1} \left[(\tau s + 1) + \tau^2 s^2 p_A^{-1}|_{s=-\frac{1}{\tau}} - s(\tau s + 1) \left[\frac{dp_A}{ds} \Big|_{s=0} + \tau \right] \right]$$

$$= p_M^{-1} \left[s^2 \tau^2 (p_A^{-1}|_{s=-\frac{1}{\tau}} - \frac{1}{\tau} \cdot \frac{dp_A}{ds} \Big|_{s=0} - 1) - s \cdot \frac{dp_A}{ds} \Big|_{s=0} + 1 \right] \tag{5.2-20}$$

Note that for $\tau = 0$ (5.2–20) reduces to the corresponding entry in the last row of Table 4.1-1.

5.2.3   Minimum Error Norm for Step Inputs to Stable Systems

Theorem 4.1-3 states that the contributions of different NMP elements to the ISE are additive.

Theorem 4.1-3. *Let the plant be given by* $p = p_1 p_2 p_M$, *where both* p_1 *and* p_2 *are allpass and* $p_1(0) = p_2(0) = 1$. *Then for step inputs*

$$\min_{\tilde{q}} \|e\|_2^2 = \|(1 - p_1 p_2)s^{-1}\|_2^2 = \|(1 - p_1)s^{-1}\|_2^2 + \|(1 - p_2)s^{-1}\|_2^2 \qquad (5.2-21)$$

The proof uses similar arguments as the proof of Thm. 5.2-1.

Proof. We know from the last section that the optimal $\tilde{q} = p_M^{-1}$. Thus

$$\min_{\tilde{q}} \|e\|_2^2 = \frac{1}{2\pi i} \int_{-i\infty}^{i\infty} |1 - p_1 p_2|^2 |s|^{-2} ds$$

$$= \frac{1}{2\pi i} \int_{-i\infty}^{i\infty} |p_1^{-1} - p_2|^2 |s|^{-2} ds$$

$$= \frac{1}{2\pi i} \int_{-i\infty}^{i\infty} |(p_1^{-1} - 1)s^{-1} + (1 - p_2)s^{-1}|^2 ds$$

The first term is strictly unstable, the second term strictly stable. Because of the orthogonality we can rewrite the last expression as

$$\frac{1}{2\pi i} \left[\int_{-i\infty}^{i\infty} |(p_1^{-1} - 1)s^{-1}|^2 ds + \int_{-i\infty}^{i\infty} |(1 - p_2)s^{-1}|^2 ds \right]$$

$$= \frac{1}{2\pi i} \left[\int_{-i\infty}^{i\infty} |(1 - p_1)s^{-1}|^2 ds + \int_{-i\infty}^{+i\infty} |(1 - p_2)s^{-1}|^2 ds \right]$$

$\square$

5.2.4   Two-Degree-of-Freedom Controller

In Sec. 3.3.2 we derived for the two-degree-of-freedom controller in Fig. 3.3-1 the closed-loop relationship

$$e = (1 - pq_d)d - (1 - pq_r)r \qquad (3.3-7)$$

when there is no modelling error ($p = \tilde{p}$). In the case of open-loop unstable plants, the structure in Fig. 3.3-1 cannot be implemented. The feedback structure of Fig. 2.4-2 has to be used instead. Equations (3.3-6) and (3.3-7), however, still describe the relation between d, r, and e provided that

$$q_d = \frac{c_1 c_3}{1 + \tilde{p} c_1 c_3} \qquad (5.2-22)$$

$$q_r = \frac{c_1 c_2}{1 + \tilde{p} c_1 c_3} \qquad (5.2-23)$$

In this section we want to determine what limitations plant RHP poles place on disturbance rejection and setpoint tracking performance.

We found earlier (Sec. 5.2.2) that for MP systems the optimal controller is

$$\tilde{q} = p^{-1} \qquad (5.2-17)$$

which implies that perfect control ($\epsilon = 0$) can be achieved for *any* MP system (stable or unstable) in the limit as improper controllers are used.

For NMP systems the controller choice (5.2–17) is not allowed because p^{-1} is unstable. In addition to stability, q_d has to satisfy condition ii) of Thm. 5.1-1. That is, q_d has to be chosen such that $(1 - p q_d)$ has zeros at the unstable plant poles. *The unstable plant poles restrict the choice of q_d and thus they impose limitations on the disturbance rejection performance which can be achieved.*

Let us first design $\tilde{q}_d$ via (5.2–6) for $v = d$. Then we can use (5.2–22) to obtain $\hat{c} \triangleq c_1 c_3$

$$\hat{c} = \frac{\tilde{q}_d}{1 - \tilde{p} \tilde{q}_d} \qquad (5.2-24)$$

where the individual factors c_1 and c_3 are yet to be determined.

Next, let us consider the design of c_2 for setpoint tracking.

$$e = y - r = -(1 - \frac{p c_1}{1 + p c_1 c_3} c_2) r \qquad (5.2-25)$$

If we define a new plant

$$\hat{p} = \frac{p c_1}{1 + p c_1 c_3} \qquad (5.2-26)$$

and a new controller

$$\hat{q} = c_2 \qquad (5.2-27)$$

we can rewrite (5.2–25) in the familiar form

$$e = -(1 - \hat{p}\hat{q}) r \qquad (5.2-28)$$

The RHP zeros of $\hat{p}$ are the RHP zeros of p and c_1 and the RHP poles of c_3. In order to minimize the effect of RHP zeros on e, c_1 should be selected MP and

c_3 stable. Then the only RHP zeros of $\hat{p}$ are those of p. Obviously, for internal stability there must not be any RHP pole-zero cancellations between c_1 and c_3. Generally, the decomposition of $\hat{c}$ into c_1 and c_3 satisfying these requirements is not unique.

With these choices of c_1 and c_3 we can find the H_2-optimal $\hat{q}$ from (5.2–6) with $\hat{p} = \hat{p}_A \hat{p}_M$, $\hat{p}_A = p_A$ and $\hat{b}_p = b_v = 1$:

$$c_2 = \hat{q} = (\hat{p}_M r_M)^{-1} \left\{ p_A^{-1} r_M \right\}_* \qquad (5.2-29)$$

and for $d = 0$

$$y = \hat{p}\hat{q}r = p_A r_M^{-1} \left\{ p_A^{-1} r_M \right\}_* r \qquad (5.2-30)$$

which shows explicitly that the setpoint response is affected by the RHP zeros of p only. *In the two-degree-of-freedom control structure unstable plant poles impose no limitations on the nominal setpoint tracking performance.*

For special cases the number of control blocks can be reduced to two without affecting internal stability. For example, when c_3 is MP it can be combined with c_1 and c_2. Also, when c_1 is stable it can be incorporated into c_2 and c_3.

Of course, when *robust* performance is the objective the designs of q_r and q_d are not independent and the RHP plant poles affect the achievable performance for both disturbance rejection and setpoint tracking (see Secs. 3.4.2 and 5.5).

5.3 The IMC Filter

The philosophy behind the IMC filter is the same as for stable systems. The objective is robust stability and robust performance. The filter structure is fixed a priori and a few parameters (usually just one) are adjusted to meet the robustness objectives.

5.3.1 Filter Form

The IMC filter $f(s)$ is chosen to be a rational function that satisfies the following requirements:

(i) Pole-zero excess. The controller $q = \tilde{q}f$ must be proper. Assume that the designer has specified a pole-zero excess of m for the filter $f(s)$.

(ii) Internal stability, i.e., $\tilde{q}f$ and $(1 - p\tilde{q}f)p$ have to be stable.

(iii) Asymptotic tracking of inputs — i.e., $(1 - p\tilde{q}f)v$ has to be stable.

We assume that p and v have k distinct poles in the closed RHP at locations $\pi_1, \ldots, \pi_k$ (counting poles which appear in both p and v only once) and that $\tilde{q}$ has been designed such that

$$1 - p\tilde{q}(s) = 0 \quad \text{at} \quad s = \pi_1, \ldots, \pi_k \qquad (5.3-1)$$

Then, for simple poles, requirements (ii) and (iii) translate to

$$f(s) = 1 \quad \text{at} \quad s = \pi_1, \ldots, \pi_k \qquad (5.3-2)$$

A filter of the form

$$f(s) = \frac{a_{k-1}s^{k-1} + \ldots + a_1 s + a_0}{(\lambda s + 1)^{m+k-1}} \qquad (5.3-3)$$

automatically satisfies requirement (i) as well as (with the k coefficients $a_{k-1}, \ldots, a_0$ properly chosen) requirements (ii) and (iii). Furthermore, the closed-loop system will be internally stable and will track inputs asymptotically as time $\rightarrow \infty$ for all $\lambda > 0$. This will allow the adjustable filter time constant λ to be selected freely without any concern for nominal stability.

In the following a procedure is presented to obtain the coefficients of the filter numerator. The k equalities (5.3–2) expand to the following system of linear equations:

$$\begin{pmatrix} \pi_1^{k-1} & \cdots & \pi_1 & 1 \\ \vdots & & \vdots & \vdots \\ \pi_k^{k-1} & \cdots & \pi_k & 1 \end{pmatrix} \begin{pmatrix} a_{k-1} \\ \vdots \\ a_0 \end{pmatrix} = \begin{pmatrix} (\lambda\pi_1 + 1)^{m+k-1} \\ \vdots \\ (\lambda\pi_k + 1)^{m+k-1} \end{pmatrix} \qquad (5.3-4)$$

The matrix in (5.3–4) is of the Vandermonde form. Using a Lagrange-type interpolation formula it is quite simple to develop the expression for the filter explicitly. Define

$$b_j(s) = \frac{1}{(\lambda s + 1)^n} \prod_{\substack{i=1 \\ i \neq j}}^{k} (s - \pi_i), \qquad j = 1, \ldots, k \qquad (5.3-5)$$

where $n = m + k - 1$ and note that

$$b_j(\pi_i) = 0 \qquad i = 1, \ldots k; \quad i \neq j \qquad (5.3-6)$$

$$b_j(\pi_j) = \frac{1}{(\lambda\pi_j + 1)^n} \prod_{\substack{i=1 \\ i \neq j}}^{k} (\pi_j - \pi_i) \qquad (5.3-7)$$

The stabilizing filter is easily constructed by additively combining b_j's (5.3-5) with the appropriate scaling factors (5.3-7):

$$f(s) = \frac{1}{(\lambda s + 1)^n} \sum_{j=1}^{k} (\lambda\pi_j + 1)^n \prod_{\substack{i=1 \\ i \neq j}}^{k} \frac{s - \pi_i}{\pi_j - \pi_i} \qquad (5.3-8)$$

This formula is not valid for repeated poles. For a pole of multiplicity ℓ at π_i the filter has to satisfy

$$f(\pi_i) = 1 \qquad\qquad (5.3-9)$$

and

$$\frac{d^j f(s)}{ds^j}\Big|_{s=\pi_i} = 0 \qquad j = 1, \ldots, \ell-1 \qquad\qquad (5.3-10)$$

For each multiple pole the system of equations (5.3–4) has to be augmented by the linear equalities (5.3–10) to determine the coefficients $a_0, \ldots, a_{k-1}$. Most common are repeated poles at the origin. For k distinct poles at $\pi_1, \ldots, \pi_k$ in the open RHP and a double pole at the origin the formula (5.3–8) is modified as follows

$$f(s) = \frac{1}{(\lambda s + 1)^n}\left[\sum_{j=1}^{k}(\lambda\pi_j + 1)^n \frac{s^2}{\pi_j^2}\prod_{\substack{i=1 \\ i \neq j}}^{k}\frac{s-\pi_i}{\pi_j-\pi_i} + (\alpha s + 1)\prod_{i=1}^{k}\frac{s-\pi_i}{-\pi_i}\right] \qquad (5.3-11)$$

where

$$\alpha = \lambda n + \sum_{i=1}^{k}\pi_i^{-1} \qquad\qquad (5.3-12)$$

Example 5.3-1. Assume that p and v have a common RHP pole at $s = \pi (k = 1)$.

Then the filter form for $m = 2$ is

$$f(s) = \frac{a_0}{(\lambda s + 1)^2} \qquad\qquad (5.3-13)$$

and (5.3–2), or equivalently (5.3–4), becomes

$$a_0 = (\lambda\pi + 1)^2$$

resulting in the IMC filter

$$f(s) = \frac{(\lambda\pi + 1)^2}{(\lambda s + 1)^2} \qquad\qquad (5.3-14)$$

If $\pi = 0$, (5.3–14) reduces to the standard filter for stable systems with Type 1 performance: $f = (\lambda s + 1)^{-2}$.

Note that the filter (5.3–14) was not designed to satisfy any specific low-frequency performance requirement. This could be accomplished by imposing the additional interpolation constraint $f(0) = 1$.   □

Example 5.3-2. Assume that the same pole-zero excess requirements as in Ex. 5.3-1 apply but that p and v have a double pole at the origin. The filter form is

$$f(s) = \frac{a_1 s + a_0}{(\lambda s + 1)^3} \qquad\qquad (5.3-15)$$

(5.3–2) and (5.3–10) become respectively

$$\frac{a_1 \cdot 0 + a_0}{(\lambda \cdot 0 + 1)^3} = 1 \Rightarrow a_0 = 1 \qquad (5.3 - 16)$$

$$\frac{a_1 \cdot (\lambda \cdot 0 + 1)^3 - (a_1 \cdot 0 + a_0)3\lambda(\lambda \cdot 0 + 1)^2}{(\lambda \cdot 0 + 1)^6} = 0 \Rightarrow a_1 = 3\lambda \qquad (5.3 - 17)$$

and 5.3–14 yields

$$f(s) = \frac{3\lambda s + 1}{(\lambda s + 1)^3} \qquad (5.3 - 18)$$

This is the filter we proposed for stable systems for Type 2 performance.   □

5.3.2   Qualitative Interpretation of the Filter Function

With the filter f and the H_2-optimal controller $\tilde{q}$ the complementary sensitivity function becomes

$$\tilde{\eta} = p\tilde{q}f \qquad (5.3 - 19)$$

In Sec. 5.2.2. we noted that $p\tilde{q} = 1$ for MP systems. Thus for MP systems the finite closed-loop poles are the poles of the filter f. Adjusting the filter time constant λ is equivalent to adjusting the speed of the closed-loop response.

The filter is always designed such that the closed-loop system is stable for all values of $\lambda > 0$. For stable systems the filter form is unconstrained except at $s = 0$ where Type 1 performance requires $f(0) = 1$ and Type 2 requires, in addition, $\frac{df}{ds}|_{s=0} = 0$. For unstable systems the filter has to be unity at all unstable system poles. This constraint limits the range of filter time constants λ which can be chosen for reasonable performance as we will show next.

Consider the MP system

$$p = p_M = \frac{1}{-s + \beta} \qquad \beta > 0 \qquad (5.3 - 20)$$

and assume that a step disturbance acts at the process input

$$v = v_M = \frac{1}{s(-s + \beta)} \qquad (5.3 - 21)$$

The H_2-optimal controller is $\tilde{q} = p^{-1}$. For internal stability and asymptotically error-free disturbance compensation we have to require that the filter satisfies

$$f(0) = f(\beta) = 1 \qquad (5.3 - 22)$$

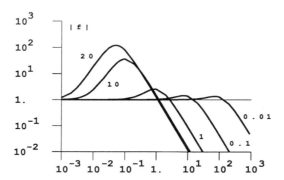

Figure 5.3-1. Filter magnitude for different values of λ; system has one unstable pole at (1,0).

We will assume that the filter is second order $(n = 2)$ and use (5.3–8) to find its transfer function

$$f(s) = \frac{1}{(\lambda s + 1)^2} \left[\frac{-s + \beta}{\beta} + \frac{(\lambda\beta + 1)^2 s}{\beta} \right] = \frac{1}{(\lambda s + 1)^2} [\lambda s(\lambda\beta + 2) + 1]$$

$$(5.3 - 23)$$

The following approximate expressions are valid for large and small filter time constants respectively

$$\lambda\beta >> 2 \qquad f(s) \cong \frac{\lambda^2\beta s + 1}{(\lambda s + 1)^2} \qquad\qquad (5.3 - 24)$$

$$\lambda\beta << 2 \qquad f(s) \cong \frac{2\lambda s + 1}{(\lambda s + 1)^2} \qquad\qquad (5.3 - 25)$$

For large λ's the zero $(-\lambda^2\beta)^{-1}$ of (5.3–24) will occur at much smaller frequencies than the pole $-\lambda^{-1}$. This will result in a large peak of the complementary sensitivity function η and is therefore undesirable. For small λ's the zero $(-2\lambda)^{-1}$ is close to the pole $-\lambda^{-1}$ and therefore the peak in η will be moderate.

In conclusion, for unstable systems *it is undesirable from a performance and robustness point of view to make the filter time constant much larger than the inverse of the unstable pole* β^{-1} (Fig. 5.3-1). The closed loop response cannot be made arbitrarily slow, otherwise the performance deteriorates drastically. This observation is very reasonable from a physical point of view: If a system has a very fast unstable pole — i.e., the pole is far out in the RHP, rapid control action is necessary to get acceptable performance.

For NMP systems, in general, the optimal $\tilde{\eta}$ (5.3–19) includes as poles the mirror images of the plant RHP zeros and poles. Increasing the filter time constant —

i.e., moving the filter pole closer to the origin will affect the closed-loop response noticeably only when the filter poles become of the same order of magnitude as the poles of $p\tilde{q}$.

Difficulties arise when the unstable pole and the unstable zero are located close to each other. Because $\epsilon(\zeta) = 1$ at the zero ζ, and $\epsilon(\pi) = 0$ at the pole π, the sensitivity function has to vary rapidly which leads to high peaks in ϵ as we will demonstrate in Sec. 5.7.2. These control difficulties are inherent in the plant itself and little can be done via sophisticated control system design.

5.4   Robust Stability

5.4.1   Norm-Bounded Uncertainty Regions

For IMC ($\tilde{\eta} = \tilde{p}\tilde{q}f$) Thm. 2.5-1 becomes Cor. 5.4-1.

Corollary 5.4-1 (Robust Stability). *Assume that all plants p in the family Π*

$$\Pi = \left\{ p : \left| \frac{p(i\omega) - \tilde{p}(i\omega)}{\tilde{p}(i\omega)} \right| < \bar{\ell}_m(\omega) \right\} \qquad (5.4-1)$$

have the same number of RHP poles. Then the system is robustly stable if and only if the IMC filter satisfies

$$|f| < \frac{1}{|\tilde{p}\tilde{q}\bar{\ell}_m|} \qquad \forall \omega \qquad (5.4-2)$$

where $\tilde{q}$ is a stabilizing controller for the nominal plant $\tilde{p}$.

For stable systems there are no constraints on f. Therefore there always exists a filter f which satisfies (5.4–2) regardless how large the uncertainty $\bar{\ell}_m$. For unstable systems f is constrained to be unity at the RHP system poles. Thus, depending on $\bar{\ell}_m$, there may not exist any filter time constant λ for which the constraint (5.4–2) is met. Indeed, there may not exist any filter — however complicated — which satisfies (5.4–2). A minimum amount of information is necessary or equivalently a maximum amount of uncertainty is allowed to stabilize an unstable system. The necessary information at $\omega = 0$ can be characterized easily.

Corollary 5.4-2. *Assume that a filter f is to be designed for a system with poles at the origin — i.e., $f(0) = 1$. There exists an f such that the closed-loop system is robustly stable for the family Π described by (5.4–1) only if $\bar{\ell}_m(0) < 1$.*

Note that contrary to Cor. 4.3-2, Cor. 5.4-2 is only necessary. For unstable systems, in general, the filter has to satisfy other constraints in addition to the one at $s = 0$.

The filter design procedure is again to increase the filter time constant until (5.4–2) is satisfied. In Sec. 5.3.2. we found that making the filter poles slower than the mirror images of the RHP system poles leads to peaks in the complementary sensitivity function and consequently to increased robustness problems. Thus the range over which the time constant λ can be varied to improve robustness is limited.

5.4.2 General Uncertainty Regions

The problem formulation is identical to the one for stable systems presented in Sec. 4.3.2. The next corollary follows directly from Thm. 4.3-1.

Corollary 5.4-3. *Assume that all plants in the family Π have n poles in the open RHP. The closed-loop system is robustly stable for Π if and only if the Nyquist band $\pi(\omega)c(i\omega)$ makes n net counterclockwise encirclements of the point (-1,0) without covering it.*

The discussion of Sec. 4.3.2 carries over to the case of unstable systems with minor obvious modifications.

5.5   Robust Performance

The key results, Cor. 4.4-1 and Thm. 4.4-1, hold for unstable systems if it is assumed that all plants in the family Π have the same number of RHP poles and if the controller $\tilde{q}$ and filter f are stabilizing for the nominal plant $\tilde{p}$. The remarks of Sec. 4.4. can be interpreted in the context of open-loop unstable systems with minor self evident changes. They are not repeated here for brevity.

For the two-degree-of-freedom structure, it is necessary to use the Structured Singular Value (used in the chapters on MIMO systems) in order to obtain non-conservative conditions for robust performance. An alternative is to use condition (3.4–20) (which is only sufficient) and the following procedure. First design f_d ($q_d = \tilde{q}_d f_d$) according to the one-degree-of-freedom design method. Then design f_r ($q_r = \tilde{q}_r f_r$) to satisfy (3.4–20). When this is accomplished, the control system performance will be robust for both disturbance rejection and setpoint tracking. However, the fact that (3.4–20) is only sufficient may make the task of designing a satisfactory f_r impossible. Clearly, the above procedure emphasizes robustness of disturbance rejection rather than of setpoint tracking as is appropriate for process control applications.

After q_d and q_r have been designed, the controllers for the feedback structure in Fig. 3.3-1 can be obtained through

$$c_1 c_3 = \frac{q_d}{1 - \tilde{p} q_d} \qquad (5.5-1)$$

$$c_2 = q_r \frac{1 + c_1 c_3 \tilde{p}}{c_1} \qquad (5.5-2)$$

where c_1 is MP, c_3 is stable and there are no RHP pole-zero cancellations between c_1 and c_3. Otherwise c_1 and c_3 are arbitrary.

5.6 Summary of the IMC Design Procedure

The required information for the IMC design is the same as that for stable systems: process model, input type, performance specifications and uncertainty information. The input specification requires some care. If the physical disturbance enters at the plant input, the input v used in the design procedure has to include the unstable system poles for the resulting controller to yield offset free performance.

Design Procedure
  Step 1: Nominal Performance
  The stabilizing H_2-optimal controller $\tilde{q}$ is determined which minimizes

$$\|(1 - \tilde{p}\tilde{q})\|_2$$

for the specified input v. After the unstable plant and input poles and zeros have been determined the optimal controller $\tilde{q}$ can be found explicitly from:

$$\tilde{q} = b_p (p_M b_v v_M)^{-1} \left\{ (b_p p_A)^{-1} b_v v_M \right\}_* \qquad (5.2-6)$$

If all the unstable plant/input poles are at the origin, the entries in Table 4.1-1, 2 and 3 can be used to find $\tilde{q}$ for typical inputs.

  Step 2: Robust Stability and Robust Performance
  The controller $\tilde{q}$ is augmented by the IMC filter f

$$q = \tilde{q} f$$

In order for q to be stabilizing, f has to be unity at all unstable plant/input poles $\pi_1, \ldots, \pi_k$. For distinct poles the one-parameter filter has the form

$$f(s) = \frac{1}{(\lambda s + 1)^n} \sum_{j=1}^{k} (\lambda \pi_j + 1)^n \prod_{\substack{i=1 \\ i \neq j}}^{k} \frac{s - \pi_i}{\pi_j - \pi_i} \qquad (5.3-8)$$

For repeated unstable poles, in general, a system of linear equations has to be solved to determine the numerator coefficients of the filter (Sec. 5.3.1). The order

n is selected large enough for q to be proper. The filter poles are a subset of the system closed-loop poles. If they are made much slower than the mirror images of the system RHP poles undesirable performance and robustness properties result. For MP systems and in the absence of modelling error, λ is the closed-loop time constant. For NMP systems λ becomes the dominant closed-loop time constant when it is made large enough.

Robust Stability. Check if

$$|\tilde{p}\tilde{q}f\bar{\ell}_m| < 1 \quad \text{for} \quad \omega = 0 \tag{5.6 - 1}$$

This condition is necessary for a filter time constant $\lambda > 0$ to exist for which the system is robustly stable.

Robust Performance. Increase λ just enough to meet the condition

$$|\tilde{p}\tilde{q}f\bar{\ell}_m| + |(1 - \tilde{p}\tilde{q}f)w| < 1 \quad \forall\omega \tag{4.4 - 4}$$

i.e., choose λ to make (4.4–4) (almost) an equality for some specific value(s) of ω.

If the two-degree-of-freedom structure is used, then, at least for the nominal case, q_d for disturbance rejection and $\hat{q}$ for setpoint tracking can be designed sequentially (Sec. 5.2.4). Recall that with the two-degree-of-freedom structure the RHP system poles impose limitations on disturbance rejection only but not on setpoint tracking performance.

5.7 Applications

5.7.1 Distillation Column Base Level Control

When the steam input to a distillation column is increased, the liquid level in the column base can exhibit inverse response behavior. The following type of model

$$\frac{y}{u} = \frac{1}{s}(1 - 2e^{-s\theta}) \tag{5.7 - 1}$$

was found to describe the behavior of many industrial columns adequately. The response of y to a step change in u is shown in Fig. 5.7-1. The deadtime θ is equal to the sum of the hydraulic time constants τ of the n individual trays ($\theta = n\tau$). Actually, the time delay in (5.7–1) is used to approximate a large number of first order lags modelling the tray behavior

$$e^{-sn\tau} \cong \frac{1}{(\tau s + 1)^n} \tag{5.7 - 2}$$

The following numbers would be typical: $\tau = 5$ sec, $n = 18$ and $\theta = 1.5$ min.

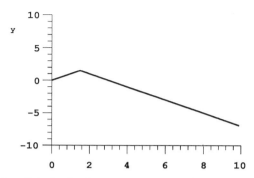

Figure 5.7-1. Reboiler level response to a step change in steam input.

For the purpose of this example we will assume that the "real" column behaves like a sequence of lags. The model uses the time delay approximation (5.7–2). The absolute value of the multiplicative error

$$\ell_m = \frac{p - \tilde{p}}{\tilde{p}} = \frac{e^{-1.5s} - (0.083s + 1)^{-18}}{0.5 - e^{-1.5s}} \qquad (5.7-3)$$

as well as the smooth bound $\bar{\ell}_m = |2.6s(1.3s + 1)^{-1}|$ are shown in Fig. 5.7-2 as a function of frequency. Because here we are concerned only with *one* model and *one* plant it is clearly conservative to define a *set* of plants bounded by $\bar{\ell}_m(\omega)$. Nevertheless, we will design the control system to meet the performance specifications for all plants in this set, assuming that some of the unknown model errors are captured with this uncertainty description.

All major disturbances are step-like and enter at the plant input. Thus for the control system design we will assume $v = s^{-2}$. As a robust performance specification we will require the maximum peak of ϵ not to exceed $w^{-1} = 2.5$.

Design Procedure
  Step 1: Nominal Performance
  For (5.7–1) $b_p = b_v = 1, v_M = s^{-2}$. Because (5.7–1) has an infinite number of RHP zeros

$$\zeta_k = \frac{1}{\theta}(\ln 2 + 2k\pi i) \quad -\infty < k \text{ (integer) } < +\infty \qquad (5.7-4)$$

it is somewhat more complicated to determine the allpass portion p_A of the model. The allpass must have the poles at the mirror images of the zeros

$$\pi_k = \frac{1}{\theta}(-\ln 2 + 2k\pi i) \quad -\infty < k \text{ (integer) } < +\infty \qquad (5.7-5)$$

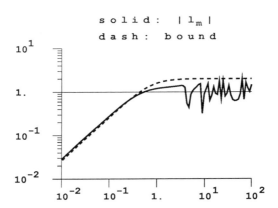

Figure 5.7-2. Model error caused by delay approximation.

and therefore

$$p_A = \frac{1 - 2e^{-s\theta}}{-2 + e^{-s\theta}} \tag{5.7 - 6}$$

$$p_M = \frac{1}{s}(-2 + e^{-s\theta}) \tag{5.7 - 7}$$

The H_2-optimal controller is

$$\tilde{q} = (-2 + e^{-s\theta})^{-1} s^3 \left\{ \frac{-2 + e^{-s\theta}}{1 - 2e^{-s\theta}} \frac{1}{s^2} \right\}_* = (-2 + e^{-s\theta})^{-1} s^3 \left[\frac{1}{s^2} + \frac{3\theta}{s} \right]$$

$$= (-2 + e^{-s\theta})^{-1} s(3\theta s + 1) \tag{5.7 - 8}$$

$$\bar{\eta} = \tilde{p}\tilde{q} = \frac{1 - 2e^{-s\theta}}{-2 + e^{-s\theta}}(3\theta s + 1) \tag{5.7 - 9}$$

The error resulting from a ramp disturbance is shown in Fig. 5.7-3. $\bar{\eta}$ and $\bar{\epsilon}$ can be seen in Fig. 5.7-4. Note that because of the NMP components the achievable performance measured either in terms of time domain behavior (Fig. 5.7-3) or bandwidth (Fig. 5.7-4B) is quite poor.

Step 2: Robust Stability and Robust Performance

Because the only two unstable poles are at the origin, a Type 2 filter can be used to improve the systems robustness. The order has to be three to make q proper.

$$f(s) = \frac{3\lambda s + 1}{(\lambda s + 1)^3} \tag{5.7 - 10}$$

The effect of λ on robustness ($|\bar{\eta}|$) and performance ($|\bar{\epsilon}|$) is shown in Fig. 5.7-4. Robust stability for the set of plants bounded by $\bar{\ell}_m$ requires $\lambda > 5$.

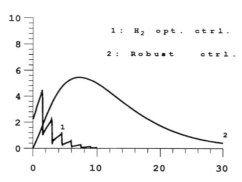

Figure 5.7-3. Closed-loop response for ramp disturbance; ISE-optimal controller (1) and "robust" controller (2).

The robust performance constraints for different λ's are seen in Fig. 5.7-5. A time constant $\lambda = 7$ is necessary for robust performance when $w^{-1} = 2.5$. The response of the "real" plant for this controller is compared to the ideal response in Fig. 5.7-3.

Because the open-loop system is unstable, the controller must be implemented via the classic feedback structure. The optimal controller is calculated as

$$c = \frac{\tilde{q}}{1 - \tilde{p}\tilde{q}} = \frac{(3\theta s + 1)s}{3\theta s(-1 + 2e^{-s\theta}) - 3(1 - e^{-s\theta})} \qquad (5.7 - 11)$$

Note that c has a double pole at the origin. One is canceled by the zero at the origin, the other one provides the integral action required for the error free tracking of ramp inputs.

Mainly because of the delay, the expression (5.7–11) for the controller is quite complex. It becomes even more complex when the Type 2 filter is added. In the next chapter we will show that approximating c by a PID controller brings about only a minor loss of performance.

5.7.2 NMP Unstable Systems

In this section we will demonstrate the design difficulties arising when the plant contains both RHP poles and zeros. Consider the general example

$$p = \frac{-s + \alpha}{-s + \beta} \qquad \alpha, \beta > 0 \qquad (5.7 - 12)$$

so that

$$p_A = \frac{-s + \alpha}{s + \alpha} \qquad (5.7 - 13)$$

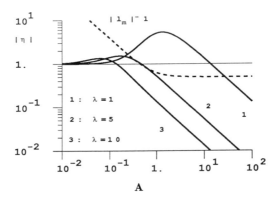

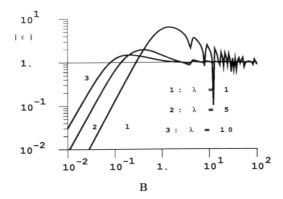

Figure 5.7-4. $|\tilde{\eta}|$ and $|\tilde{\varepsilon}|$ for different filter parameters; $\lambda > 5$ is required for robust stability.

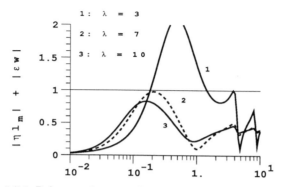

Figure 5.7-5. Robust performance bound for different filter parameters.

$$p_M = \frac{s + \alpha}{-s + \beta} \qquad (5.7 - 14)$$

$$b_p = \frac{-s + \beta}{s + \beta} \qquad (5.7 - 15)$$

A step disturbance is assumed to pass through the unstable system portion:

$$v_M = \frac{1}{s(-s + \beta)} \qquad (5.7 - 16)$$

$$b_v = \frac{-s + \beta}{s + \beta} \qquad (5.7 - 17)$$

Substituting (5.7–13 to 17) into (5.2–6) yields after some algebra

$$\tilde{q} = \frac{(-s + \beta)(2s + \alpha - \beta)}{(\alpha - \beta)(s + \alpha)} \qquad (5.7 - 18)$$

The appropriate filter satisfying (5.3–2) for $\pi_2 = 0$ and $\pi_2 = \beta$ is

$$f(s) = \frac{\beta^{-1}\left[(\lambda\beta + 1)^3 - 1\right]s + 1}{(\lambda s + 1)^3} \qquad (5.7 - 19)$$

The sensitivity and complementary sensitivity functions are shown in Fig. 5.7-6 A and B for $\alpha = 1, \beta = 10$. One notices from the behavior of ϵ as a function of λ that the bandwidth is limited to about 1 rad min^{-1} because of the RHP zero at $s = 1$. Also, the complementary sensitivity cannot be forced to drop below one at frequencies below 10 rad min^{-1} because of the RHP pole at $s = 10$. This pole is also the cause of the high peaks in ϵ and η when λ is decreased much below 0.05 min.

When the pole moves closer to the zero ($\alpha = 1, \beta = 2$) the maximum peak in both ϵ and η cannot be reduced much below 13 regardless of how λ is chosen (Fig. 5.7-7). This reflects the control difficulties inherent in systems with RHP poles and zeros close to each other.

5.8   References

5.1.2. A similar controller parametrization was proposed by Zames & Francis (1983). It was modified here to allow poles at the origin. The parametrization is equivalent to that proposed by Youla et al. (1976).

5.2.3. The additivity is pointed out in the book by Frank (1974).

5.2.4. The limitations imposed by RHP zeros and poles in the context of the two-degree-of-freedom structure are discussed by Vidyasagar (1985).

5.3.1. An efficient and robust algorithm for solving systems of linear equations in the Vandermonde form is reported by Golub & VanLoan (1983; pp. 119-124). When the conditions (5.3–10) are added, the system is of a confluent Vandermonde form. Efficient solution algorithms for related systems are studied by Bjork and Elfring (1973).

5.7.1. This control problem was proposed to the authors by J. Shunta of DuPont. The model is derived in the book by Buckley, Luyben & Shunta (1985) where other control schemes for this system are also discussed.

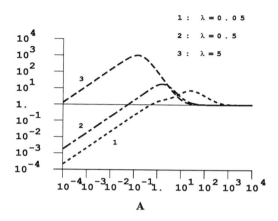

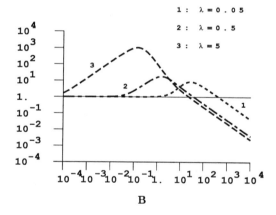

Figure 5.7-6. Sensitivity (*A*) and complementary sensitivity (*B*) for $\alpha = 1, \beta = 10$ for different values of the filter parameter λ.

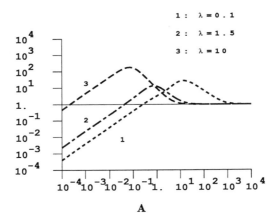

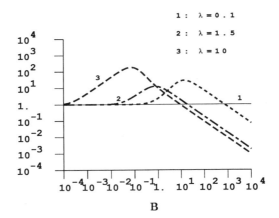

Figure 5.7-7. Sensitivity (A) and complementary sensitivity (B) for $\alpha = 1, \beta = 2$ for different values of the filter parameter λ.

Chapter 6

ISSUES IN SISO IMC DESIGN

6.1 Implications of IMC for Classic Feedback Controllers

By far the most widely used controllers in the process industries are the two-term PI and the three-term PID controller. Though IMC is clearly more general and therefore more powerful it is worthwhile to explore the relationships between IMC, PI and PID in order to gain insight into the tuning of these simpler controllers, their performance, robustness and limitations.

6.1.1 General Relationships

In Chap. 3 we derived the general relationship between the classic feedback controller c and the IMC controller q when the process model is $\tilde{p}$:

$$c = \frac{q}{1 - \tilde{p}q} \qquad (6.1-1)$$

In Chaps. 4 and 5 we proposed to choose as the IMC controller q the controller $\tilde{q}$, which is H_2 optimal for the system $\tilde{p}$ for a particular input v, and to augment it by a lowpass filter f.

$$q = \tilde{q}f \qquad (6.1-2)$$

The order of q obtained in this fashion is generally higher than the order of the model $\tilde{p}$. The complexity of q and similarly, the complexity of the equivalent classic controller c is determined by the complexity of the model. Simple models give rise to simple controllers. In particular, for a large number of simple models used commonly in the process industries the IMC design procedure yields classic PI and PID controllers as we will show in the next sections. We will compare these IMC-tuning rules with the rules proposed by Ziegler & Nichols and Cohen & Coon in terms of performance and robustness.

6.1.2   PID Settings for Simple Models

The IMC controllers shown in Table 6.1-1 were designed via the standard procedure developed in Chaps. 4 and 5. We find that IMC leads to PID controllers for virtually all models common in industrial practice. Note that the table includes systems with pure integrators and RHP zeros. Occasionally, the PID controllers are augmented by a first-order lag with time constant τ_F.

For Case O the IMC design procedure would yield a controller of too high an order to be interpreted as a PID controller with lag. Therefore a slightly modified procedure was used resulting in a controller with the same attractive properties as the standard IMC controller. A few remarks regarding Table 6.1-1 are appropriate.

- The controllers have a single adjustable parameter λ. For MP systems λ is the closed-loop time constant. For NMP systems λ is the dominant closed-loop time constant when $\lambda > \beta$, where β^{-1} is the location of the zero in the RHP.

- When the PID controller of the specified form is applied to the model $\tilde{p}$ the closed-loop system is stable for all values of $\lambda \geq 0$.

- For the specified inputs the controllers yield offset free performance for all $\lambda \geq 0$ and are H_2 optimal for $\lambda = 0$.

- For MP plants the controllers with $\lambda = 0$ are H_2 optimal for *all* inputs which behave asymptotically $(t \to \infty)$ like the specified v_M.

- For Cases A, B, C, G, and I the controller gain is inversely proportional to λ, thus demonstrating that on-line PID controller adjustment is effectively achieved by manipulating k_c. These are MP models, for which the model itself imposes no limitations on the bandwidth.

  For Cases D, E, F, K, L, and N the controller gain k_c and the filter constant τ_F are the only parameters dependent on λ, but because of the presence of a RHP zero, there is a maximum gain which cannot be surpassed no matter how small λ is.

  For Cases H, J, M, and O, λ appears in *all* the parameters of the classic feedback controller. It is not surprising that for such processes trial and error tuning of PID controllers is notoriously difficult. However, the IMC parametrization shows how all the controller parameters may be adjusted *simultaneously* in an effective manner.

- In practice the ideal PID controller does not exist. An additional lag is always present in the controller to provide roll-off at high frequencies. Cases D, E, F, and K through O yield "practical" PID controllers with "optimal" roll-off elements $(\tau_F s + 1)^{-1}$.

- No systems with LHP zeros are listed in Table 6.1-1. LHP zeros translate into lags in the feedback controller structure when the IMC design procedure is used. Therefore, for models with LHP zeros, the PID controller from Table 6.1-1 should be augmented with the corresponding lags.

- Controller complexity, as stated in the introduction, depends on the model and the control system objectives. Consider the cases G (a pure integrator), A (a first-order model) and B (a second-order noninteracting model), for which the input is a step. Only a proportional controller is necessary for G, a PI-controller must be used for A, while a PID controller is needed for B. Likewise, consider Cases N and O where the process model is the same: as the inputs become more demanding (ramps instead of steps) the complexity of the controller increases.

- Table 6.1-1 can also be used for systems with delays by approximating the deadtime with a Padé element; then the entry for the rational approximate model provides the controller parameters. This design approach leads to simpler controllers than when the time delay is considered exactly. The use of the Padé approximation, however, introduces modelling error, which consequently limits the achievable bandwidth and the minimum value for λ. This loss of performance is usually insignificant. Corollary 4.3-1 provides the condition

$$\bar{\ell}_m < \frac{1}{|\tilde{\eta}|} = \frac{1}{|\tilde{p}\tilde{q}f|} \tag{6.1-3}$$

which guarantees robust stability when the multiplicative model error is bounded by $\bar{\ell}_m$. The Padé approximation does not generate a family of plants but generates a specific error ℓ_m. If we set $\bar{\ell}_m = |\ell_m|$, (6.1-3) is only sufficient and therefore conservative. Nevertheless it is useful for estimating a lower bound on λ which assures closed-loop stability.

Table 6.1-1. IMC controllers for simple models interpreted as PID controllers with filter:

$$c = k_c\left(1+\tau_D s+\frac{1}{\tau_I s}\right)\frac{1}{\tau_F s+1}; \qquad \gamma = \frac{2\beta\tau_2}{\beta+\tau_2}$$

	Model $\tilde{p}$	Input v_M	$\tilde{\eta} = \tilde{p}q = \tilde{p}\tilde{q}f$	$k_c k$	τ_I	τ_D	τ_F
A	$\dfrac{k}{\tau s+1}$	$\dfrac{1}{s}$	$\dfrac{1}{\lambda s+1}$	$\dfrac{\tau}{\lambda}$	τ	—	—
B	$\dfrac{k}{(\tau_1 s+1)(\tau_2 s+1)}$	$\dfrac{1}{s}$	$\dfrac{1}{\lambda s+1}$	$\dfrac{\tau_1+\tau_2}{\lambda}$	$\tau_1+\tau_2$	$\dfrac{\tau_1\tau_2}{\tau_1+\tau_2}$	—
C	$\dfrac{k}{\tau^2 s^2+2\varsigma\tau s+1}$	$\dfrac{1}{s}$	$\dfrac{1}{\lambda s+1}$	$\dfrac{2\varsigma\tau}{\lambda}$	$2\varsigma\tau$	$\dfrac{\tau}{2\varsigma}$	—
D	$k\dfrac{-\beta s+1}{\tau s+1}$	$\dfrac{1}{s}$	$\dfrac{-\beta s+1}{(\beta s+1)(\lambda s+1)}$	$\dfrac{\tau}{2\beta+\lambda}$	τ	—	$\dfrac{\beta\lambda}{2\beta+\lambda}$
E	$k\dfrac{-\beta s+1}{\tau_1 s+1}$	$\dfrac{1}{\tau_2 s+1}\dfrac{1}{s}$	$\dfrac{(-\beta s+1)(\gamma s+1)}{(\beta s+1)(\lambda s+1)}$	$\dfrac{\gamma+\tau_1}{2\beta-\gamma+\lambda}$	$\gamma+\tau_1$	$\dfrac{\gamma\tau_1}{\gamma+\tau_1}$	$\dfrac{\beta(\gamma+\lambda)}{2\beta-\gamma+\lambda}$
F	$k\dfrac{-\beta s+1}{\tau^2 s^2+2\varsigma\tau s+1}$	$\dfrac{1}{s}$	$\dfrac{-\beta s+1}{(\beta s+1)(\lambda s+1)}$	$\dfrac{2\varsigma\tau}{2\beta+\lambda}$	$2\varsigma\tau$	$\dfrac{\tau}{2\varsigma}$	$\dfrac{\beta\lambda}{2\beta+\lambda}$
G	$\dfrac{k}{s}$	$\dfrac{1}{s}$	$\dfrac{1}{\lambda s+1}$	$\dfrac{1}{\lambda}$	—	—	—

H	$\frac{k}{s}$	$\frac{1}{s^2}$	$\frac{2\lambda s+1}{(\lambda s+1)^2}$	$\frac{2}{\lambda}$	2λ	—	—
I	$\frac{k}{s(\tau s+1)}$	$\frac{1}{s}$	$\frac{1}{\lambda s+1}$	$\frac{1}{\lambda}$	—	τ	—
J	$\frac{k}{s(\tau s+1)}$	$\frac{1}{s^2}$	$\frac{2\lambda s+1}{(\lambda s+1)^2}$	$\frac{2\lambda+\tau}{\lambda^2}$	$2\lambda+\tau$	$\frac{2\lambda\tau}{2\lambda+\tau}$	—
K	$k\frac{-\beta s+1}{s}$	$\frac{1}{s}$	$\frac{-\beta s+1}{(\beta s+1)(\lambda s+1)}$	$\frac{1}{2\beta+\lambda}$	—	—	$\frac{\beta\lambda}{2\beta+\lambda}$
L	$k\frac{-\beta s+1}{s}$	$\frac{1}{\tau_2 s+1}\frac{1}{s}$	$\frac{(-\beta s+1)(\gamma s+1)}{(\beta s+1)(\lambda s+1)}$	$\frac{1}{2\beta-\gamma+\lambda}$	—	γ	$\frac{\beta(\gamma+\lambda)}{2\beta-\gamma+\lambda}$
M	$k\frac{-\beta s+1}{s}$	$\frac{1}{s^2}$	$\frac{(-\beta s+1)(2\beta s+1)(2\lambda s+1)}{(\beta s+1)(\lambda s+1)^2}$	$\frac{2(\beta+\lambda)}{2\beta^2+\lambda^2}$	$2(\beta+\lambda)$	$\frac{2\beta\lambda}{\beta+\lambda}$	$\frac{\beta\lambda^2+4\beta^2\lambda}{2\beta^2+\lambda^2}$
N	$k\frac{-\beta s+1}{s(\tau s+1)}$	$\frac{1}{s}$	$\frac{-\beta s+1}{(\beta s+1)(\lambda s+1)}$	$\frac{1}{2\beta+\lambda}$	—	τ	$\frac{\beta\lambda}{2\beta+\lambda}$
O	$k\frac{-\beta s+1}{s(\tau s+1)}$	$\frac{1}{s^2}$	$\frac{(-\beta s+1)(2(\beta+\lambda)s+1)}{(\beta s+1)(\lambda s+1)^2}$	$\frac{2(\beta+\lambda)+\tau}{2\beta^2+4\beta\lambda+\lambda^2}$	$2(\beta+\lambda)+\tau$	$\frac{2\tau(\beta+\lambda)}{2(\beta+\lambda)+\tau}$	$\frac{\beta\lambda^2}{2\beta^2+4\beta\lambda+\lambda^2}$

Example 6.1-1. The system

$$p(s) = \frac{1 - k_1 e^{-\theta s}}{s} \tag{6.1 - 4}$$

was studied in Sec. 5.7.1. Using a first-order Padé approximation we find

$$\tilde{p}(s) = \frac{1 - k_1 + \frac{\theta}{2}(1 + k_1)s}{s(\frac{\theta}{2}s + 1)} \tag{6.1 - 5}$$

If $k_1 > 1$ then (6.1–5) has a RHP zero and a controller from Case N or O can be selected. If $k_1 < 1$ the resulting LHP zero should be removed by a simple lag, as explained above. PID parameters can then be obtained from Case I or J.

For the specific distillation column model studied in Sec. 5.7.1, (6.1–5) becomes

$$\tilde{p}(s) = \frac{-1 + 2.25s}{s(0.75s + 1)} \tag{6.1 - 6}$$

Case O provides the PID parameters

$$k = -\frac{5.25 + 2\lambda}{10.125 + 9\lambda + \lambda^2} \tag{6.1 - 7}$$

$$\tau_I = 5.25 + 2\lambda \tag{6.1 - 8}$$

$$\tau_D = \frac{3.375 + 1.5\lambda}{5.25 + 2\lambda} \tag{6.1 - 9}$$

$$\tau_F = \frac{2.25\lambda^2}{10.125 + 9\lambda + \lambda^2} \tag{6.1 - 10}$$

With this controller

$$|\tilde{\eta}| = \left| \frac{2(2.25 + \lambda)s + 1}{(\lambda s + 1)^2} \right| \tag{6.1 - 11}$$

In Fig. 6.1-1 the norm of the model error

$$\ell_m = \frac{p - \tilde{p}}{\tilde{p}} = \frac{-e^{-\theta s} + \frac{-\theta s + 2}{\theta s + 2}}{\frac{1}{k_1} - \frac{-\theta s + 2}{\theta s + 2}} \tag{6.1 - 12}$$

$$= \frac{2 - 1.5s - (1.5s + 2)e^{-1.5s}}{-1 + 2.25s} \tag{6.1 - 13}$$

is compared with $|\tilde{\eta}|^{-1}$ for $\lambda = 2$. Condition (6.1–3) is satisfied and stability is assured for all $\lambda \geq 2$.

The error resulting from a ramp input (s^{-2}) is shown in Fig. 6.1-2 for $\lambda = 2$. It should be compared with the error obtained with the H_2-optimal controller. With regard to the maximum deviation from steady state there is little incentive

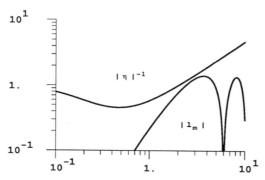

Figure 6.1-1. Error ℓ_m caused by Padé approximation and complementary sensitivity $\tilde{\eta}$ for $\lambda = 2$.

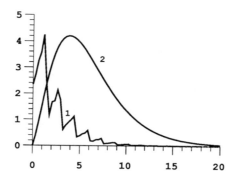

Figure 6.1-2. Error resulting from ramp input for ISE-optimal controller (1) and PID controller (2).

for using anything more sophisticated than a PID controller. If settling time or ISE is critical, the H_2-optimal controller is preferred. Note however, that the H_2-optimal controller used in the simulation is improper and would have to be detuned in practice for robustness reasons. Then its advantages over a properly tuned PID controller might become quite small. □

Example 6.1-2. Most commonly in process control the time delay appears as a multiplicative term in the transfer function

$$p = p_M e^{-\theta s} \qquad (6.1-14)$$

For the zeroth-order Padé approximation $(e^{-\theta s} \cong 1)$

$$|\ell_m| = |1 - e^{-\theta s}| \qquad (6.1-15)$$

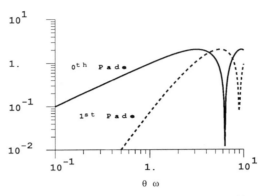

Figure 6.1-3. Error ℓ_m caused by zeroth and first order Padé approximation.

and for the first-order Padé approximation

$$|\ell_m| = \left| \frac{-\frac{\theta}{2}s + 1}{\frac{\theta}{2}s + 1} - e^{-\theta s} \right| \qquad (6.1-16)$$

Both multiplicative errors are plotted in Fig. 6.1-3. For the first-order Padé approximation, $|\ell_m| = 1$ at $\omega \cong 3/\theta$ and thus a sufficient condition for stability is to choose $\lambda > \theta/3$. For the zeroth-order approximation $|\ell_m| = 1$ at $\omega \cong 1/\theta$ and therefore $\lambda > \theta$ is required. Because the bandwidth is inherently limited by the NMP nature of the system, one can expect that using the first-order Padé approximation will yield designs very close to optimal (i.e., as if no approximation were present). The zeroth-order approximation will be adequate, however, when small bandwidths and low frequency inputs are involved.

For the popular first-order model with deadtime

$$p(s) = \frac{ke^{-\theta s}}{\tau s + 1} \qquad (6.1-17)$$

the zeroth-order Padé approximation yields

$$\tilde{p}(s) = \frac{k}{\tau s + 1} \qquad (6.1-18)$$

Entry A in Table 6.1-1 provides a PI controller for this structure. A first-order Padé approximation yields

$$\tilde{p}(s) = \frac{k(-\frac{\theta}{2}s + 1)}{(\tau s + 1)(\frac{\theta}{2}s + 1)} \qquad (6.1-19)$$

Entry F (PID controller with first-order lag) is applicable here. This problem is discussed in more detail in the next subsection.   □

Table 6.1-2. IMC-based "real" PID parameters for $p(s) = ke^{-\theta s}(\tau s + 1)^{-1}$ and practical recommendations for λ.

Controller	kk_c	τ_I	τ_D	τ_F	Recommended λ/θ ($\lambda > 0.2\tau$ always)
PID	$\frac{2\tau+\theta}{2(\lambda+\theta)}$	$\tau + \frac{\theta}{2}$	$\frac{\tau\theta}{2\tau+\theta}$	$\frac{\lambda\theta}{2(\lambda+\theta)}$	> 0.25
PI	$\frac{\tau}{\lambda}$	τ	–	–	> 1.7
Improved PI	$\frac{2\tau+\theta}{2\lambda}$	$\tau + \frac{\theta}{2}$	–	–	> 1.7

6.1.3   PID Settings for a First-Order System with Deadtime

The important role of the first-order lag/deadtime model in process control mandates a more detailed discussion of Ex. 6.1-2. Our attention is directed to a further understanding of the PI and PID rules generated by cases A and F.

The IMC-based controllers obtained with first- and zeroth-order Padé approximations for the time delay are shown in the first two rows of Table 6.1-2. Inspection of the closed-loop transfer functions for system (6.1–17) with these controllers indicates a number of advantages:

- The closed-loop response is independent of the system time constant τ (the process lag $(1 + \tau s)^{-1}$ is cancelled by the controller).

- Time is scaled by θ.

- The shape of the response depends on λ/θ only.

In other words, specifying one value of λ/θ for any first-order-lag-with-deadtime model results in an identical response when time is scaled by θ, regardless of k, θ,

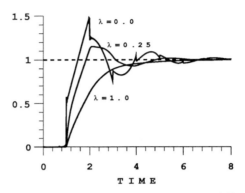

Figure 6.1-4. Step responses with IMC-PID controller for different values of λ.

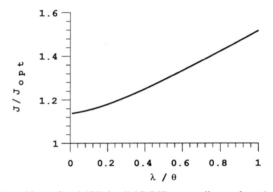

Figure 6.1-5. Normalized ISE for IMC-PID controller as function of λ/θ.

and τ. The choice of the "best" ratio λ/θ must be based on performance and robustness considerations.

PID Controller

For the PID controller Fig. 6.1-4 demonstrates the dependence of the step response on λ/θ. A ratio of $\lambda/\theta = 0.25$ offers a good compromise between rise time and overshoot. Because of the scaling properties of the closed-loop transfer function a convenient design plot can be made (Fig. 6.1-5). The performance measure J, the ISE for a step input change, has been plotted as a function of λ/θ. J is normalized by J_{opt}, the ISE corresponding to the optimum response $y/r = e^{-\theta s}$. In theory, a Smith predictor with infinite gain ($k_c = \infty$) accomplishes this response (see Sec. 6.2). Very small values of λ/θ lead to robustness problems. Based on Figs. 6.1-4 and 6.1-5, $\lambda/\theta = 0.25$ is recommended.

Figure 6.1-5 also confirms that the first-order Padé approximation leads to

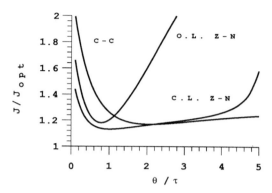

Figure 6.1-6. Comparison of normalized ISE for PID controllers tuned according to the closed-loop and open-loop Ziegler-Nichols rules and according to Cohen-Coon.

relatively little performance deterioration. For $\lambda/\theta = 0.25$ the result is a PID controller that performs with only 20% greater ISE than the *optimal* Smith predictor, while retaining favorable robustness characteristics.

The recommendation $\lambda/\theta = 0.25$ is made under the assumption of no plant uncertainty apart from the "uncertainty" introduced by the Padé approximation. Significant plant uncertainty within the bandwidth of the controller will require the designer to select a larger value of λ. This consideration is of particular concern when $\lambda/\tau \ll 1$. Because for the process industries the closed-loop bandwidth can rarely exceed ten times the open-loop bandwidth $(10/\tau)$, a practical requirement is to always select $\lambda > \tau/10$. For the IMC-PID parameters this inequality is dominant for $\theta/\tau < 0.4$.

Next we compare IMC-PID with "real" PID controllers

$$c(s) = k_c \left[1 + \frac{1}{\tau_I s} + \frac{\tau_D s}{0.1\tau_D s + 1} \right] \qquad (6.1 - 20)$$

tuned according to the classic Ziegler-Nichols and Cohen-Coon tuning rules (Fig. 6.1-6). The first notable difference between these rules and those from IMC is that J depends strongly on θ/τ, while for the IMC rules the performance is independent of $\theta/\tau (J/J_{\text{opt}} = 1.2)$. For small and large values of θ/τ the IMC rules clearly lead to superior performance. For $\theta/\tau > 6$ the closed-loop system is even *unstable* with the closed-loop Ziegler-Nichols parameters. Figures 6.1-7A to C show that in the range of $\theta/\tau \approx 1$ the smaller ISE obtained with the non-IMC settings is obtained at the cost of excessive overshoot: 50% are rarely acceptable in practice.

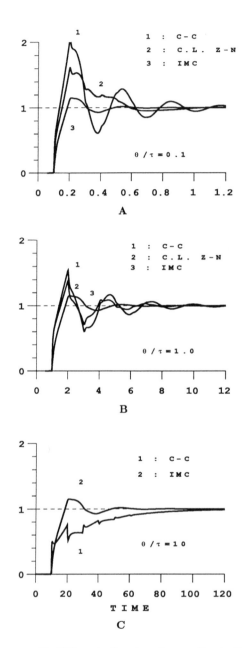

Figure 6.1-7. Step responses with PID controllers tuned according to IMC ($\lambda/\theta = 0.25$), Ziegler-Nichols and Cohen-Coon for different ratios θ/τ. A: $\theta/\tau = 0.1$, B: $\theta/\tau = 1.0$, C: $\theta/\tau = 10$.

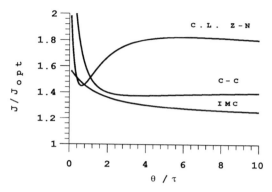

Figure 6.1-8. Normalized ISE for PI controllers tuned according to IMC, Ziegler-Nichols and Cohen-Coon.

PI Controller

Overall the IMC-PI rule is not superior to the Ziegler-Nichols and Cohen-Coon expressions because the zeroth-order Padé approximation neglects deadtime altogether. This can be remedied by incorporating the deadtime in the internal model through other means and leads to the "Improved PI rule" shown in Table 6.1-2. The performance of the IMC-PI controller is now dependent on θ/τ, but for $\lambda/\theta = 1.7$ it is superior to Ziegler-Nichols and Cohen-Coon for almost all values of θ/τ (Fig. 6.1-8).

Robustness to Deadtime Errors

It is important to emphasize that the improved performance obtained through the IMC settings has not been obtained at the cost of robustness. It can be verified that both the PID and the PI controller allow at least 140% deadtime error for all values of θ/τ before the closed-loop system becomes unstable.

6.1.4 Summary

The IMC design procedure was used to derive PID controller parameters for a variety of models commonly used in the process industries (Tab. 6.1-1). The adjustable parameter λ corresponds approximately to the closed-loop time constant and is selected by the designer to achieve the appropriate compromise between performance and robustness and to keep the action of the manipulated variable within bounds. The closed-loop system is stable for all $\lambda \geq 0$. The performance is ISE optimal for $\lambda = 0$.

It is recommended that for time delays a first-order Padé approximation is used $(\theta \cong (-\theta s + 2)/(\theta s + 2))$. The PID parameters for the rational approximate model can then be found again in Table 6.1-1. Because of the approximation ("model error") there is a lower bound on λ for stability. When the delay appears in the transfer function as a factor, $\lambda > \theta/3$ is recommended. For the specific case of a first order process with delay the results are summarized in Table 6.1-2.

6.2 IMC Interpretation of Smith Predictor Controller

The "Smith predictor" was proposed in the 1950s to improve the closed-loop performance for systems with time delay. The opinions on its merits expressed in dozens of papers which have appeared in the last three decades vary widely. Our discussion aims to clarify some of the misconceptions and to make some concrete statements regarding the robustness properties.

6.2.1 General Relationships

The feedback control structure with the Smith predictor in place is shown in Fig. 6.2.1A. Here

$$p = p^* e^{-s\theta} \tag{6.2 - 1}$$

where p^* does not contain any delay terms. The block marked with double lines is called "predictor." The reason for the name "predictor" can be understood from the following arguments. Without the predictor the feedback signal z is equal to the output y

without predictor:

$$y = pu \tag{6.2 - 2}$$

with predictor:

$$z = p^* u \tag{6.2 - 3}$$

From (6.2–2) and (6.2–3) the feedback signal z with the predictor in place is related to the output y by

$$z = e^{+s\theta} y \tag{6.2 - 4}$$

Thus the block $(p - p^*)$ *predicts the effect of the manipulated variable u on the output y and modifies the feedback signal z accordingly.*

A simple rearrangement of the block diagram in Fig. 6.2-1A which leaves all input-output relationships unaffected makes the similiarity with the IMC structure (Fig. 6.2-1B) apparent.

$$q = \frac{c}{1 + p^* c} \tag{6.2 - 5}$$

The IMC structure in Fig. 6.2-1B is internally stable if and only if p and q are stable (Thm. 3.2-1). This implies first that the Smith-predictor controller implemented in the way indicated in Fig. 6.2-1B can be used for *stable* plants

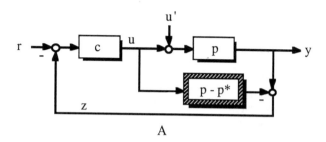

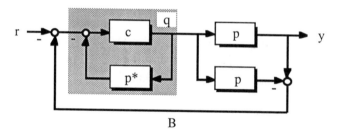

Figure 6.2-1. Smith predictor controller in standard (A) and IMC (B) form.

only. The alternative implementation indicated in Fig. 6.2-1A is also internally unstable for unstable plants. To prove this consider the effect of the external input u' on the output y

$$\frac{y}{u'} = \frac{p^* e^{-s\theta}}{1 + p^* c}[1 - (e^{-s\theta} - 1)p^* c] \qquad (6.2-6)$$

The transfer function (6.2–6) has RHP poles for all choices of the controller c: If c cancels the RHP poles of p^* the first part of expression (6.2–6) is unstable. If it does not, the term in brackets is unstable.

The second implication of Thm. 3.2-1 is that q defined by (6.2–5) has to be stable for internal stability of the Smith-predictor control structure. q is stable if the closed-loop system with the controller c applied to the *delay-free* plant p^* is stable. This is one of the main motivations for using the Smith-predictor: stability of the *nominal* system can be studied in terms of the delay-free plant.

While the Smith-predictor structure is in many aspects similar to the IMC structure, it does not utilize the same controller parametrization and thus loses some of the main advantages of IMC: (1) The set of c's which yields closed-loop stable systems is difficult to define. With the IMC parametrization any stable IMC controller q guarantees closed loop stability. (2) Performance measured in terms of the sensitivity function is a nonlinear function of c while it is a linear function of q.

6.2.2   Some Myths about the Tuning of Smith-Predictor Controllers

The literature abounds with myths regarding the properties, advantages, and drawbacks of Smith-predictor controllers. We hope to diffuse some of them in the following.

Myth #1. With the predictor in place the controller c can be tuned just like for a system p^* without time delay. If stability is the only issue then the statement is correct. If the designer is also concerned about *performance* then it becomes a myth which is clearly not true. For Fig. 6.2-1 the sensitivity function which is indicative of performance is

$$\tilde{\epsilon} = 1 - pq = 1 - p\frac{c}{1 + p^* c} \qquad (6.2-7)$$

It depends on the plant *with* delay and this dependence has to be taken into consideration when adjusting c. Let us assume for example that p^* is MP. If we design c for p^* then it would be optimal (with respect to any objective function) to increase the gain of the controller c to infinity leading to

$$q = \frac{1}{p^*} \qquad (6.2-8)$$

and

$$\tilde{\epsilon} = 1 - e^{-s\theta} = 1 - p_A \qquad (6.2-9)$$

Tables 4.1-1 and 4.1-2 indicate that the sensitivity function (6.2–9) is H_2-optimal for step inputs *only*. For any other input, tuning the controller c for good performance on the delay-free plant p^* will *not* lead to good performance when c is applied to the real plant p (with delay) in the context of the Smith-predictor structure.

Myth #2. Smith-predictor controllers work well for setpoint changes but not for disturbances. Implicit in this statement is usually the assumption that both the setpoint changes and the disturbances are steps but that the disturbances go through a lag (e.g., the plant itself) before acting on the plant output. Myth #1 indicates that unknowingly c is usually tuned for an H_2-optimal response to step inputs. It is indeed true that these controller parameters will not result in good performance for slowly changing disturbances. However, this is not a fault of the Smith-predictor structure but rather a consequence of incorrect tuning of the controller c. Using the results of Chaps. 4 and 5, q and thus c can be found easily for optimal disturbance rejection. However, these parameters again will generally not yield a good setpoint step response. This is the basic trade-off problem the designer faces when tuning a controller for two very different inputs and it has nothing to do with the Smith predictor. It can be easily resolved by the two-degree-of-freedom structure discussed on several occasions in this book.

Myth #3. Smith-predictor controllers are sensitive to model errors. Let the norm of the multiplicative model error be bounded by $\bar{\ell}_m(\omega)$. Then we know from Thm. 2.5-1 that the system is robustly stable if the complementary sensitivity function of the model satisfies

$$|\tilde{\eta}| < \frac{1}{\bar{\ell}_m} \qquad (6.2-10)$$

Because $|p| = |p^*|$, (6.2–10) becomes for the Smith-predictor structure

$$\left| \frac{p^*c}{1 + p^*c} \right| < \frac{1}{\bar{\ell}_m} \qquad (6.2-11)$$

The same robust stability condition would have to be satisfied if the model did not include a time delay in the first place. The Smith-predictor does not in any way increase the sensitivity to modelling error. Problems arise in practice because c tends to be adjusted without attention to the limitations imposed by model uncertainty. The ideas presented in Chap. 4 can be easily adapted to design high performance robust controllers for Smith-predictor control systems as is demonstrated in the next subsection.

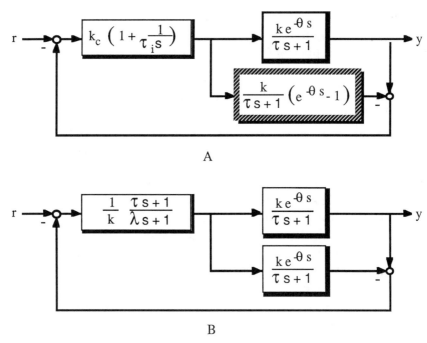

Figure 6.2-2. Smith-predictor/PI controller for first-order deadtime system in standard (A) and IMC (B) form.

6.2.3   Robust Tuning of Smith-Predictor Controller for First-Order System with Deadtime

For a first-order system with deadtime a Smith predictor with a PI controller (Fig. 6.2-2A) is equivalent to IMC (Fig. 6.2-2B) if

$$k_c = \frac{\tau}{\lambda k} \qquad\qquad (6.2-12)$$

$$\tau_I = \tau \qquad\qquad (6.2-13)$$

If the integral time is set equal to the system time constant, adjusting the PI controller gain is equivalent to adjusting the IMC filter parameter λ. Because IMC is stable for all $\lambda \geq 0$ the Smith predictor with PI controller is stable for all $k_c \geq 0, \tau_I = \tau$. Model uncertainty imposes an upper (lower) limit on $k_c(\lambda)$.

Deadtime Uncertainty. Let us first consider uncertainty in the process deadtime only. This was discussed in detail in Sec. 4.6.1. We found that the rule of thumb $\lambda = \bar{\delta}$ or

$$k_c = \frac{\tau}{\bar{\delta}k} \tag{6.2 - 14}$$

yields acceptable robust performance for a maximum deadtime error of $\bar{\delta}$.

General Uncertainty. When all three model parameters have error bounds associated with them the complete robust design procedure demonstrated in Sec. 4.6.2 has to be carried out to determine the appropriate choice of k_c. The integral time is always set equal to the model time constant regardless of model uncertainty.

6.2.4 Summary

We related the Smith-predictor control structure to the IMC structure (Fig. 6.2-1) and concluded that IMC has the same predictor capabilities. For the design of the feedback control the IMC design procedure is preferred because it deals optimally with different input forms (steps, ramps, etc.) and addresses robustness explicitly. If designed properly — e.g., via the IMC design procedure, the Smith-predictor controller can offer performance advantages over classic feedback control. These advantages are small (or negligible) if model uncertainty requires significant detuning.

6.3 Feedforward Control

6.3.1 Objectives and Structure

Feedforward control can be used to compensate for measurable disturbances. It is discussed here for stable systems only. The combined classic feedforward-feedback structure is shown in Fig. 6.3-1.

The effect of the disturbance on the output is described by

$$y = \frac{p_d - c_f p}{1 + pc} d' \tag{6.3 - 1}$$

Ideally one should select

$$c_f = \frac{p_d}{p} \tag{6.3 - 2}$$

which leads to perfect disturbance compensation. If (6.3–2) is noncausal or unstable then perfect feedforward control is not possible, and c_f should be chosen to minimize the effect of d' on y. It is apparent from (6.3–1) that the optimal choice of c_f depends both on the feedback controller c and the nature of the

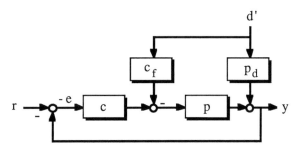

Figure 6.3-1. Classic feedforward/feedback control scheme.

disturbance d'. In the context of IMC the structure shown in Fig. 6.3-2 is recommended. Through some block diagram manipulations it can be shown that the IMC controllers are related to the classic controllers by

$$c = \frac{q}{1 - \tilde{p}q} \qquad (6.3 - 3)$$

$$c_f = \frac{q_f - \tilde{p}_d q}{1 - \tilde{p}q} \qquad (6.3 - 4)$$

The advantage of the IMC parametrization is that it separates the effect of the feedforward controller q_f and the feedback controller q and allows them to be designed independently. When there is no model error ($p = \tilde{p}, p_d = \tilde{p}_d$)

$$e = (p_d - pq_f)d' - (1 - pq)r \qquad (6.3 - 5)$$

If

$$q_f = \frac{p_d}{p} \qquad (6.3 - 6)$$

is causal and stable it can be implemented for perfect disturbance compensation. If it is not then it is natural to select q_f such that

$$\|(p_d - pq_f)d'\|_2 \qquad (6.3 - 7)$$

is minimized.

6.3.2 Design

Theorem 6.3-1. *Assume that p and p_d are stable and that d' is bounded. The H_2-optimal feedforward controller $\tilde{q}_f$ minimizes*

$$\frac{1}{2\pi} \int_{-\infty}^{\infty} |(p_d - p\tilde{q}_f)d'|^2 d\omega \qquad (6.3 - 8)$$

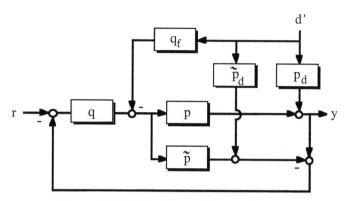

Figure 6.3-2. IMC feedforward/feedback scheme.

and is given by

$$\tilde{q}_f = (p_M d'_M)^{-1} \left\{ p_A^{-1} p_d d'_M \right\}_* \qquad (6.3-9)$$

where the operator { }∗ denotes that after a partial fraction expansion of the operand all terms involving the poles of p_A^{-1} are omitted.

Proof. The proof follows the same arguments as the proof of Thm. 5.2-1.

It is instructive to compare the feedforward design problem which aims to minimize (6.3–7) with the standard design problem discussed in Chap. 5 which has as its objective to minimize

$$\|(1 - p\tilde{q})p_d d'\|_2^2 \qquad (6.3-10)$$

In the latter case $p\tilde{q}$ is to approximate identity, in the former case $p\tilde{q}_f$ is to approximate p_d. If p_d and p happen to have the same NMP elements then $\tilde{q}_f$ can be chosen to make the approximation perfect (see 6.3–6). On the other hand, perfect control of NMP systems is not feasible through feedback control. *Thus, feedforward control can only be more effective than feedback if both p and p_d are NMP.* This is seen more clearly by comparing the error norms obtained with the two optimal controllers (6.3–9) and (4.1–7).

Feedforward:

$$\left\| p_d d'_M - p_A \left\{ p_A^{-1} p_d d'_M \right\}_* \right\|_2 \qquad (6.3-11)$$

Feedback:

$$\left\| p_{dM} d'_M - p_A \left\{ p_A^{-1} p_{dM} d'_M \right\}_* \right\|_2 \qquad (6.3-12)$$

If the process is MP ($p_A = 1$) then both (6.3–11) and (6.3–12) are zero. If the disturbance transfer function p_d is MP ($p_d = p_{dM}$) then (6.3–11) and (6.3–12) are identical and as far as nominal performance is concerned there is no incentive to use feedforward. Evaluation of (6.3–11) and (6.3–12) allows one to determine the potential improvements via feedforward control in the general case.

Example 6.3-1.

$$p = p_A = e^{-2s}, \quad p_d = e^{-s}, \quad d' = \frac{1}{s}$$

Feedforward:

$$q_f = s\{e^{2s} \cdot e^{-s} \cdot s^{-1}\}_*$$

$$= s\{e^s \cdot s^{-1}\}_* = s \cdot s^{-1} = 1$$

$$\|e\|_2 = \|e^{-s}s^{-1} - e^{-2s}s^{-1}\|_2 = \|(1 - e^{-s})s^{-1}\|_2 = 1$$

Feedback:

$$\|e\|_2 = \|(1 - e^{-2s})s^{-1}\|_2 = 2$$

$\square$

Example 6.3-2.

$$p = p_A = \frac{-s + 1}{s + 1}, \quad p_d = e^{-2s}, \quad d' = \frac{1}{s}$$

Feedforward:

$$q_f = s\left\{\frac{(s+1)}{(-s+1)}e^{-2s}\frac{1}{s}\right\}_* = s\left[\frac{(s+1)e^{-2s}}{(-s+1)s} - \frac{2e^{-2}}{-s+1}\right] = \frac{s(e^{-2s} - 2e^{-2}) + e^{-2s}}{-s+1}$$

$$\|e\|_2 = \left\|e^{-2s}\frac{1}{s} - \frac{-s+1}{(s+1)s}\left[\frac{s(e^{-2s} - 2e^{-2}) + e^{-2s}}{-s+1}\right]\right\|_2 = \left\|\frac{2e^{-2}}{s+1}\right\|_2 = 2e^{-2} \cdot \frac{1}{\sqrt{2}} = 0.19$$

Feedback: From Thm. 4.1-4

$$\|e\|_2 = \sqrt{2} = 1.41$$

$\square$

All the discussions about the advantages of feedforward over feedback in this section neglect model error. On one hand, model error forces the feedback controller to be detuned for robustness and nominal performance to be sacrificed. On the other hand, the performance of a feedforward controller is more sensitive to model mismatch than that of a feedback controller. A comparison of feedforward and feedback in the presence of model mismatch is a topic of current research.

6.3.3 Summary

If perfect feedforward compensation is possible then the IMC feedforward/feedback structure does not offer any specific advantage over the classic structure. If perfect compensation is not possible then the feedforward and the feedback designs "interact" in the classic structure

$$\frac{y}{d'} = \frac{p_d - c_f p}{1 + pc} \tag{6.3 - 1}$$

but are *independent* for IMC. The H_2-optimal IMC feedforward controller is computed from

$$\tilde{q}_f = (p_M d'_M)^{-1} \left\{ p_A^{-1} p_d d'_M \right\}_* \tag{6.3 - 9}$$

The achievable performance with feedforward is superior to that with feedback only if *both* the plant and the disturbance transfer function are NMP.

6.4 Cascade Control

6.4.1 Objectives, Structure, and Design

Cascade control (Fig. 6.4-1) is a specific form of a two-degree-of-freedom control structure which can be used if in addition to the output to be controlled (y_2) another process output (y_1) can be measured. It is popular to design c_1 first with the outer loop open and then to design c_2 for the new "plant" consisting of the closed inner loop combined with p_2. Four questions arise:

1. Consider the error e_2 in response to the inputs r_2, d_1 and d_2. Under what conditions is the performance which is achievable by the 2-loop structure superior to that which can be obtained with one controller alone — i.e., with the inner loop open?

2. Does the two step-design procedure, where the inner loop is designed first, limit the performance which is achievable with the outer loop?

3. How should c_1 and c_2 be designed?

4. How does model uncertainty affect design and performance?

1. Advantages of the Two-Controller Structure
   If we assume that only a single loop (the outer one) is used, then the three inputs affect the output in the following manner

$$e_2 = (1 - p_1 p_2 q)(d_2 + p_2 d_1 - r_2) \tag{6.4 - 1}$$

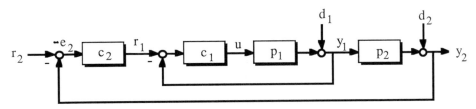

Figure 6.4-1. Cascade control system.

Here we have employed the IMC parametrization for the controller. If two controllers are employed then (6.4–1) becomes

$$e_2 = (1 - p_1 q_1 p_2 q_2)(d_2 - r_2) + (1 - p_1 q_1)p_2 d_1 \qquad (6.4 - 2)$$

If we set $q_1 q_2 = q$ and compare (6.4–1) and (6.4–2) we note that tracking of r_2 and suppression of d_2 are equally effective with a one and a two-controller structure. We rewrite the response to d_1

$$1 - \text{loop}: \quad e_2 = (1 - p_1 p_2 q)p_2 d_1 \qquad (6.4 - 3)$$

$$2 - \text{loop}: \quad e_2 = (1 - p_1 q_1)p_2 d_1 \qquad (6.4 - 4)$$

The suppression of d_1 with the 2-loop structure is superior if q_1 can be designed such that

$$\min_{q_1} \|(1 - p_1 q_1)p_2 d_1\|_2 << \min_{q} \|(1 - p_1 p_2 q)p_2 d_1\|_2 \qquad (6.4 - 5)$$

i.e., if the inner loop alone compensates for the disturbance $p_2 d_1$ much better than the outer loop by itself ever could. It should be clear from our discussions of H_2-optimal control that (6.4–5) is an equality if p_2 is MP and stable. *Thus, a cascade controller is only useful if p_2 has RHP zeros or a time delay.*

2. Performance Limitations Imposed on Outer Loop

If the inner loop is effective ((6.4–5) holds), then the outer loop should be designed primarily for r_2 and d_2. The question is whether performance of the outer loop has been sacrificed by the design of the inner loop (q_1). Let us assume that the controller for the inner loop has been implemented in the form shown in Fig. 6.4-2 where

$$c_1 c_3 = \frac{q_1}{1 - p_1 q_1} \qquad (6.4 - 6)$$

We require that c_1 is MP and c_3 stable and that there are no RHP pole-zero cancellations between c_1 and c_3. Otherwise c_1 and c_3 are arbitrary. We showed in

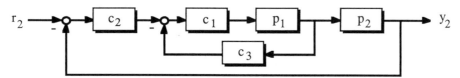

Figure 6.4-2. Cascade control system with additional controller block c_3.

Sec. 5.2.4 that with this choice the inner loop imposes no performance limitations on the outer loop. Furthermore the outer loop can be designed without regard for any RHP poles in p_1. When c_3 is MP or c_1 is stable then the number of control blocks can be reduced from three to two (Fig. 6.4-1) without effect on closed-loop performance or internal stability.

3. Design

Usually only two controller blocks (Fig. 6.4-1) are used for the implementation. Let us assume that (6.4–5) is satisfied as a strong inequality or in other words, that p_2 has significant NMP character. Then a two-step design procedure is reasonable where q_1 is designed first to solve

$$\min_{q_1} \|(1 - p_1 q_1)p_2 d_1\|_2 \qquad (6.4-7)$$

Then q_2 is determined from

$$\min_{q} \|(1 - p_1 p_2 q)d_2\|_2 \qquad (6.4-8)$$

$$q_2 = \frac{q}{q_1} \qquad (6.4-9)$$

(This assumes that q_1 is MP, otherwise more than the two blocks in Fig. 6.4-1 are necessary.) If $p_1 q_1 \cong 1$ over the expected bandwidth of the outer loop then it is even justifiable to assume for the outer loop design that the inner loop is under perfect control ($p_1 q_1 = 1$) and to obtain q_2 from

$$\min_{q_2} \|(1 - p_2 q_2)d_2\|_2 \qquad (6.4-10)$$

4. Uncertainty

The preceding discussions dealt with the cases when model uncertainty was negligible. To date we have not explored cascade control under uncertainty in any depth. It is without question that very often the inner loop is introduced primarily to remove model uncertainty. For example, a flow control loop can correct for uncertain and possibly nonlinear valve characteristics.

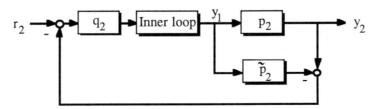

Figure 6.4-3. IMC implementation of cascade control system to avoid reset windup.

6.4.2   Implementation

The cascade controller implementation shown in Fig. 6.4-1 or 6.4-2 can lead
to serious reset windup problems. Reset windup can be avoided by informing
the controller when its output, the manipulated variable, is saturating. Upon
saturation the controller turns off the integral action as an "anti-reset windup"
measure.

Difficulties arise because the output of controller c_2 is the setpoint r_1 of the
inner loop rather than the true manipulated variable u. Reset windup occurs
when r_1 continues to change after u has saturated.

The IMC implementation (Fig. 6.4-3) avoids this windup problem. The inner
loop can be implemented in the IMC or the classic form. By feeding the true
output y_1 of the inner loop to the model $\tilde{p}_2$, the outer loop will recognize any
saturation of u as it is reflected in y_1 and windup does not occur (compare the
discussion on classic feedback via IMC implementation in Sec. 3.2.3).

6.4.3   Summary

Cascade control in its basic form shown in Fig. 6.4-1 is useful under two circumstances:

(i) Disturbance d_1 is significant and p_2 is NMP.

(ii) Plant p_1 has much uncertainty associated with it — e.g., poorly known
nonlinear valve characteristics — and the inner loop serves to remove this
uncertainty.

Disregarding model uncertainty we arrived at the following conclusions:

(i) For the extended structure in Fig. 6.4-2 the achievable performance (regarding control of d_2 and r_2) with the outer loop is unaffected by the inner loop. If $c_3 = q_1$ is MP the same is true for the simple structure in Fig. 6.4-1.

(ii) When p_2 has significant NMP character then a simple two step design procedure is recommended:

(1) q_1 is designed to solve

$$\min_{q_1} \|(1 - p_1q_1)p_2d_1\|_2 \qquad (6.4-7)$$

(2) q_2 is determined from

$$\min_{q_2} \|(1 - p_2q_2)d_2\|_2 \qquad (6.4-10)$$

(This assumes that $p_1q_1 \cong 1$ over the expected bandwidth of the outer loop).

6.5 References

6.1. The ideas in this section were expressed originally by Morari, Skogestad & Rivera (1984). The presentation here follows closely that by Rivera, Skogestad & Morari (1986). They also studied the performance of the ideal (improper) PID controller for a first-order system with deadtime.

6.1.3. Similar settings for the PID controller were proposed by Smith & Corripio (1985). It should be noted that on commercial hardware different PID implementations can be found (see Smith & Corripio, 1985). The PID parameters proposed here have to be adapted accordingly. The form (6.1–20) was mentioned in the book by Åström & Wittenmark (1984). The tuning rules by Ziegler & Nichols were published in 1942. The rules by Cohen & Coon (1953) are limited to first-order deadtime models.

In case the process model is not of a form which leads naturally to PI or PID controller, model reduction has to be applied first or simultaneously with the controller design procedure (Rivera, 1987). This technique has been implemented in the ROBEX software (Lewin et al., 1987).

6.2. The predictor was first proposed by Smith (1957). Palmor & Shinnar (1981) discuss its optimality and robustness. They show that gain margin and phase margin are useless robustness measure for this case. Åström (1977) points out that the Smith predictor is an extreme form of lead compensator. The insights presented in a paper by Horowitz (1983) are similar to the ones in this section. The tuning rule (6.2–14) was mentioned first by Brosilow (1979).

Part II

SAMPLED DATA SINGLE-INPUT SINGLE-OUTPUT SYSTEMS

Chapter 7

FUNDAMENTALS OF SAMPLED-DATA SYSTEMS CONTROL

The chapters in this book dealing with the control of sampled-data systems are not self-contained. It is assumed that the reader has mastered the preceding chapters addressing the same topics for continuous-time systems. Indeed, some issues which are essentially identical (for example, the design of two-degree-of-freedom controllers) are completely omitted. In other cases only those features which distinguish sampled-data systems from continuous systems are emphasized. The equivalence of the classic feedback structure with the IMC structure was firmly established for continuous systems. Therefore, rather than deriving all stability and performance conditions first for the classic feedback structure (Chap. 2) and then translating them to the IMC structure (Chap. 3) we will proceed directly with the IMC structure after some general definitions and results for sampled-data systems control.

Out treatment of sampled-data systems is different from that in many other books in that we define performance in terms of the *continuous* plant output – i.e., we pay close attention to the intersample behavior.

7.1 Sampled-Data Feedback Structure

The block diagram of a typical computer-controlled system is shown in Fig. 7.1-1A. Thick lines are used to represent the paths along which the signals are continuous (analog). The sampling switch is used to describe the A/D converter which is modelled as an impulse modulator. When a signal $a(t)$ is fed to a switch with a sampling time T, it yields as an output the impulse sequence $a^*(t)$

$$a^*(t) = \sum_{k=0}^{\infty} a(kT)\delta(t - kT) \qquad (7.1-1)$$

The Laplace transform of $a^*(t)$ is

$$\mathcal{L}\{a^*(t)\} = a^*(e^{sT}) = \sum_{k=0}^{\infty} a(kT)e^{-skT} \qquad (7.1-2)$$

Alternatively we can represent the impulse sequence by its Fourier series

$$a^*(t) = \frac{1}{T} \sum_{k=-\infty}^{\infty} a(t)e^{ik\omega_s t} \qquad (7.1-3)$$

where ω_s is the sampling frequency

$$\omega_s = \frac{2\pi}{T} \qquad (7.1-4)$$

From (7.1–3) we obtain a different representation of $\mathcal{L}\{a^*(t)\}$

$$\mathcal{L}\{a^*(t)\} = a^*(e^{sT}) = \frac{1}{T} \sum_{k=-\infty}^{\infty} a(s + ik\omega_s) \qquad (7.1-5)$$

The transformation

$$z = e^{sT} \qquad (7.1-6)$$

will be used throughout the book. Then $a^*(z)$ is the z-transform of the signal $a(t)$. The following notation describes (7.1–2) and (7.1–5).

$$a^*(z) = \mathcal{ZL}^{-1}\{a(s)\} \qquad (7.1-7)$$

It is clear from (7.1–5) that $a^*(e^{i\omega T})$ is periodic in ω with period ω_s. It is also important to note that for a rational function $a^*(z)$ we have $a^*(z)^H = a^*(z^H)$, where the superscript ^{H} indicates complex conjugate, and therefore for $\pi/T < \omega < 2\pi/T$ we have:

$$a^*(e^{i\omega T})^H = a^*(e^{-i\omega T}) = a^*(e^{i(\omega_s-\omega)})^T \qquad (7.1-8)$$

Hence, in addition to periodicity, a rational z-transform $a^*(z)$ has the property that its values for frequencies larger than π/T are uniquely determined by those for $0 \le \omega \le \pi/T$.

For the signals in Fig. 7.1-1A we have

$$r^*(z) = \mathcal{ZL}^{-1}\{r(s)\} \qquad (7.1-9)$$

$$d^*(z) = \mathcal{ZL}^{-1}\{d(s)\} \qquad (7.1-10)$$

$$y^*(z) = \mathcal{ZL}^{-1}\{y(s)\} \qquad (7.1-11)$$

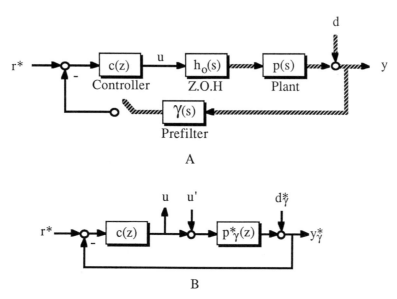

Figure 7.1-1. Block diagram of computer controlled system. A: Sampled-data structure with thick lines indicating analog signals. B: Discrete structure with all signals discrete.

The controller z-transfer function $c(z)$ represents a difference equation, which models the computer program. The zero-order hold $h_0(s)$ models the D/A converter, which constructs the piecewise-constant input to the plant from the impulse sequence described by $u(z)$. We have

$$h_0(s) = \frac{1 - e^{-sT}}{s} \qquad (7.1-12)$$

The block $\gamma(s)$ represents an analog anti-aliasing prefilter. Briefly one can understand the problem of aliasing from (7.1–5). After substitution of s with $i\omega$, it follows that the value of a^* at a frequency ω is the sum of the values of the continuous signal a at the frequencies $\omega + k\omega_s$ divided by T. The result is that after sampling, a high-frequency disturbance or measurement noise cannot be distinguished from an equivalent low frequency one. The objective of the prefilter is to cut off high frequency components from the analog signals before sampling, when that is necessary. Its transfer function is stable.

Note that no measurement device block is included in Fig. 7.1-1. When the dynamics of the measurement device function are significant they can be included in the prefilter $\gamma(s)$.

When the continuous output y is not observed directly but after the prefilter and only at the sampling intervals, then Fig. 7.1-1A can be simplified to Fig.

7.1-1B.

$$d^*_\gamma(z) = \mathcal{Z L}^{-1}\{\gamma(s)d(s)\} \qquad (7.1-13)$$

$$y^*_\gamma(z) = \mathcal{Z L}^{-1}\{\gamma(s)y(s)\} \qquad (7.1-14)$$

Here all signals are impulse sequences. The block $p^*_\gamma(s)$ is the pulse transfer function representing the zero-order hold equivalent of $p(s)\gamma(s)$. We define

$$p^*_\gamma(z) = \mathcal{Z L}^{-1}\{h_0(s)p(s)\gamma(s)\} \qquad (7.1-15)$$

and similarly

$$p^*(z) = \mathcal{Z L}^{-1}\{h_0(s)p(s)\} \qquad (7.1-16)$$

Pulse transfer functions are always rational in z, although the continuous transfer functions may include time delays. Time delays appear as poles at $z = 0$. It should also be noted that in the case of pulse transfer functions, the definitions of *proper* and *causal* in the spirit of Sec. 2.1 coincide.

Definition 7.1-1. *A system $g(z)$ is proper or causal if $\lim_{z\to\infty}g(z)$ is finite. A proper system is strictly proper if $\lim_{z\to\infty}g(z) = 0$ and semiproper if $\lim_{z\to\infty}|g(z)| > 0$. All systems which are not proper are called improper or noncausal.*

A system $g(z)$ is improper if the order of the numerator polynomial exceeds the order of the denominator polynomial and proper otherwise. An improper system is not physically realizable because it requires prediction.

It is useful to understand the relationship between the poles and the zeros of a continuous-time system and of the corresponding discrete-time system. Poles are mapped in a simple manner: if π_i is a pole of the continuous system then $e^{\pi_i T}$ is a pole of the corresponding discrete system (zero order hold included). It is not possible to give a simple formula for the mapping of the zeros. The zeros of the discrete system depend on the sampling period. In particular, it is possible for a discrete-time system to have zeros outside the unit circle (UC) even when the corresponding continuous system is MP. The converse can also happen.

It is well known that poles of a continuous system can become unobservable by sampling. We will assume throughout the book that the sampling rate has been chosen such that all unstable poles of the continuous system $p(s)$ appear in the pulse transfer function $p^*_\gamma(z)$. With this assumption the internal stability of the system in Fig. 7.1-1A can be assessed in terms of the system in Fig. 7.1-1B.

Theorem 7.1-1. *The sampled-data system in Fig. 7.1-1A is internally stable if and only if the transfer matrix in (7.1-17)*

$$\begin{pmatrix} y^*_\gamma \\ u \end{pmatrix} = \begin{pmatrix} \frac{p^*_\gamma c}{1+p^*_\gamma c} & \frac{p^*_\gamma}{1+p^*_\gamma c} \\ \frac{c}{1+p^*_\gamma c} & \frac{-p^*_\gamma c}{1+p^*_\gamma c} \end{pmatrix} \begin{pmatrix} r^* \\ u' \end{pmatrix} \qquad (7.1-17)$$

is stable — i.e., if and only if all poles of the four pulse-transfer functions are strictly inside the unit circle (UC).

7.2 IMC Structure

The block diagram of the sampled-data IMC loop is shown in Fig. 7.2-1A. The block $\tilde{p}_\gamma^*(z)$ is the pulse transfer function representing the zero order hold equivalent of $\tilde{p}(s)\gamma(s)$, where $\tilde{p}(s)$ is the continuous plant model. We define

$$\tilde{p}_\gamma^*(z) = \mathcal{ZL}^{-1}\left\{h_0(s)\tilde{p}(s)\gamma(s)\right\} \qquad (7.2-1)$$

and similarly

$$\tilde{p}^*(z) = \mathcal{ZL}^{-1}\left\{h_0(s)\tilde{p}(s)\right\} \qquad (7.2-2)$$

The same block manipulations as in the continuous case can be used here to derive the relations between the feedback controller $c(z)$ and the IMC controller $q(z)$:

$$c = \frac{q}{1 - \tilde{p}_\gamma^* q} \qquad (7.2-3)$$

$$q = \frac{c}{1 + \tilde{p}_\gamma^* c} \qquad (7.2-4)$$

When c and q are related through (7.2–3) or (7.2–4), $u(z)$ and $y(s)$ react to inputs $r^*(z)$ and $d(s)$ in exactly the same way for both the classic feedback and the IMC structure.

In Fig. 7.2-1B a different configuration is drawn for the sampled-data IMC structure. This configuration is equivalent to that of Fig. 7.2-1A, but is not suitable for computer implementation because of the presence of the continuous model $\tilde{p}(s)$. However Fig. 7.2-1B demonstrates the properties of the IMC structure, that were discussed in Sec. 3.1, in a clearer way.

If only the sampled signals are of interest, then Fig. 7.2-1A and B are equivalent to Fig. 7.2-1C where all signals are digital.

Finally, it should be noted that the implicit assumption has been made throughout this section that an exact model is available for the anti-aliasing prefilter $\gamma(s)$. The simplicity of this control-loop element makes this assumption valid and allows us to avoid unnecessary complications.

7.3 Formulation of Control Problem

For the design of a discrete controller the same items have to be specified as in the continuous case:

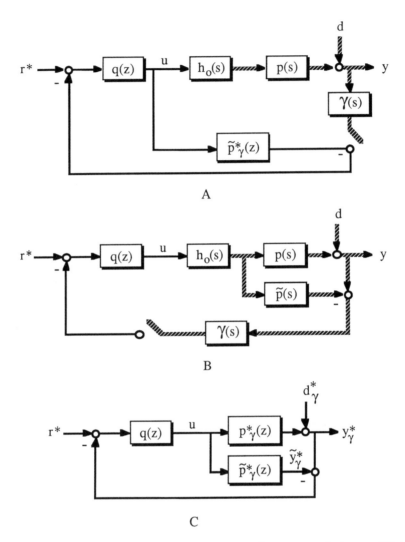

Figure 7.2-1. IMC structure. A: Sampled-data structure; B: Structure equivalent to (A) but not implementable; C: Discrete structure (all signals discrete).

- process model

- model uncertainty bounds

- type of inputs

- performance objectives

The process model can be continuous or discrete. There are advantages to starting with a continuous model (see Sec. 7.3.1). The inputs of interest, in particular the disturbances, are continuous in nature. Therefore the same input specifications (specific inputs, sets of inputs) as discussed in Sec. 2.2.3 are relevant here. In terms of performance, one is usually interested in the behavior of tne *continuous* rather than the *sampled* output. The fact that only the sampled output is available to the controller leads to some complications in the specification of a meaningful design objective which will be addressed in Sec. 7.5.

7.3.1 Process Model

Most popular identification schemes generate pulse transfer function models. Such models are sufficient for control system design but do not allow the analysis of the intersample behavior which can be significantly worse than the behavior at the sampling points, as we will show later in this chapter. Furthermore, model uncertainty is more naturally described in terms of the continuous system. Thus, it is desirable that a continuous system model be available. The system itself will be assumed to be linear and time invariant but not necessarily finite dimensional. Systems with time delays do not cause any problems for the design of discrete controllers.

7.3.2 Model Uncertainty Description

In Sec. 2.2.2 the additive and multiplicative uncertainty descriptions were presented, which assume that for each frequency ω, the actual plant $p(i\omega)$ lies in a disk-shaped region of known radius around the model $\tilde{p}(i\omega)$. For sampled data systems we also need to know how far $p_\gamma^*(e^{sT})$ lies from the known $\tilde{p}_\gamma^*(e^{sT})$. This information can be obtained from the information on $p(s)$. Let $p(s)$ belong to the family Π of plants defined by

$$\Pi = \{p : |p(i\omega) - \tilde{p}(i\omega)| \leq \bar{\ell}_a(\omega)\} \qquad (7.3-1)$$

or equivalently

$$p(i\omega) = \tilde{p}(i\omega) + \ell_a(i\omega) \qquad (7.3-2)$$

$$|\ell_a(i\omega)| \leq \bar{\ell}_a(\omega) \qquad \forall p \in \Pi \qquad (7.3-3)$$

From the definitions (7.1–15) and (7.2–1) we find

$$p_\gamma^*(e^{sT}) - \tilde{p}_\gamma^*(e^{sT}) = \mathcal{ZL}^{-1}\{h_0(s)\gamma(s)(p(s) - \tilde{p}(s))\} = \mathcal{ZL}^{-1}\{h_0(s)\gamma(s)\ell_a(s)\}$$

$$(7.3-4)$$

and by using (7.1–5)

$$p_\gamma^*(e^{i\omega T}) - \tilde{p}_\gamma^*(e^{i\omega T}) = \frac{1}{T}\sum_{k=-\infty}^{\infty} h_0\gamma\ell_a(i\omega + ik\omega_s) \qquad (7.3-5)$$

With (7.3–3) we obtain the following bound

$$|p_\gamma^*(e^{i\omega T}) - \tilde{p}_\gamma^*(e^{i\omega T})| \leq \frac{1}{T}\sum_{k=-\infty}^{\infty} |h_0\gamma(i\omega + ik\omega_s)|\bar{\ell}_a(\omega + k\omega_s) \triangleq \bar{\ell}_a^*(\omega) \qquad (7.3-6)$$

The above sum converges because $|h_0\gamma(i\omega)|\bar{\ell}_a(\omega) \to 0$ faster than $1/\omega$ as $\omega \to \infty$. This happens because $|h_0\gamma(i\omega)| \to 0$ at least as fast as $1/\omega$ as $\omega \to \infty$, even if $\gamma(s) = 1$. Also a bound $\bar{\ell}_a(\omega)$ such that $\bar{\ell}_a(\omega) \to 0$ as $\omega \to \infty$ can always be found since any physical system $p(s)$ and its model $\tilde{p}(s)$ are strictly proper and therefore $\ell_a(\omega) \to 0$ as $\omega \to \infty$. Note that if a prefilter $\gamma(s)$ is used, the property $\ell_a(\omega) \to 0$ as $\omega \to \infty$ is not needed for convergence.

Let us now define the family Π^* of plants $p(s)$ as follows

$$\Pi^* = \{p(s) : |p_\gamma^*(e^{i\omega T}) - \tilde{p}_\gamma^*(e^{i\omega T})| \leq \bar{\ell}_a^*(\omega)\} \qquad (7.3-7)$$

Clearly Π^* depends on the choice of T and $\gamma(s)$. However, the steps to arrive at (7.3–6) imply that if a plant $p(s)$ belongs to Π, then it also belongs to Π^*.

Because of the step from (7.3–5) to (7.3–6) the description (7.3–7) is conservative but not much so. The reason is that the sum in (7.3–5) and (7.3–6) has only a few dominant terms: $|\gamma(i\omega)|$ is designed to be small for $\omega > \pi/T$ in order to cut off high frequency components. Also $h_0(i\omega)/T$ is small for $\omega > \pi/T$. Therefore, the only dominant term in (7.3–5) and (7.3–6) is the one for which $-\pi/T \leq \omega + k\omega_s \leq \pi/T$. Hence for $0 \leq \omega \leq \pi/T$, the dominant term corresponds to $k = 0$. Computationally it is rare that more than two or three terms are significant.

7.4 Internal Stability

Assuming that the sampling time T has been chosen to avoid unobservable unstable poles in p_γ^* we only need to study the internal stability of the system in Fig. 7.2-1C where all signals are digital. The internal stability conditions can be stated in terms of pulse transfer functions and the arguments of Section 3.2.1 carry over directly.

Theorem 7.4-1. *Assume that the model is perfect* $(p(s) = \tilde{p}(s))$; *then the IMC system in Fig. 7.2.1A is internally stable if and only if both the plant* $p(s)$ *and the controller* $q(z)$ *are stable.*

7.5 Nominal Performance

The objective is to keep the *continuous* error e between the plant output y and the reference r small when the overall system is affected by external signals r and d. Contrary to the continuous case, there is no transfer function between d and e but the relationship is time varying. We will explain the problem in Sec. 7.5.1 and suggest meaningful approximations.

7.5.1 Sensitivity and Complementary Sensitivity Function

From the IMC structure of Fig. 7.2-1A or B we can easily obtain for $p = \tilde{p}$

$$y(s) = h_0(s)\tilde{p}(s)q(e^{sT})(r^*(e^{sT}) - d_\gamma^*(e^{sT})) + d(s) \qquad (7.5-1)$$

We are interested in finding transfer functions relating the external inputs $r(s)$ and $d(s)$ to the error

$$e(s) = y(s) - r(s) \qquad (7.5-2)$$

where $r(s)$ is the Laplace transform of the continuous time function we wish the plant output to follow. The signal $r(s)$ is related to $r^*(e^{sT})$ through (7.1-9) but it does not appear in the block diagrams since no hardware (A/D converter modelled by the sampling switch) is actually used to obtain $r^*(e^{sT})$.

Simple inspection of (7.5-1) indicates that it is not possible to obtain transfer functions relating $r(s)$ and $d(s)$ to $e(s)$. Let us first consider the relation between $r(s)$ and $e(s)$. Equations (7.5-1) and (7.5-2) yield

$$e(s) = h_0(s)\tilde{p}(s)q(e^{sT})r^*(e^{sT}) - r(s) \qquad (7.5-3)$$

Clearly there is no transfer function relating $r(s)$ to $e(s)$. The reason is that the relation is time-varying — i.e., the response of $e(s)$ to $r(s)$ depends on the time relative to the sampling instant at which the signal $r(s)$ is applied. A transfer function can be obtained in the special case when $\mathcal{L}^{-1}\{r(s)\}$ remains constant between sampling instants. In this case we have $r(s) = h_0(s)r^*(e^{sT})$ and then (7.5–3) yields

$$\frac{-e(s)}{r(s)} = \frac{-e(s)}{h_0(s)r^*(e^{sT})} = 1 - \tilde{p}(s)q(e^{sT}) \triangleq \tilde{\epsilon}_r(s) \qquad (7.5-4)$$

The complementary sensitivity function $\tilde{\eta}_r(s)$ relating $y(s)$ to $r(s)$ can be obtained by subtracting the sensitivity function $\tilde{\epsilon}_r(s)$ from unity.

$$\frac{y(s)}{h_0(s)r^*(e^{sT})} = \tilde{p}(s)q(e^{sT}) \triangleq \tilde{\eta}_r(s) \qquad (7.5-5)$$

Let us now consider the relation between $d(s)$ and $e(s)$ or equivalently $d(s)$ and $y(s)$. From (7.5–1) we have

$$y(s) = d(s) - h_0(s)\tilde{p}(s)q(e^{sT})d_\gamma^*(e^{sT}) \qquad (7.5-6)$$

Again the relation is time varying and there is no transfer function connecting $d(s)$ to $y(s)$. If, of course, $\gamma(s) = 1$ and $\mathcal{L}^{-1}\{d(s)\}$ remained constant between the sampling instants, then we could proceed in a manner similar to that for $r(s)$ and obtain the same expression for the sensitivity function as in (7.5–4). The assumption, however, that $\mathcal{L}^{-1}\{d(s)\}$ is constant between the sampling instants is not realistic.

There are three possible approaches to deal with this problem:

1. The time varying sensitivity operator can be bounded by a "conic sector."

2. The bandwidth of the disturbance signal $d(s)$ can be assumed to be limited and an approximate sensitivity function can be defined.

3. The plant output can be studied at the sampling instants only and an appropriate pulse-transfer function can be derived.

We will discuss the latter two approaches in the following.

Approximate sensitivity function for bandlimited disturbance signal. We will assume the disturbance to be approximately limited to the frequency band up to π/T.

From (7.1–5) we find

$$d_\gamma^*(e^{sT}) = \frac{1}{T}\sum_{k=-\infty}^{\infty} d(s + ik\omega_s)\gamma(s + ik\omega_s) \qquad (7.5-7)$$

Because d is band limited and because γ is designed to attenuate signals at frequencies larger than π/T, (7.5–7) can be approximated by

$$d_\gamma^*(e^{i\omega T}) \cong \frac{1}{T}d(i\omega)\gamma(i\omega) \qquad 0 \leq \omega \leq \frac{\pi}{T} \qquad (7.5-8)$$

With this approximation (7.5–6) becomes

$$y(i\omega) \cong \left[1 - \frac{1}{T}h_0(i\omega)\tilde{p}(i\omega)q(e^{i\omega T})\gamma(i\omega)\right]d(i\omega) \qquad 0 \leq \omega \leq \frac{\pi}{T} \qquad (7.5-9)$$

Defining the new "controller"

$$\hat{q}(s) = \frac{1}{T}h_0(s)q(e^{sT})\gamma(s) \tag{7.5 - 10}$$

(7.5–9) can be rewritten as

$$y(i\omega) \cong (1 - \tilde{p}(i\omega)\hat{q}(i\omega))d(i\omega) \triangleq \tilde{\epsilon}_d(i\omega)d(i\omega) \qquad 0 \leq \omega \leq \frac{\pi}{T} \tag{7.5 - 11}$$

which is identical in *structure* with what can be obtained for the continuous system. For continuous systems the expression is exact, while for sampled-data systems it represents an approximation of a time-varying relationship.

Sensitivity pulse-transfer function. Sampling of (7.5–1) yields

$$y^*(z) = \tilde{p}^*(z)q(z)(r^*(z) - d_\gamma^*(z)) + d^*(z) \tag{7.5 - 12}$$

Then by assuming $\gamma(s) = 1$, we can obtain sensitivity and complementary sensitivity pulse-transfer functions, connecting $e^*(z)$ to $r^*(z)$ and $d^*(z)$, where

$$e^*(z) = \mathcal{ZL}^{-1}\{e(s)\} \tag{7.5 - 13}$$

$$\tilde{\epsilon}^*(z) \triangleq 1 - \tilde{p}^*(z)q(z) \tag{7.5 - 14}$$

$$\tilde{\eta}^*(z) \triangleq \tilde{p}^*(z)q(z) \tag{7.5 - 15}$$

However, disregarding the intersample behavior of the plant output may lead to serious problems as will be illustrated in Sec. 7.5.3.

7.5.2   Asymptotic Properties of Closed-Loop Response

"System types" were defined in Sec. 2.4.3 to classify the asymptotic closed-loop behavior. A "Type m" system, where m is a non-negative integer is defined as a system which tracks perfectly, as time $\rightarrow \infty$, inputs $r(s)$ and $d(s)$ with all the poles in the LHP except m or less poles at $s = 0$. The conditions that have to be satisfied in order for this to happen impose certain requirements on the controller $q(z)$ and the anti–aliasing prefilter $\gamma(s)$, described by the following theorem (see for comparison Sec. 3.3.3).

Theorem 7.5-1. *Provided that the closed-loop system is stable, the necessary and sufficient conditions for the system to be "Type m" ($m > 0$) are the following:*

$$\lim_{z \to 1} \frac{d^k}{dz^k}(1 - \tilde{p}^*(z)q(z)) = 0, \quad 0 \leq k < m \tag{7.5 - 16}$$

$$\lim_{s \to 0} \frac{d^k}{ds^k}(1 - \gamma(s)) = 0, \quad 0 \leq k < m \tag{7.5 - 17}$$

Proof. The disturbance $d(s)$ goes through $\gamma(s)$ before it is sampled and therefore we clearly need

$$\lim_{\text{time}\to\infty} \mathcal{L}^{-1}\{d(s) - \gamma(s)d(s)\} = \lim_{s\to 0}(s(1 - \gamma(s))d(s)) = 0 \qquad (7.5 - 18)$$

Since (7.5–18) must be satisfied for all $d(s)$ with m or less poles at $s = 0$, $(1-\gamma(s))$ has to have m zeros at $s = 0$, which will be the case if and only if (7.5–17) holds.

We obtain from Fig. 7.2-1A or B

$$e^*(z) = \frac{p^*(z)q(z)}{1 + q(z)(p_\gamma^*(z) - \tilde{p}_\gamma^*(z))}(r^*(z) - d_\gamma^*(z)) + (d^*(z) - r^*(z)) \qquad (7.5 - 19)$$

Condition (7.5–17) implies that

$$\lim_{\text{time}\to\infty} \mathcal{Z}^{-1}\{d^*(z) - d_\gamma^*(z)\} = \lim_{z\to 1}(1 - z^{-1})(d^*(z) - d_\gamma^*(z)) = 0 \qquad (7.5 - 20)$$

Hence for tracking considerations, $d_\gamma^*(z)$ can be replaced by $d^*(z)$ in (7.5–19)

$$e^*(z) = \frac{1 + q(p_\gamma^* - \tilde{p}_\gamma^*) - p^*q}{1 + q(p_\gamma^* - \tilde{p}_\gamma^*)}v^* \qquad (7.5 - 21)$$

where $v^* = d^* - r^*$. (7.5–17) also implies that

$$\lim_{\text{time}\to\infty} \mathcal{Z}^{-1}\{(p_\gamma^* - p^*)v^*\} = 0 \qquad (7.5 - 22)$$

Thus, for tracking considerations p_γ^* can be replaced by p^* and similarly $\tilde{p}_\gamma^*$ by $\tilde{p}^*$. Then (7.5–21) becomes

$$e^*(z) = \frac{1 - \tilde{p}^*q}{1 + q(p^* - \tilde{p}^*)}v^* \qquad (7.5 - 23)$$

Assume that v^* has at most m poles at $z = 1$ and apply the final value theorem to (7.5–23). Condition (7.5–16) follows directly. □

The implications of (7.5–16) for the design of $q(z)$ will be considered in Chap. 8. Let us discuss briefly the design of the prefilter $\gamma(s)$, whose objective is to cut off high-frequency components. Most digital control books discuss different types of anti-aliasing prefilters, which satisfy (7.5–16) for $m = 1$, like Butterworth and Bessel filters. For the case of $m > 1$ a simple modification can be used. Let us write

$$\gamma(s) = \gamma_1(s)\gamma_m(s) \qquad (7.5 - 24)$$

where

$$\gamma_m(s) = \frac{c_{m-1}s^{m-1} + \ldots + c_1 s + 1}{(\tau s + 1)^{m-1}} \qquad (7.5 - 25)$$

and $\gamma_1(s)$ is an appropriate prefilter for $m = 1$. Then for a specified τ, (7.5–17) can be used to compute the coefficients $c_1, \ldots, c_{m-1}$. Qualitatively it is clear that the use of $\gamma_m(s)$ to satisfy (7.5–17) should not change the behavior of $\gamma_1(s)$ significantly. Condition (7.5–17) simply adds some properties at $\omega = 0$ and this can be done without affecting the high-frequency properties of $\gamma_1(s)$. A large τ should be used to push the effect of $\gamma_m(s)$ toward $\omega = 0$. Indeed for a usual second-order filter $\gamma_1(s) = \omega_0^2/(s^2 + 2\omega_0\zeta s + \omega_0^2)$ and for $m = 2$ (ramp inputs), (7.5–17) yields $c_1 = \tau + 2\zeta/\omega_0$ and therefore for a sufficiently large $\tau, \gamma_m(s)$ does not affect the high-frequency performance of $\gamma(s)$ significantly.

7.5.3 Limitations on Achievable Performance

In Sec. 3.3.4 the concept of "perfect control" was discussed and three sources of limitations on the achievable closed-loop performance were given, namely the NMP characteristics of the plant, constraints on the inputs and model uncertainty. In this section some additional sources, particular to sampled-data control systems will be discussed.

(i) *Intersample rippling*

To demonstrate the problem we shall assume that $p(s) = \tilde{p}(s)$ and $d(s) = 0$. Let us consider the system

$$p(s) = \frac{2}{(s^2 + 1.2s + 1)(s + 2)} \qquad (7.5 - 26)$$

and choose a sampling time $T = 1.8$. Then

$$p^*(z) = 0.483 \frac{z^2 + 1.01z + 0.0597}{z^3 - 0.116z^2 + 0.118z - 0.00315} \qquad (7.5 - 27)$$

The behavior of two control algorithms will be examined:

$$q_1(z) = (zp^*(z))^{-1} \qquad (7.5 - 28)$$

$$q_2(z) = 1.001 \frac{z^3 - 0.116z^2 + 0.118z - 0.00315}{z^3} \qquad (7.5 - 29)$$

The response to a step change in the setpoint $r(s)$ is shown for both algorithms in Fig. 7.5-1. Clearly $q_1(z)$ produces an unacceptable response. However if one concentrated only at the sampling instants, which is equivalent to using (7.5–12) instead of (7.5–1), then it would seem that $q_1(z)$ produces a perfect response which reaches the setpoint in one sampling interval and remains there. On the other hand, although $q_2(z)$ produces an excellent response, if one looked only at the sample points it would seem inferior to that of $q_1(z)$ since it takes three sampling intervals to reach the setpoint.

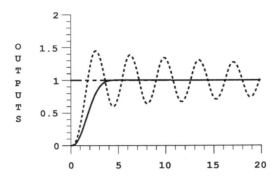

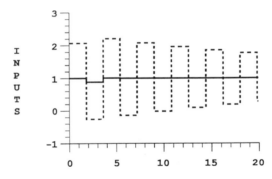

Figure 7.5-1. Demonstration of intersample rippling. Dash: q_1; Solid: q_2.

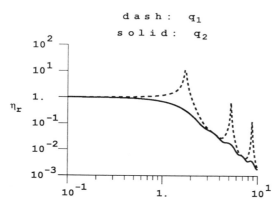

Figure 7.5-2. Bode plot of complementary sensitivity η_r.

The cause of the problem is the pole of $q_1(z)$ at $z = -0.94$. From (7.5–1), and (7.5–12) we obtain for $p(s) = \tilde{p}(s)$:

$$y(s) = p(s)q(e^{sT})h_0(s)r^*(e^{sT}) \qquad (7.5 - 30)$$

$$y^*(z) = p^*(z)q(z)r^*(z) \qquad (7.5 - 31)$$

In (7.5–31) this pole cancels with the zero of $p^*(z)$ and its bad effect does not show up in $y^*(z)$. This does not happen in (7.5–30), however, as is shown on the Bode plots of $\eta_r(s)$ in Fig. 7.5-2, where the pole of $q_1(z)$ at $z = -0.94$ causes a peak in $|\eta_r(i\omega)|$. Figure 7.5-1 clearly indicates that the problem appears because $q_1(z)$ produces an oscillatory output $u(z)$ with a period that matches the sampling period and whose effect does not show up in the sampled output $y^*(z)$. This is a characteristic of poles near (-1,0) on the z-plane. Hence, to avoid such hidden oscillations (intersample rippling) one should use an IMC controller $q(z)$ which has no poles near (-1,0) or in general no poles with negative real part. A controller $q(z)$ which inverts the model $\tilde{p}(z)$ cannot be used when $\tilde{p}(z)$ has zeros close to (-1,0).

(ii) *Effect of sampling on performance*

From a qualitative point of view, sampling clearly puts a limitation on the achievable performance since one can obtain information on the system output and change the control action only at every sampling point. We can demonstrate this fact quantitatively by looking at $\eta_r(s)$, given by (7.5–5) for $p(s) = \tilde{p}(s)$. In Fig. 7.5-3 a typical Bode plot of $p(s)$ is shown. For perfect performance $\eta_r(s) = 1$ – i.e., $q(e^{sT})$ should be equal to the inverse of $p(s)$. However, as shown by (7.1–8),

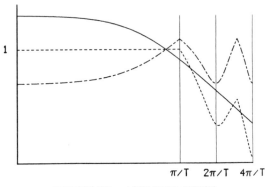

Figure 7.5-3. Effect of sampling on performance (logarithmic plot). Solid line: $|\tilde{p}(i\omega)|$. Dash and dot line: $|q(e^{i\omega T})|$. Dashed line: $|\tilde{p}(i\omega)q(e^{i\omega T})|$. (Reprinted with permission from Int. J. Control, **44**, 716(1986), Taylor & Francis Ltd.)

$q(e^{i\omega T})$ is periodic in ω with period ω_s and its values for frequencies larger than π/T are uniquely determined by those for $\omega \leq \pi/T$. In Fig. 7.5-3 an ideal q is plotted which inverts $p(s)$ for ω up to π/T. In order for this to be accomplished, q has to be of infinite order. Even for this q, it is clear from Fig. 7.5-3 that the closed-loop transfer function $p(s)q(e^{sT})$ cannot have a bandwidth larger than π/T.

7.5.4   Discrete Linear Quadratic (H_2^*-) Optimal Control

In the continuous case, the objective of H_2-optimal control theory is to minimize the integral of the squared error — i.e., the H_2 norm of the error — for a particular input. The H_2^* norm for a discrete signal $e^*(z)$ is given by

$$\|e^*\|_2^2 = \sum_{k=0}^{\infty} e_k^2 \qquad (7.5-32)$$

where the sequence $\{e_k\}$ is defined from

$$\{e_k\} = \mathcal{Z}^{-1}\{e^*(z)\} \qquad (7.5-33)$$

The objective of the H_2^*-optimal controller $\tilde{q}_H$ is to minimize (7.5–32) resulting from a particular reference and/or disturbance change. Recall that a discrete controller is not effective in rejecting disturbances in the frequency range $\omega > \pi/T$. Thus for the computation of the H_2^*-optimal control law it is not meaningful to specify disturbances with large high-frequency components. Therefore $\tilde{q}_H^*$ should be designed for the filtered disturbance d_γ^* rather than d^*. Then we find from Sec. 7.5.1

$$e^*(z) = \epsilon^*(z)(d_\gamma^*(z) - r^*(z)) \qquad (7.5-34)$$

Let us define the combined inputs

$$v^*(z) = d_\gamma^*(z) - r^*(s) \tag{7.5 - 35}$$

With the help of Parseval's theorem we can rewrite the objective (7.5–32)

$$\|e^*\|_2^2 = \frac{1}{2\pi} \int_{-\pi}^{\pi} |e^*(e^{i\theta})|^2 d\theta \tag{7.5 - 36}$$

and upon substitution of (7.5–34) and (7.5–35)

$$\|e^*\|_2^2 = \frac{1}{2\pi} \int_{-\pi}^{\pi} |\tilde{\epsilon}^*(e^{i\theta})v^*(e^{i\theta})|^2 d\theta \tag{7.5 - 37}$$

Thus, the H_2^*-optimal control problem becomes

$$\min_{q(z)} \|\tilde{\epsilon}^* v^*\|_2 = \min_{q(z)} \|(1 - \tilde{p}^* q) v^*\|_2 \tag{7.5 - 38}$$

Note that in this formulation no attention is paid to intersample behavior. Experience has shown that the H_2^*-optimal controller can lead to unacceptable intersample rippling.

7.5.5 H_∞ Performance Objective

In Sec. 2.4.5 we introduced the H_∞ performance objective

$$\|\epsilon w\|_\infty < 1 \tag{2.4 - 20}$$

We found it particularly relevant for disturbance rejection because rather than restricting the disturbance to a specific function it assumes the disturbance to belong to a set. Usually this is more realistic. Because of the time varying nature of the sampling operation there exists no sensitivity function for sampled-data systems and thus (2.4–20) cannot be defined. We can, however, state an objective similar to (2.4–20) for the approximate relation (7.5–11).

We assume the disturbance to be approximately limited to the frequency band up to π/T. This implies for the weight w

$$|w(\omega)| << 1, \quad \omega > \frac{\pi}{T} \tag{7.5 - 39}$$

Therefore (2.4–20) can be approximated by

$$|\epsilon w| < 1, \quad 0 \le \omega \le \frac{\pi}{T} \tag{7.5 - 40}$$

Using (7.5–11), (7.5–40) can be expressed as

$$|(1 - \tilde{p}(i\omega)\hat{q}(i\omega))w(\omega)| < 1, \quad 0 \le \omega \le \frac{\pi}{T} \tag{7.5 - 41}$$

Note that (7.5–41) is exactly equal to (2.4–20) if the disturbance d has no components at frequencies higher than π/T.

Let us assume that the performance specification for disturbance rejection has been stated in the form of (2.4–20) in terms of the continuous input and output signals. Then (7.5–41) can be used to assess if these specifications can be met with the digital controller $q(z)$.

7.6 Robust Stability

We wish to derive a condition that guarantees stability of the control loop for all plants in the family Π^* defined by (7.3–7). The Nyquist stability criterion as applied to discrete systems can be used to obtain such a condition in exactly the same way as for continuous systems. Hence in the same way as in Sec. 2.5 we can derive the following theorem, where for consistency with the continuous case we define

$$\bar{\ell}_m^*(\omega) = \bar{\ell}_a^*(\omega)/|\tilde{p}^*(e^{i\omega T})| \tag{7.6 – 1}$$

Theorem 7.6-1 (Robust Stability). *Assume that all plants $p(s)$ in the family*

Π^*

$$\Pi^* = \{p(s) : |p_\gamma^*(e^{i\omega T}) - \tilde{p}_\gamma^*(e^{i\omega T})| \leq \bar{\ell}_a^*(\omega)\} \tag{7.3 – 7}$$

have the same number of RHP poles and that these poles do not become unobservable after sampling. Let $c(z)$ be a controller that stabilizes the system in Fig. 7.1-1A for the nominal plant $\tilde{p}(s)$. Then the system is robustly stable with the controller c if and only if the complementary sensitivity function $\tilde{\eta}^*(z)$ for $\tilde{p}(s)$ satisfies

$$|\tilde{\eta}^*(e^{i\omega T})|\bar{\ell}_m^*(\omega) < 1, \quad 0 \leq \omega \leq \pi/T \tag{7.6 – 2}$$

(Note that the periodicity and (7.1–8) imply that (7.6–2) holds for all ω if it holds for $0 \leq \omega \leq \pi/T$.)

The IMC structure can be used for control system implementation only when the plant is stable. Then the robust stability condition is described by the following theorem.

Theorem 7.6-2 (Robust Stability). *Assume that all plants $p(s)$ in the family Π^* are stable, that $q(z)$ is stable, and that $c(z)$ is related to $q(z)$ through (7.2–3). Then the systems in Figs. 7.1-1A and 7.2-1A are robustly stable if and only if*

$$|\tilde{\eta}^*(e^{i\omega T})|\bar{\ell}_m^*(\omega) < 1, \quad 0 \leq \omega \leq \pi/T \tag{7.6 – 3}$$

7.7 Robust Performance

In a similar manner as in Sec. 7.5.5 we will develop an approximate sensitivity function on the basis of which we will assess robust performance. It follows from Figs. 7.2-1A or B that $y(s)$ and $d(s)$ are related by the time-varying expression

$$y(s) = d(s) - \frac{h_0(s)p(s)q(e^{sT})}{1 + q(e^{sT})(p_\gamma^*(e^{sT}) - \tilde{p}_\gamma^*(e^{sT}))}d_\gamma^*(e^{sT}) \qquad (7.7-1)$$

When d is bandlimited we can use the approximation

$$d_\gamma^*(e^{i\omega T}) \cong \frac{1}{T}d(i\omega)\gamma(i\omega) \qquad 0 \le \omega \le \frac{\pi}{T} \qquad (7.5-8)$$

derived in Sec. 7.5.1. We can use the arguments of Sec. 7.3.2 to justify

$$p_\gamma^*(e^{i\omega T}) - \tilde{p}_\gamma^*(e^{i\omega T}) \cong \frac{1}{T}h_0(i\omega)\gamma(i\omega)\ell_a(i\omega) \quad 0 \le \omega \le \frac{\pi}{T} \qquad (7.7-2)$$

With (7.5–8) and (7.7–2), (7.7–1) becomes

$$y(i\omega) \cong \frac{1 - h_0(i\omega)\tilde{p}(i\omega)q(e^{i\omega T})\gamma(i\omega)/T}{1 + \ell_a(i\omega)q(e^{i\omega T})h_0(i\omega)\gamma(i\omega)/T}d(i\omega) \quad 0 \le \omega \le \frac{\pi}{T} \qquad (7.7-3)$$

With the "controller"

$$\hat{q}(s) = \frac{1}{T}q(e^{sT})h_0(s)\gamma(s) \qquad (7.5-10)$$

equation (7.7–3) can be rewritten as

$$y(i\omega) \cong \frac{1 - \tilde{p}(i\omega)\hat{q}(i\omega)}{1 + \ell_a(i\omega)\hat{q}(i\omega)}d(i\omega) \quad 0 \le \omega \le \frac{\pi}{T} \qquad (7.7-4)$$

Equation (7.7–4) is identical in structure with what can be obtained for the continuous system. For continuous systems the expression is exact, while for sampled-data systems it represents an approximation of the time-varying relationship between y and d. We can take advantage of this structural similarity and restate approximate conditions for sampled-data systems which were derived for continuous systems in Sec. 2.6.

7.7.1 H_2 Performance Objective

To estimate the worst error that can occur when a specific controller is used for a family Π of plants we can use the expression derived from (7.7–4)

$$\max_{p\in\Pi} \|e\|_2^2 \cong \frac{1}{2\pi} \int_{-\pi/T}^{\pi/T} \left| \frac{(1 - \tilde{p}(i\omega)\hat{q}(i\omega))d(i\omega)}{1 - \bar{\ell}_a(\omega)|\hat{q}(i\omega)|} \right|^2 d\omega \qquad (7.7-5)$$

Because of the robust stability condition (7.6–2)

$$\bar{\ell}_a(\omega)|\hat{q}(i\omega)| < \bar{\ell}_a^*(\omega)|q^*(e^{i\omega T})| < 1 \tag{7.7 – 6}$$

and the integrand in (7.7–5) is always bounded. Because of the approximations made to arrive at (7.7–4), (7.7–5) is valid only when the bandwidth of d is limited to π/T. Furthermore, it is optimistic — i.e., the error bound is underestimated because (7.7–2) underestimates the uncertainty.

7.7.2 H_∞ Performance Objective

Based on the approximation (7.7–4), the H_∞ objective (2.4–20) for robust performance can be stated

$$\left| \frac{1 - \tilde{p}(i\omega)\hat{q}(i\omega)}{1 + \ell_a(i\omega)\hat{q}(i\omega)} \right| < \frac{1}{w(\omega)}, \quad \forall \ell_a \ni |\ell_a(i\omega)| \le \bar{\ell}_a(\omega), \quad 0 \le \omega \le \pi/T \tag{7.7 – 7}$$

where the weight $w(\omega)$ is designer specified. Note however that $w(\omega)$ cannot be chosen arbitrarily large because even for $\ell_a = 0$, the left hand side of (7.7–7) may be nonzero. The selection of $w(\omega)$ will be discussed in Sec. 8.4.1.

The following conditions are completely equivalent to (7.7–7)

$$\frac{|1 - \tilde{p}(i\omega)\hat{q}(i\omega)|w(\omega)}{1 - \bar{\ell}_a(\omega)|\hat{q}(i\omega)|} < 1, \quad 0 \le \omega \le \pi/T \tag{7.7 – 8}$$

$$|\hat{q}(i\omega)|\bar{\ell}_a(\omega) + |1 - \tilde{p}(i\omega)\hat{q}(i\omega)|w(\omega) < 1, \quad 0 \le \omega \le \pi/T \tag{7.7 – 9}$$

Condition (7.7–9) is identical in structure with the result for continuous systems. While it is exact for continuous systems, (7.7–9) is generally optimistic because of the approximation (7.5-8) and (7.7–4).

7.8 Summary

The basic IMC concepts carry over to the discrete case without major modifications. Because of the sampling operation an anti-aliasing prefilter $\gamma(s)$ has to be included in the control system (Fig. 7.2-1) and the process model $\tilde{p}_\gamma^*(z)$ (7.2–1) has to be defined accordingly. When the classic feedback controller $c(z)$ and the IMC controller $q(z)$ are related through

$$c = \frac{q}{1 - \tilde{p}_\gamma^* q} \tag{7.2 – 3}$$

$$q = \frac{c}{1 + \tilde{p}_\gamma^* c} \tag{7.2 – 4}$$

then the input-output behavior of the IMC structure and the classic feedback structure is the same. The IMC structure can be used for implementation only if $\tilde{p}$ and q are stable. The classic feedback system is internally stable for $p = \tilde{p}$ if and only if q defined by (7.2–4) is stable (Thm. 7.4-1).

For the design of a discrete controller the following has to be specified:

- process model

- model uncertainty bounds

- type of inputs

- performance objectives

In order to account for the intersample performance the availability of a continuous plant model is essential. Model uncertainty bounds for the discrete model can be obtained from the bounds $\bar{\ell}_a$ for the continuous model:

$$|p_\gamma^*(e^{i\omega T}) - \tilde{p}_\gamma^*(e^{i\omega T})| \leq \frac{1}{T} \sum_{k=-\infty}^{\infty} |h_0 \gamma(i\omega + ik\omega_s)| \bar{\ell}_a(\omega + k\omega_s) \triangleq \bar{\ell}_a^*(\omega) \quad (7.3 - 6)$$

If the reference trajectory r is assumed to be constant between samples ($r(s) = h_0(s)r^*(e^{sT})$) then the sensitivity and complementary sensitivity can be defined in the usual manner:

$$\tilde{\epsilon}_r(s) = 1 - \tilde{p}(s)q(e^{sT}) \qquad\qquad\qquad (7.5 - 4)$$

$$\tilde{\eta}_r(s) = \tilde{p}(s)q(e^{sT}) \qquad\qquad\qquad (7.5 - 5)$$

The disturbance d is usually *not* constant between samples. Then the relationship between d and y is time varying and a transfer function cannot be defined. However, if d is band limited up to π/T then approximately

$$y(i\omega) \cong (1 - \tilde{p}(i\omega)\hat{q}(i\omega))d(i\omega) \qquad 0 \leq \omega \leq \frac{\pi}{T} \qquad (7.5 - 11)$$

where

$$\hat{q}(s) = \frac{1}{T}h_0(s)q(e^{sT})\gamma(s) \qquad\qquad\qquad (7.5 - 10)$$

For asymptotically error-free response to polynomial inputs (i.e., Type m behavior) the controller q and anti-aliasing filter γ must have the following properties

$$\lim_{z \to 1} \frac{d^k}{dz^k}(1 - \tilde{p}^*(z)q(z)) = 0, \quad 0 \leq k < m \qquad (7.5 - 16)$$

$$\lim_{s \to 0} \frac{d^k}{ds^k}(1 - \gamma(s)) = 0, \qquad 0 \leq k < m \qquad (7.5 - 17)$$

Apart from the factors which limit the closed loop performance of continuous systems (NMP characteristics, constraints and model uncertainty) two more limitations arise for discrete systems: intersample rippling caused by poles of the controller q close to (-1,0) and a limitation of the effective closed loop bandwidth to π/T caused by the sampling operation.

The two performance objectives discussed in this book for discrete systems are the sum of the squared errors

$$H_2^* : \quad ||e^*||_2^2 = \frac{1}{2\pi} \int_{-\pi}^{\pi} |\tilde{\epsilon}^*(e^{i\theta})v^*(e^{i\theta})|^2 d\theta \qquad (7.5-37)$$

and a bound on the sensitivity utilizing the approximate relationship (7.5-11):

$$H_\infty : \quad |(1 - \tilde{p}(i\omega)\hat{q}(i\omega))w(\omega)| < 1, \qquad 0 \leq \omega \leq \frac{\pi}{T} \qquad (7.5-41)$$

The robust stability condition for discrete systems is formally similar to that for continuous systems

$$|\tilde{\eta}^*(e^{i\omega T})||\bar{\ell}_m^*(\omega) < 1, \qquad 0 \leq \omega \leq \frac{\pi}{T} \qquad (7.6-3)$$

An approximate (somewhat optimistic) condition for robust performance in the H_∞ sense can be derived for (7.5–11):

$$|\hat{q}(i\omega)||\bar{\ell}_a(\omega) + |1 - \tilde{p}(i\omega)\hat{q}(i\omega)|w(\omega) < 1, \quad 0 \leq \omega \leq \frac{\pi}{T} \qquad (7.7-9)$$

7.9   References

7.1. A detailed discussion of z-transforms can be found in any book on digital control. Two very good such books are Åström and Wittenmark (1984) and Kuo (1980). The effect of sampling time on the location of the zeros of discrete systems was analyzed by Åström, Hagander and Sternby (1984).

7.3.1. Åström and Wittenmark (1984, Chap. 12) provide a concise discussion of identification techniques for discrete systems and further references.

7.5. Another well known class of digital control algorithms, besides the H_2 and H_∞ types, are the so-called deadbeat controllers. A brief but good discussion on these algorithms can be found in Kuo (1980), Chap. 10. For more details see Isermann (1981). For a comparison of the deadbeat and H_2 controllers to the Dahlin and Vogel-Edgar controllers see Zafiriou and Morari (1985).

7.5.1. Dailey (1987) and Thompson (1982) propose design approaches that consider the continuous plant output by bounding appropriate time varying operators, like the sensitivity operator or the sampling switch, with "conic sectors."

7.5.2. For a list of anti-aliasing prefilters for "Type 1" systems, see Åström and Wittenmark (1984, p. 28).

Chapter 8

SISO IMC DESIGN FOR STABLE SAMPLED-DATA SYSTEMS

As in the continuous case the IMC design procedure consists of two steps.

STEP 1: Nominal Performance
The controller $\tilde{q}(z)$ is selected to yield a "good" system response for the input(s) of interest, without regard for constraints and model uncertainty.

STEP 2: Robust Stability and Performance
The controller $\tilde{q}(z)$ is augmented by a lowpass filter $f(z)$ $(q(z) = \tilde{q}(z)f(z))$ to achieve robust stability and robust performance.

8.1 Nominal Performance

In the continuous case $\tilde{q}$ is designed so that it minimizes the integral of the squared error for a particular input. The analogous approach in the discrete case would be to design the controller to minimize the sum of the squared errors for some external setpoint or disturbance input. Although such a controller may suffer from the problem of intersample rippling as exhibited in Sec. 7.5.3, it can be used as a starting point for the design of $\tilde{q}(z)$. In Sec. 8.1.1 the design of the discrete linear quadratic optimal controller will be discussed and in Sec. 8.1.2 an appropriate simple modification of this controller will be introduced to avoid intersample rippling.

8.1.1 H_2^*-Optimal Control

The H_2^*-optimal controller $\tilde{q}_H(z)$ is designed by solving the following minimization problem

$$\min_{\tilde{q}_H(z)} \|e^*\|_2 = \min_{\tilde{q}_H(z)} \|(1 - \tilde{p}^*(z)\tilde{q}_H(z))v^*(z)\|_2 \qquad (7.5-38)$$

subject to the constraint that $\tilde{q}_H(z)$ be stable and causal.

The following theorem which provides the solution of (7.5–38) will be proven in Chapter 9 for the general case of unstable plants.

Theorem 8.1-1. *Assume that $\tilde{p}$ is stable. Factor the model $\tilde{p}^*(z)$ into an allpass part $\tilde{p}_A^*(z)$ and $\tilde{p}_M^*(z)$*

$$\tilde{p}^*(z) = \tilde{p}_A^*(z)\tilde{p}_M^*(z) \qquad (8.1-1)$$

where

$$\tilde{p}_A^*(z) = z^{-N} \prod_{j=1}^{h} \frac{(1 - (\zeta_j^H)^{-1})(z - \zeta_j)}{(1 - \zeta_j)(z - (\zeta_j^H)^{-1})} \qquad (8.1-2)$$

and ζ_j, $j = 1, \ldots, h$ are the zeros of $\tilde{p}^*(z)$ which are outside the UC. The positive integer N is chosen such that $\tilde{p}_M^*(z)$ is semi-proper — i.e., its numerator and denominator have the same degree, which is equivalent to saying that N is such that $z^N \tilde{p}^*(z)$ is semi-proper.

Factor the input $v^*(z)$ similarly — i.e.,

$$v^*(z) = v_A^*(z)v_M^*(z) \qquad (8.1-3)$$

$$v_A^*(z) = z^{-N_v} \prod_{j=1}^{h_v} \frac{(1 - (\zeta_{vj}^H)^{-1})(z - \zeta_{vj})}{(1 - \zeta_{vj})(z - (\zeta_{vj}^H)^{-1})} \qquad (8.1-4)$$

where ζ_{vj}, $j = 1, \ldots, h_v$ are the zeros of $v^*(z)$ outside the UC and N_v is such that $z^{N_v} v^*(z)$ is semi-proper. The H_2^*- optimal controller $\tilde{q}_H(z)$ is given by

$$\tilde{q}_H(z) = z(\tilde{p}_M^* v_M^*)^{-1} \left\{ z^{-1} \tilde{p}_A^{*-1} v_M^* \right\}_* \qquad (8.1-5)$$

where the operator $\{\cdot\}_*$ denotes that after a partial fraction expansion of the operand only the strictly proper and stable (including poles at $z = 1$) terms are retained.

Note that $\tilde{q}_H(z)$ is stable and causal. Also note that in order for the system to be Type m when $\tilde{q}_H(z)$ is used as the controller, the input $v(s)$ for which $\tilde{q}_H(z)$ is designed must have m poles at $s = 0$.

The evaluation of (8.1–5) for specific inputs v^* yields the results shown in Table 8.1-1. As an illustration, let us compute the H_2^*-optimal controller for two different inputs.

Example 8.1-1.

$$v^* = v_M^* = z(z - 1)^{-1} \qquad \text{(Step)}$$

$$\left\{ z^{-1} \tilde{p}_A^{*-1} v_M^* \right\}_* = \left\{ \tilde{p}_A^{*-1}(z - 1)^{-1} \right\}_* = (z - 1)^{-1}$$

$$\tilde{q}_H(z) = z(\tilde{p}_M^* v_M^*)^{-1}(z - 1)^{-1} = \tilde{p}_M^{*-1}$$

□

Table 8.1-1. H_2^*-optimal controller for some typical input forms.

Input $v(s)$	Input $v(z)$	Controller $\tilde{q}_H(z)$
$\frac{1}{s}$	$\frac{z}{z-1}$	$(\tilde{p}_M^*(z))^{-1}$
$\frac{1}{\tau s+1}$	$\frac{z/\tau}{z-e^{-T/\tau}}$	$(\tilde{p}_M^*(z))^{-1}(\tilde{p}_A^*(e^{-T/\tau}))^{-1}$
$\frac{1}{s(\tau s+1)}$	$\frac{z(1-e^{-T/\tau})}{(z-1)(z-e^{-T/\tau})}$	$(\tilde{p}_M^*(z))^{-1}\frac{(1-\tilde{p}_A^{*-1}(e^{-T/\tau})e^{-T/\tau})z+(\tilde{p}_A^{*-1}(e^{-T/\tau})-1)e^{-T/\tau}}{(1-e^{-T/\tau})z}$
$\frac{1}{s^2}$	$\frac{Tz}{(z-1)^2}$	$(\tilde{p}_M^*(z))^{-1}\frac{(N+\Xi+1)z-N-\Xi}{z}$

$$\text{where } \Xi \triangleq \frac{d}{dz}(\tilde{p}_A^{*-1}(z)z^{-N})|_{z=1}$$

$$= \sum_{j=1}^{h}\frac{(\zeta_j^H)^{-1}-\zeta_j}{(1-\zeta_j)(1-(\zeta_j^H)^{-1})}$$

Example 8.1-2.

$$v^* = v_M^* = \frac{z/\tau}{z - e^{-T/\tau}}$$

$$\left\{z^{-1}\tilde{p}_A^{*-1}v_M^*\right\}_* = \left\{\tilde{p}_A^{*-1}\frac{1/\tau}{z - e^{-T/\tau}}\right\}_* = (\tilde{p}_A^*(e^{-T/\tau}))^{-1}\frac{1/\tau}{z - e^{-T/\tau}}$$

$$\tilde{q}_H(z) = z(\tilde{p}_M^*v_M^*)^{-1}(\tilde{p}_A^*(e^{-T/\tau}))^{-1}\frac{1/\tau}{z - e^{-T/t}} = \tilde{p}_M^{*-1}(\tilde{p}_A^*(e^{-T/\tau}))^{-1}$$

□

The derivation of $\tilde{q}_H$ for the other inputs listed in Table 8.1-1 is left as an exercise.

In the case of setpoint following, one sometimes has available and supplies to the controller future values of the setpoint, which the system output is to follow after N_p time steps. By doing so, better servo-behavior is accomplished. In this case $\tilde{q}_H(z)$ can be obtained from

$$\tilde{q}_H(z) = z(\tilde{p}_M^*v_M^*)^{-1}\left\{z^{-N_p-1}\tilde{p}_A^{*-1}v_M^*\right\}_* \qquad (8.1-6)$$

8.1.2   Design of the IMC Controller $\tilde{q}(z)$

The H_2^*- optimal controller $\tilde{q}_H(z)$ obtained in Sec. 8.1.1 may exhibit intersample rippling caused by poles of $\tilde{q}_H(z)$ close to $(-1, 0)$ as explained in Sec. 7.5.3. Hence a modification is necessary to obtain $\tilde{q}(z)$ from $\tilde{q}_H(z)$. We can write

$$\tilde{q}(z) = \tilde{q}_H(z)\tilde{q}_-(z)B(z) \qquad (8.1-7)$$

where $\tilde{q}_-(z)$ cancels all the poles of $\tilde{q}_H(z)$ with negative real part and substitutes them with poles at the origin. $B(z)$ is selected to preserve the system type. The introduction of poles at the origin aims at incorporating into the design some of the advantages of a deadbeat-type response while at the same time avoiding known problems of deadbeat controllers like overshoot or undershoot.

Let κ_i, $i = 1, \ldots, \rho$ be the poles of $\tilde{q}_H(z)$ with negative real part. Then we can write

$$\tilde{q}_-(z) = z^{-\rho}\prod_{j=1}^{\rho}\frac{z - \kappa_j}{1 - \kappa_j} \qquad (8.1-8)$$

$$B(z) = \sum_{j=0}^{m-1}b_j z^{-j} \qquad (8.1-9)$$

where m is the system type and the coefficients b_j, $j = 0, \ldots, m-1$ are chosen such that $\tilde{q}(z)$ satisfies (7.5–16). By construction $\tilde{q}_H(z)$ satisfies (7.5–16). Then it follows that $\tilde{q}(z)$ satisfies (7.5–16) if and only if

$$\lim_{z \to 1}\frac{d^k}{dz^k}(1 - \tilde{q}_-(z)B(z)) = 0, \qquad k = 0, 1, \ldots, m-1 \qquad (8.1-10)$$

For the important special cases of $m = 1$ and 2 we find

$$\text{Type 1}: \quad B(z) = 1 \qquad (8.1-11)$$

$$\text{Type 2}: \quad B(z) = b_0 + b_1 z^{-1} \qquad (8.1-12)$$

with

$$b_0 = 1 - b_1 \qquad (8.1-13)$$

$$b_1 = \sum_{j=1}^{\rho} \frac{\kappa_j}{1 - \kappa_j} \qquad (8.1-14)$$

In general, use of the transformation $z = \lambda^{-1}$ in (8.1–10) leads to a system of linear equations which can be solved easily by successive substitution.

The proposed "correction scheme" might seem somewhat *ad hoc* but at least for step inputs it can be shown to lead to controllers which combine the advantages of the algorithm that minimizes the sum of squared errors and of deadbeat-type algorithms. For step inputs Table 8.1-1 shows the H_2^*-optimal controller to be $\tilde{q}_H(z) = (\tilde{p}_M^*(z))^{-1}$. In order for the system to be Type 1, $B(z) = 1$. Application of (8.1–7) leads to a controller $\tilde{q}(z)$ with the following properties:

- In the case where all the unstable zeros of $\tilde{p}^*(z)$ have negative real part, the controller is of the deadbeat type and drives the discrete output of the system to the setpoint in a finite number of time steps.

- When $\tilde{p}^*(z)$ has unstable zeros with positive real part, the controller drives the output to the setpoint asymptotically in order to avoid large overshoot or undershoot.

- When all the zeros, stable or unstable, have positive real part, the controller minimizes the sum of the squared errors of the output.

Similar desirable properties are maintained for other input types when the minimum number of coefficients b_i necessary to satisfy (8.1–10) is used. Unfortunately, unlike for the continuous case, it is impossible to state general formulas for the IMC controller $\tilde{q}(z)$ for commonly occurring process models. The reason is that the factor $\tilde{q}_-(z)$ depends on both the MP and the NMP zeros of the plant $p^*(z)$ which in turn depend on the zeros and poles of the continuous system and the sampling time. Thus, we will simply illustrate the benefits of the IMC controller with an example.

Example 8.1-3. Consider the system given by (7.5–26) in Sec. 7.5.3. Its zero-order-hold discrete equivalent for $T = 1.8$ has two zeros, both inside the UC, at $z = -0.95$ and $z = -0.06$. Hence $p_M^* = zp^*$. For a step input v, the expression

for the H_2-optimal controller given by (8.1–5) simplifies to that in Ex. 8.1-1 —
i.e.,

$$q_1 \overset{\Delta}{=} \tilde{q}_H = (\tilde{p}_M^*)^{-1}$$

This controller has a pole at $z = -0.95$ which is close enough to (-1,0) to produce
the unacceptable input ringing and output intersample rippling shown in Fig.
7.5-1.

Application of (8.1–7) yields the controller $q_2(z)$ given by (7.5–29), which produces the excellent response shown also in Fig. 7.5-1. Note that the pole at
$z = -0.06$ is so close to the origin, that it does not really make a difference
whether it is substituted with a pole at the origin or not.   □

Example 8.1-4. Consider the system

$$p(s) = \frac{1}{(10s + 1)(25s + 1)}$$

For $T = 3$ we get

$$p^*(z) = \frac{0.0157(z + 0.869)}{(z - 0.887)(z - 0.741)}$$

The discrete system has a zero at $z = -0.869$, which is close enough to (-1,0)
to produce the intersample rippling shown in Fig. 8.1-1 when it appears as a
pole of the H_2^*-optimal controller q_H. Again, application of (8.1–7) eliminates the
problem and results in a deadbeat type response for this particular example.   □

8.2   The Discrete IMC Filter

Similar to the continuous case, $\tilde{q}(z)$ is augmented by a low-pass filter $f(z)$ ($q =
\tilde{q}f$), whose structure and parameters should be determined such that an optimal
compromise between performance and robustness is reached. To simplify the
design task the filter structure is fixed and only a few adjustable parameters are
included. The simplest form is a first order one-parameter filter:

$$f_1(z) = \frac{(1 - \alpha)z}{z - \alpha} \qquad (8.2 - 1)$$

The filter should preserve the asymptotic properties of the closed-loop system —
i.e., (7.5–16) should be satisfied. The design procedure in Sec. 8.1.2 assures that
(7.5–16) is satisfied for $q(z) = \tilde{q}(z)$. Therefore, for the system to be Type m, the
filter $f(z)$ has to satisfy

$$\text{Type m}: \quad \frac{d^k}{dz^k}(1 - f(z))\Big|_{z=1} = 0, \quad 0 \leq k < m \qquad (8.2 - 2)$$

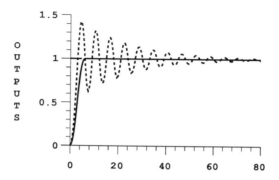

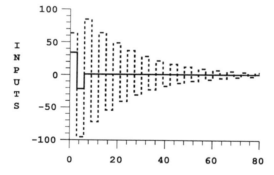

Figure 8.1-1. Dashed line: $q_1 = \tilde{q}_H$. Solid line: $q_2 = \tilde{q}_H \tilde{q}_- B$.

For a Type 1 system only $f(1) = 1$ is required and the filter given by (8.2–1) clearly meets that requirement. For $m \geq 2$ however, the filter (8.2–1) is not sufficient. In this case we postulate

$$f(z) = (\beta_0 + \beta_1 z^{-1} + \ldots + \beta_w z^{-w})\frac{(1-\alpha)z}{z-\alpha} \tag{8.2-3}$$

where the coefficients $\beta_0, \ldots, \beta_w$ are to be chosen such that $f(z)$ satisfies (8.2–2) for some specified α.

Theorem 8.2-1. *For a Type m system the coefficients β_i of the filter (8.2–3) have to satisfy*

$$\beta_0 = 1 - (\beta_1 + \ldots \beta_w) \tag{8.2-4}$$

and for $m \geq 2$, $w \geq m - 1$

$$N_w \begin{bmatrix} \beta_1 \\ \beta_2 \\ . \\ . \\ . \\ \beta_w \end{bmatrix} = \begin{bmatrix} -\alpha/(1-\alpha) \\ 0 \\ . \\ . \\ . \\ 0 \end{bmatrix} \tag{8.2-5}$$

where the elements ν_{ij} of the $(m-1) \times w$ matrix N_w are defined by

$$\nu_{ij} = \begin{cases} 0 & \text{for } i > j \\ \frac{j!}{(j-i)!} & \text{for } i \leq j \end{cases} \tag{8.2-6}$$

For the proof the following lemma will be used.

Lemma 8.2-1. *Let $h(\lambda) = \frac{1-\alpha}{1-\alpha\lambda}$. Then*

$$h^{(k)}(\lambda) = (1-\alpha)k! \, \alpha^k (1-\alpha\lambda)^{-(k+1)} \tag{8.2-7}$$

where the superscript (k) denotes k^{th} derivative.

Proof. By induction.
$k = 1$.

$$\frac{d}{d\lambda}h(\lambda) = (1-\alpha)\alpha(1-\alpha\lambda)^{-2}$$

$k = n$. Assume

$$h^{(n)}(\lambda) = (1-\alpha)n! \, \alpha^n (1-\alpha\lambda)^{-(n+1)} \tag{8.2-8}$$

$k = n + 1$. From (8.2–8) we get

$$h^{(n+1)}(\lambda) = (1-\alpha)n! \, \alpha^n \frac{d}{d\lambda}(1-\alpha\lambda)^{-(n+1)} =$$

$$= (1-\alpha)(n+1)!\,\alpha^{n+1}(1-\alpha\lambda)^{-(n+2)}$$

$\square$

Proof of Theorem 8.2-1. Equation (8.2–4) follows directly from (8.2–2) for $k = 0$. For proving (8.2–5) we define

$$\Gamma(\lambda) \triangleq \beta_0 + \beta_1\lambda + \ldots + \beta_w\lambda^w \qquad (8.2-10)$$

$$h(\lambda) \triangleq \frac{1-\alpha}{1-\alpha\lambda}$$

and express the filter (8.2–3) as

$$f(\lambda^{-1}) = \Gamma(\lambda)h(\lambda) \qquad (8.2-11)$$

Thus for $k \geq 1$ we can rewrite (8.2–2) as

$$\left.\frac{d^k}{d\lambda^k}f(\lambda^{-1})\right|_{\lambda=1} = 0, \qquad k = 1,\ldots m-1 \qquad (8.2-12)$$

For $k = 1$, (8.2–12) yields

$$\Gamma^{(1)}(1)h(1) + \Gamma(1)h^{(1)}(1) = 0 \qquad (8.2-13)$$

From Lemma 8.2-1 we find

$$h^{(k)}(1) = k!\alpha^k(1-\alpha)^{-k} \qquad (8.2-14)$$

Substituting (8.2–14) into (8.2–13) yields

$$\Gamma^{(1)}(1) = -h^{(1)}(1) = -\alpha(1-\alpha)^{-1} \qquad (8.2-15)$$

We will show next that for $k \geq 2$, (8.2–12) requires $\Gamma^{(k)}(1) = 0$. The proof will be by induction.

$k = 2$. Condition (8.2–12) becomes

$$\Gamma^{(2)}(1)h(1) + 2\Gamma^{(1)}(1)h^{(1)}(1) + \Gamma(1)h^{(2)}(1) = 0 \qquad (8.2-16)$$

Using (8.2–14) and (8.2–15), (8.2–16) yields

$$\Gamma^{(2)}(1) = 0 \qquad (8.2-17)$$

$2 \leq k \leq n < m-1$. Assume

$$\Gamma^{(k)}(1) = 0 \qquad (8.2-18)$$

$k = n+1$. Because of (8.2–18), (8.2–12) becomes

$$\Gamma^{(n+1)}(1)h(1) + (n+1)\Gamma^{(1)}(1)h^{(n)}(1) + \Gamma(1)h^{(n+1)}(1) = 0 \qquad (8.2-19)$$

or by using (8.2–14) and (8.2–15)

$$\Gamma^{(n+1)}(1) = 0 \qquad\qquad (8.2-20)$$

Hence by induction

$$\Gamma^{(k)}(1) = 0, \qquad k = 2, \ldots, m-1 \qquad\qquad (8.2-21)$$

But one can easily see that

$$\begin{bmatrix} \Gamma^{(1)}(1) \\ \cdot \\ \cdot \\ \cdot \\ \Gamma^{(m-1)}(1) \end{bmatrix} = N_w \begin{bmatrix} \beta_1 \\ \cdot \\ \cdot \\ \cdot \\ \beta_w \end{bmatrix} \qquad\qquad (8.2-22)$$

and (8.2–5) follows from (8.2–21), (8.2–22).   □

For $w > m-1$, there are several solutions to (8.2–5) and one can obtain $\beta_1, \ldots, \beta_w$ as the minimum norm solution. It can be shown that as $w \to \infty$ the norm of this solution goes to zero and from (8.2–3), (8.2–4) it follows that the properties of $f(z)$ are not significantly different from those of $f_1(z)$. Finally note that for $m = 2$, one should choose $w \geq 2$ in order to avoid the trivial solution $f(z) = 1$. Then the minimum norm solution for $m = 2$, $w \geq 2$, is found to be

$$\beta_k = \frac{-6k\alpha}{(1-\alpha)w(w+1)(2w+1)}, \qquad k = 1, \ldots, w \qquad\qquad (8.2-23)$$

Note that $\lim_{w \to \infty} \beta_0 = 1$.

8.3 Robust Stability

8.3.1 Filter Design

The robust stability condition derived in Sec. 7.6 can be stated in terms of the IMC controller $q(z)$ $(= \tilde{q}(z)f(z))$.

Corollary 8.3-1 (Robust Stability). *Assume that all plants $p(s)$ in the family Π^* are stable, that $q(z)$ is stable and that $c(z)$ is related to $q(z)$ through (7.2–3). Then the systems in Figs. 7.1-1A and 7.2-1A are robustly stable if and only if*

$$|f(e^{i\omega T})| < \left[|\tilde{p}^* \tilde{q}(e^{i\omega T})| |\bar{\ell}_m^*(\omega)| \right]^{-1} \qquad 0 \leq \omega \leq \frac{\pi}{T} \qquad\qquad (8.3-1)$$

Clearly an $f(z)$ can always be found such that (8.3–1) is satisfied. However, a small $|f|$ implies a small $|\tilde{\eta}^*|$ and thus poor performance. Hence, if performance requirements have to be met the uncertainty has to be limited. A simple performance specification is to require the closed-loop system to be Type 1 – i.e., $\tilde{p}^*(1)\tilde{q}(1) = f(1) = \gamma(0) = 1$. Then from (8.3–1) we can obtain the following corollary.

Corollary 8.3-2. *Assume that $\bar{\ell}_m^*(\omega)$ is continuous. Then there exists a filter $f(z)$ such that the closed-loop system is Type 1 and robustly stable for the family Π^* if and only if $\bar{\ell}_m(0) < 1$, where $\bar{\ell}_m(0)$ is the multiplicative steady-state error bound for the continuous system.*

Proof. All that is needed is to show that (8.3–1) is satisfied for $\omega = 0$, where $f(1) = \tilde{p}^*(1)\tilde{q}(1) = 1$. Hence we need $\bar{\ell}_m^*(0) < 1$. The steady-state gain of the zero-order hold equivalent of $\tilde{p}(s)$ is the same as the steady-state gain of $\tilde{p}(s)$ — i.e., $\tilde{p}^*(1) = \tilde{p}(0)$. Also from (7.3–6) we get $\bar{\ell}_a^*(0) = \bar{\ell}_a(0)$ since $h_0(i2\pi k/T) = 0$ for $k = \pm1, \pm2, \ldots$ and $h_0(0)/T = \gamma(0) = 1$. Thus $\bar{\ell}_m^*(0) = \bar{\ell}_a^*(0)/\tilde{p}^*(1) = \bar{\ell}_a(0)/\tilde{p}(0) = \bar{\ell}_m(0)$.   □

Note that Cor. 8.3-2 requires simply that the error between the steady-state gain of the plant and that of the model is not more than 100% of the model gain. This condition can always be satisfied by appropriate selection of the model if all the possible plants have steady-state gains with the same sign.

Note that the condition $\bar{\ell}_m(0) < 1$ is the same as the one we found for the continuous system (Cor. 4.3-2). This makes sense because a steady-state requirement should not be affected by the sampling operation.

A simple way to design the IMC filter is to use an $f(z)$ of the structure in (8.2–3) and to vary the parameter α so that (8.3–1) is satisfied. Equation (8.3–1) places a lower bound α^* on α. It can be obtained from a Bode plot of $(|\tilde{p}^*\tilde{q}(e^{i\omega T})|\bar{\ell}_m^*(\omega))^{-1}$. If this quantity is never less than 1, then $\alpha^* = 0$. If it obtains values less than 1, then α^* can be found from a Bode plot of $f(z)$, which is practically the same as that of the first-order filter $f_1(z)$ in (8.2–1) provided that the number of coefficients w in (8.2–3) is sufficiently large. For example, if $(|\tilde{p}^*\tilde{q}(e^{i\omega T})|\bar{\ell}_m^*(\omega))^{-1}$ decreases like a first-order system and reaches a value of 0.7 at $\omega = \omega_\ell$ then

$$\alpha^* \cong e^{-T\omega_\ell} \tag{8.3 – 2}$$

Note that for an open-loop stable sampled-data system a *first-order* filter $f_1(z)$ can always be designed to satisfy the robust stability condition regardless of the magnitude of the model uncertainty. For continuous systems, depending on the uncertainty a higher order filter might be required. The reason is that for sampled-data systems the frequency range over which the condition has to be met is bounded.

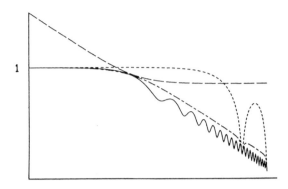

FREQUENCY

Figure 8.3-1. Effect of sampling on robust stability (logarithmic plot). Long dash: $1/\bar{\ell}_m(\omega)$. Solid: $|\tilde{p}(i\omega)\tilde{q}(e^{i\omega T})|, T = T_1$. Short dash: $|\tilde{p}(i\omega)\tilde{q}(e^{i\omega T})|, T = T_2 < T_1$. Dash and dot: $|\tilde{p}(i\omega)\tilde{q}(e^{i\omega T})f_1(e^{i\omega T})|, T = T_2$. (Reprinted with permission from Int. J. Control, **44**, 721(1986), Taylor & Francis Ltd.)

8.3.2 Effect of Sampling

As explained in Sec. 8.3-1, condition (8.3–1) can be satisfied by simply increasing the time constant of the filter, provided that $\bar{\ell}_m(0) < 1$. The increase of the filter time constant reduces the closed-loop bandwidth of the nominal system. In Sec. 7.5.3 we saw that a larger sampling time T also reduces the bandwidth. This becomes clearer if we write (8.3–1) as

$$|\tilde{p}(i\omega)\tilde{q}(e^{i\omega T})f(e^{i\omega T})| < |\tilde{p}(i\omega)|/\bar{\ell}_a^*(\omega) \qquad (8.3-3)$$

One can see that the bandwidth of the left hand side term can be reduced by either increasing α in $f(z)$ or leaving $f(z) = 1$ and increasing T. A graphical illustration of this discussion is given in Fig. 8.3-1. Note that in Fig. 8.3-1 the right-hand-side term of (8.3–3) is assumed independent of T by using the approximation $\bar{\ell}_a^*(\omega) \cong \bar{\ell}_a(\omega)$. For illustrative purposes this is a reasonable approximation for $0 \leq \omega \leq \pi/T$ but it should not be used to check (8.3–1); $\bar{\ell}_a^*(\omega)$ should be computed from (7.3–6).

8.4 Robust Performance

In Sec. 7.7.2 we derived that for robust performance the controller has to be designed such that

$$M(\omega) \triangleq |\hat{q}(i\omega)|\bar{\ell}_a(\omega) + |1 - \tilde{p}(i\omega)\hat{q}(i\omega)|w(\omega) < 1, \quad 0 \leq \omega \leq \frac{\pi}{T} \qquad (8.4-1)$$

is satisfied.

8.4.1 Filter Design

The simplest approach is to specify the structure of the filter as that in (8.2–3) and to try to satisfy (8.4–1) by varying the parameter α. Increasing α will tend to decrease the first term in $M(\omega)$ and increase the second term. Hence, depending on $\bar{\ell}_a$ and w there might be no value of α for which (8.4–1) is satisfied. Let us assume that $\tilde{q}(z)$ and $f(z)$ are selected such that the system is Type 1 or higher ($\tilde{p}^*\tilde{q}f(1) = 1$) and that $\bar{\ell}_m(0) < 1$. Then robust performance at $\omega = 0$ can be achieved for any w.

Corollary 8.4-1. *There exists a filter $f(z)$ such that (8.4–1) is satisfied at $\omega = 0$ for any weight w if and only if $\bar{\ell}_m(0) < 1$.*

Corollary 8.4-1 is similar to Cor. 4.4-2 for continuous systems and it simply states that if the system is robustly stable for a controller $c(z)$ with integral action (Type 1), then the steady-state performance is perfect even when there is modelling error.

Selection of the weight $w(\omega)$. The choice of $w(\omega)$ depends on the performance requirements set by the designer. It is consistent with the overall design philosophy to assume that an H_2-optimal controller $\tilde{q}(s)$ designed for the continuous model $\tilde{p}(s)$ would achieve the ideal performance. Hence it is reasonable to use the ideal sensitivity function $\tilde{\eta}(s) = \tilde{p}(s)\tilde{q}(s)$ as a guide for the choice of the weight:

$$w(\omega)^{-1} \geq |1 - \tilde{p}(i\omega)\tilde{q}(i\omega)| \qquad (8.4-2)$$

This sensitivity function, however, is achieved only by a non-proper controller. The properness requirement adds to (8.4–2) the condition that $1/w(\infty) \geq 1$. Also note that though for a Type m system ($m \geq 1$), (8.4–2) becomes $w(0)^{-1} \geq 0$ for $\omega = 0$, there is no need to choose $w(0) = \infty$, since $\tilde{q}(z)$ and $f(z)$ have been designed so that conditions (7.5–16) and (8.2–2) are satisfied. These conditions guarantee no steady-state offset under modelling error, provided that stability is maintained.

Computation of α. The filter parameter α has to be adjusted in an effort to satisfy (8.3–1) and (8.4–1). It was shown in Sec. 8.3.1 that (8.3–1) puts a lower bound α^* on the values of α that are allowed. Hence, to find α one must solve the following optimization problem:

$$\min_{\alpha^* \leq \alpha < 1} \max_{0 \leq \omega \leq \pi/T} M(\omega) \overset{\Delta}{=} \psi(T) \qquad (8.4-3)$$

where $M(\omega)$ is defined in (8.4–1) and the argument T has been used in ψ to indicate that the optimum value of the objective function depends on the sampling time T.

The above minimization can be carried out by computing $M(\omega)$ for a number of values for α. The computational effort is very small. It is advisable to write $\alpha = e^{-T/\tau}$ where τ is in $[\tau^*, \infty)$ with $\alpha^* = e^{-T/\tau^*}$ and minimize over τ.

8.4.2   Sampling Time Selection

A short sampling time improves the nominal performance as we have discussed in Sec. 7.5.3. However, high-frequency sampling puts a large load on the computer and for robustness nominal performance generally has to be sacrificed anyway. Thus a longer sampling time might be acceptable for robust stability and robust performance. On the other hand if the sampling time is too long, it might be impossible to meet the robust performance requirements.

As a rule, π/T should be selected larger than the bandwidth over which good performance is desired. If for a certain sampling time T^* it is found that the robust performance requirements are exceeded ($\psi(T^*) < 1$), then the specifications could be met even with a larger T. If $\psi(T^*) > 1$ then for the assumed model uncertainty and controller structure the specifications are too tight for the specific T^* and have to be relaxed.

8.4.3   Example

Let us consider the system

$$\tilde{p}(s) = \frac{3}{(s+1)(s+3)} \qquad (8.4-4)$$

A delay-type uncertainty is assumed:

$$p(s) = \tilde{p}(s)e^{-\theta s} \qquad (8.4-5)$$

where

$$0 \le \theta \le 0.05 \qquad (8.4-6)$$

Then from (2.2–2), (2.2–4)

$$\ell_m(s) = e^{-\theta s} - 1 \qquad (8.4-7)$$

from which one can easily obtain the bound $\bar{\ell}_m$ (2.2–8)

$$\bar{\ell}_m(\omega) = \begin{cases} |e^{-0.05i\omega} - 1| & 0 \le \omega \le 20\pi \\ 2 & \omega \ge 20\pi \end{cases} \qquad (8.4-8)$$

Let us examine two sampling times, different by an order of magnitude, $T_1 = 0.1$ and $T_2 = 0.01$. For the robust performance design the following weight is selected:

$$w(s)^{-1} = 0.4\frac{0.5s + 1}{0.1s + 1} \qquad (8.4-9)$$

This selection was based on the observation that at $\omega = 2$, $|\tilde{p}(i\omega)|$ is small enough ($\simeq 0.35$) to justify a relaxation of the performance requirement. Also $1/w(\infty) = 2 > 1$. It should be noted that the above choice is a rather strict performance requirement, but it is justified because the system is not inherently difficult to control and the uncertainty is small. Also note that in this simple case where $\tilde{p}(s)$ is minimum phase, the right-hand side of (8.4–2) is zero and this leaves us the freedom to select $w(\omega)$ as above.

The next step is to compute $\tilde{q}$ for the two sampling times according to the procedure of Sec. 8.1. We obtain

$$q_1(z) = \frac{40.55(z^2 - 1.64566z + 0.67032)}{z^2} \qquad (8.4 - 10a)$$

$$q_2(z) = \frac{3400(z^2 - 1.960495z + 0.960789)}{z^2} \qquad (8.4 - 10b)$$

for T_1 and T_2, respectively.

Then the quantity $\psi(T)$, which measures robust performance must be computed. For the two sampling times the solution of (8.4–3) yields

$$\psi(T_1) = 1.22 \qquad (8.4 - 11a)$$

$$\psi(T_2) = 0.90 \qquad (8.4 - 11b)$$

The corresponding optimal α's are $\alpha_1 = 0.4625$ and $\alpha_2 = 0.9363$. The optima (8.4–11) imply that for the sampling time T_2 it is possible to satisfy the tight robust performance specification set through (8.4–9), while this cannot be done for the larger T_1. Equation (8.4–11b) indicates that the specification can be met even when T is somewhat larger than T_2. Further search shows that $\psi(0.032) = 0.98 < 1$.

Let us now compare the time responses for the two controllers designed for T_1 and T_2 to see how (8.4–11) translates into the time-domain. Figure 8.4-1A shows the responses to a unit step setpoint change for the case when there is no model-plant mismatch. As expected, the controller with the smaller sampling time is somewhat better. Note that when the procedure described in this chapter is used for controller design, the use of a smaller sampling time cannot harm the nominal behavior, contrary to what could happen for some other digital algorithms, like deadbeat-type controllers. Figure 8.4-1B shows the response when the plant is $p(s) = \tilde{p}(s)e^{-0.05s}$. Again the response for T_2 is clearly better. Note that because of the robust design the faster nominal response (T_2) does not imply increased sensitivity to model uncertainty; the response for T_2 remains superior even in the presence of plant/model mismatch.

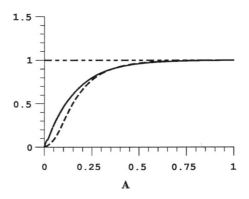

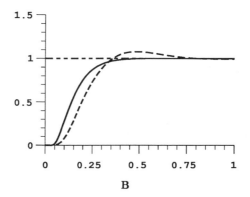

Figure 8.4-1. Step response for controllers with sampling time T_1 (dash) and T_2 (solid). A: No model error $(p = \tilde{p})$; B: Model error $p = \tilde{p}e^{-0.05s}$.

Roughly speaking, both controllers produce acceptable responses. This is not surprising since the ψ's for the two controllers are similar. This simple example demonstrates, however, that the frequency domain based quantity $\psi(T)$ captures the time domain behavior in an excellent manner and even small differences in $\psi(T)$ translate into noticeable differences in the time responses.

8.5 Summary

In the *first step* of the IMC design procedure the controller is designed to yield a "good" system response for the input(s) of interest without regard for constraints or uncertainty. The starting point is the H_2^*-optimal controller which is calculated from

$$\tilde{q}_H(z) = z(\tilde{p}_M^* v_M^*)^{-1} \left\{ z^{-1} \tilde{p}_A^{*-1} v_M^* \right\}_*, \qquad (8.1-5)$$

Here the operator $\{\cdot\}_*$ denotes that after a partial fraction expansion of the operand only the strictly proper and stable (including poles at $z = 1$) terms are retained. The allpass and MP portions of the model are denoted by $\tilde{p}_A^*$ and $\tilde{p}_M^*$ respectively (8.1–1 and 8.1–2); v_M^* is defined similarly (8.1–3 and 8.1–4). Table 8.1-1 lists formulas for $\tilde{q}_H(z)$ for some typical inputs v_M^*.

Because the H_2^*-optimal controller can lead to undesirable intersample rippling it is modified to

$$\tilde{q}(z) = \tilde{q}_H(z)\tilde{q}_-(z)B(z) \qquad (8.1-7)$$

where

$$\tilde{q}_-(z) = z^{-\rho} \prod_{j=1}^{\rho} \frac{z - \kappa_j}{1 - \kappa_j} \qquad (8.1-8)$$

$$B(z) = \sum_{j=0}^{m-1} b_j z^{-j} \qquad (8.1-9)$$

Here κ_i, $i = 1, \ldots \rho$ are the poles of $\tilde{q}_H(z)$ with negative real part, m is the system type and b_j are coefficients to be chosen to satisfy the type requirements.

In the *second step* of the IMC design procedure the controller $\tilde{q}(z)$ is augmented by a filter $f(z)$ for robustness

$$q(z) = \tilde{q}(z)f(z)$$

Recommended one-parameter filters are

$$\text{Type 1}: \ f_1(z) = \frac{(1 - \alpha)z}{z - \alpha} \qquad (8.2-1)$$

$$\text{Type 2}: \ f_2(z) = (\beta_0 + \beta_1 z^{-1} + \ldots + \beta_w z^{-w})\frac{(1 - \alpha)z}{z - \alpha} \qquad (8.2-3)$$

where

$$\beta_k = \frac{-6k\alpha}{(1-\alpha)w(w+1)(2w+1)}, \quad k = 1,\ldots,w; \quad w \geq 2 \qquad (8.2-23)$$

For robust stability the filter parameter α is increased until

$$|f(e^{i\omega T})| < [|\tilde{p}^* \tilde{q}(e^{i\omega T})|\bar{\ell}^*_m(\omega)]^{-1} \qquad 0 \leq \omega \leq \frac{\pi}{T} \qquad (8.3-1)$$

For robust performance

$$\min_{\alpha^* \leq \alpha < 1} \max_{0 \leq \omega \leq \pi/T} M(\omega) \qquad (8.4-3)$$

where

$$M(\omega) \triangleq |\hat{q}(i\omega)|\bar{\ell}_a(\omega) + |1 - \tilde{p}(i\omega)\hat{q}(i\omega)|w(\omega) < 1, \quad 0 \leq \omega \leq \frac{\pi}{T} \qquad (8.4-1)$$

and α^* is the minimum filter parameter needed to assure robust stability. Instead of increasing the filter parameter α, the sampling time T can be increased with a similar effect on robustness.

8.6   References

8.1.1. For a discussion of the state-space approach to discrete Linear Quadratic Control see Kwakernaak and Sivan (1972). For a brief discussion, which also includes deadbeat controllers, see Kucera (1972).

8.1.2. The reasons behind the "correction scheme" were presented by Zafiriou and Morari (1985).

8.3. The original formulation of the discrete IMC controller and filter and some early results on robust stability can be found in Garcia and Morari (1982, 1985).

8.4.2. For a more detailed discussion of the procedure for sampling time selection see Zafiriou and Morari (1986a). Additional material on signal sampling and reconstruction is available from Åström and Wittenmark (1984, Chap. 2).

Chapter 9

SISO DESIGN FOR UNSTABLE SAMPLED-DATA SYSTEMS

In Chap. 5 it was pointed out that the IMC structure is unsuitable for implementation when the plant is open-loop unstable. However, the IMC controller parametrization remains a valuable tool that simplifies the controller design and greatly clarifies the robustness problems of open-loop unstable plants.

9.1 Parametrization of All Stabilizing Controllers

9.1.1 Internal Stability

For open-loop unstable systems the classic feedback structure shown in Fig. 7.1-1 has to be used for implementation. For internal stability the system described by (7.1–17) has to be stable. We can express (7.1–17) in terms of the IMC controller $q(z)$ by substituting (7.2–3) into (7.1–17). We obtain (for $p = \tilde{p}$)

$$\begin{pmatrix} y_\gamma^* \\ u \end{pmatrix} = \begin{pmatrix} p_\gamma^* q & (1 - p_\gamma^* q) p_\gamma^* \\ q & -p_\gamma^* q \end{pmatrix} \begin{pmatrix} r^* \\ u' \end{pmatrix} \qquad (9.1-1)$$

All four transfer functions in (9.1–1) have to be stable. Note that since the prefilter $\gamma(s)$ is stable, the only unstable poles of $p_\gamma^*(z)$ are the unstable poles of $p^*(z)$. By using the same arguments as in Sec. 5.1.1, we can derive the following theorem.

Theorem 9.1-1. *Assume that the model is perfect* $(p = \tilde{p})$ *and that* $p^*(z)$ *has* k *unstable poles at* $\pi_1, \ldots, \pi_k$, *and that* $p(s)$ *has also* k *unstable poles (i.e., that none of the unstable poles of* $p(s)$ *become unobservable after sampling). Then the feedback system in Fig. 7.1-1 with the controller* $c = q(1 - p_\gamma^* q)^{-1}$ *is internally stable if and only if*

(i) $q(z)$ *is stable.*

(ii) $(1 - p_\gamma^* q)$ has zeros at $\pi_1, \ldots, \pi_k$.

Theorem 9.1-1 reduces to Thm. 7.4-1 when p is stable.

9.1.2   Controller Parametrization

A parametrization of *all* q's that satisfy the conditions of Thm. 9.1-1 will be found in this section. Define the allpass z-transfer function

$$b_p^*(z) = \prod_{j=1}^{k} \frac{(1 - (\pi_j^H)^{-1})(z - \pi_j)}{(1 - \pi_j)(z - (\pi_j^H)^{-1})} \qquad (9.1-2)$$

where π_j, $j = 1, \ldots, k$ are the poles of $p^*(z)$ strictly outside the UC.

Theorem 9.1-2. *Assume that $p^*(z) = \tilde{p}^*(z)$ has k poles $\pi_1, \ldots, \pi_k$ strictly outside the UC and ℓ poles at $z = 1$, and that $p(s)$ has no unstable poles that become unobservable after sampling. Also assume that there exists a causal $q_0(z)$ such that $c = q_0(1 - p_\gamma^* q_0)^{-1}$ stabilizes the system in Fig. 7.1-1. Then all causal controllers that stabilize the system are parametrized by*

$$c = q(1 - p_\gamma^* q)^{-1} \qquad (7.2-3)$$

$$q = q_0 + (b_p^*)^2 (1 - z^{-1})^{2\ell} q_1 \qquad (9.1-3)$$

where $q_1(z)$ is any arbitrary causal stable z-transfer function.

Proof. The proof is similar to that of Thm. 5.1-2, and uses the fact that $p^*(z)$ and $p_\gamma^*(z)$ have the same unstable poles.   □

Note that Thm. 9.1-2 assumes the existence of a stabilizing $q_0(z)$. The construction of the H_2^*-optimal controller in Sec. 9.2.1 will serve as proof that such a controller always exists. Also note that for an open-loop stable system we have $b_p = 1$, $\ell = 0$ and by choosing the stabilizing $q_0 = 0$, we obtain $q = q_1$, which is the IMC parametrization for stable systems.

9.2   Nominal Performance

The design procedure for unstable systems is the same as for stable ones. First the H_2^*-optimal controller $\tilde{q}_H(z)$ is designed and then a modification is introduced to avoid the problem of intersample rippling. Subsequently $\tilde{q}(z)$ is augmented by a low-pass filter to achieve robust stability and performance. In this section we shall derive the formulas for the design of $\tilde{q}(z)$.

9.2.1 H_2^*-Optimal Controller

As explained in Sec. 8.1.1, the H_2^*-optimal controller solves the problem defined by (7.5–38), subject to the constraint that $\tilde{q}$ is a stabilizing controller. The external system input $v(s)$ can be either a setpoint ($v = r$, $d = 0$) or a disturbance ($r = 0$, $v = d$ or $v = \gamma d$). Then the following theorem holds:

Theorem 9.2-1. *Let* $p^*(z) = \tilde{p}^*(z)$ *have* k *poles at* $\pi_1, \ldots, \pi_k$ *strictly outside the UC and a pole of multiplicity* ℓ *at* $z = 1$. *Define*

$$b_p^* = \prod_{j=1}^{k} \frac{(1 - (\pi_j^H)^{-1})(z - \pi_j)}{(1 - \pi_j)(z - (\pi_j^H)^{-1})} \qquad (9.1 - 2)$$

and factor the plant into an allpass portion $p_A^*(z)$ *and a semi-proper MP portion* $p_M^*(z)$.

$$p^*(z) = p_A^*(z)p_M^*(z) \qquad (9.2 - 1)$$

Factor the input v *similarly*

$$v^*(z) = v_A^*(z)v_M^*(z) \qquad (9.2 - 2)$$

Assume without loss of generality that the unstable poles of $v^*(z)$ *strictly outside the UC are the first* k' *poles* π_i *of the plant[1] and define accordingly*

$$b_v^* = \prod_{j=1}^{k'} \frac{(1 - (\pi_j^H)^{-1})(z - \pi_j)}{(1 - \pi_j)(z - (\pi_j^H)^{-1})} \qquad (9.2 - 3)$$

Assume further that $v^*(z)$ *has at least* ℓ *poles at* $z = 1$.[2] *Then the* H_2^*-optimal *controller* $\tilde{q}_H(z)$ *is given by*

$$\tilde{q}_H = zb_p^*(p_M^* b_v^* v_M^*)^{-1}\{(zb_p^* p_A^*)^{-1} b_v^* v_M^*\}_* \qquad (9.2 - 4)$$

where the operator $\{\cdot\}_*$ *denotes that after a partial fraction expansion only the strictly proper terms are retained except for those corresponding to poles of* $(p_A^*)^{-1}$.

Proof. Some preliminary definitions and facts are necessary. Let L_2^* be the Hilbert space of complex-valued functions defined on the unit circle ($UC = \{e^{i\theta} : -\pi \leq \theta < \pi\}$) and square integrable with respect to θ. The inner product on L_2^* is

$$< f, g > = \frac{1}{2\pi} \int_{-\pi}^{\pi} f(e^{i\theta})g(e^{i\theta})^H d\theta \qquad (9.2 - 5)$$

---

[1]If this assumption were not made the problem would not be meaningful. Unbounded controller action would be necessary for the error to vanish as $t \to \infty$. See also the discussion at the end of Sec. 2.2.3.

[2]This assumption is made so that disturbances occurring at the plant *input* can be rejected with vanishing error as $t \to \infty$.

The closed subspace of L_2^* of functions having analytic continuations inside the UC is defined as $(H_2^*)^\perp$; its orthogonal complement is denoted by H_2^*.[3] Note that with the above definitions a constant function is in $(H_2^*)^\perp$. $(H_2^*)^\perp$ also includes all rational z-transfer functions that are strictly unstable — i.e., which have all their poles strictly outside the UC [including poles at $z = \infty$ (improper transfer functions)]. All strictly proper, stable rational z-transfer functions are in H_2^*. Any rational $f(z)$ with no poles on the UC, can be uniquely decomposed into a strictly proper, stable part $\{f\}_+$ in H_2^* and a strictly unstable part $\{f\}_-$ in $(H_2^*)^\perp$:

$$f = \{f\}_- + \{f\}_+ \qquad (9.2-6)$$

For a rational z-transfer function, such a decomposition can be obtained by simply taking a partial fraction expansion. Note that any constant and improper terms belong in $\{f\}_-$.

If $f_1(z) \in H_2^*$ and $f_2(z) \in (H_2^*)^\perp$, then

$$< f_1, f_2 > = \frac{1}{2\pi} \int_{-\pi}^{\pi} f_1(e^{i\theta}) f_2(e^{i\theta})^H d\theta = 0 \qquad (9.2-7)$$

Hence the L_2^*-norm (H_2^*-norm) of f, defined by the inner product (9.2–5) can be computed as

$$\|f\|_2^2 = < f, f > = < \{f_-\}, \{f_-\} > + < \{f_+\}, \{f_+\} > = \|\{f\}_-\|_2^2 + \|\{f\}_+\|_2^2 \qquad (9.2-8)$$

According to (7.5–38) we define the objective function ϕ:

$$\phi \triangleq \|(1 - p^*(z)\tilde{q}(z))v^*(z)\|_2^2 \qquad (9.2-9)$$

Since multiplication of a function by an allpass does not change its L_2^*-norm, we get:

$$\phi = \|(zb_p^* p_A^* v_A^*)^{-1} b_v^* (1 - p^* \tilde{q}) v^*\|_2^2 \qquad (9.2-10)$$

Use of (9.1–3), (9.2–1), (9.2–2) yields:

$$\phi = \|(zb_p^* p_A^*)^{-1}(1 - \tilde{p}^* q_0) b_v^* v_M^* - z^{-1} p_M^* b_p^* (1 - z^{-1})^{2\ell} b_v^* v_M^* q_1\|_2^2$$

---

[3]This definition of H_2^* and $(H_2^*)^\perp$ is exactly the opposite of the one encountered in the mathematics literature, where H_2 corresponds to the L_2-functions with analytic continuations inside the UC. Our definitions have been chosen to be consistent with the common definitions of $H_2, H_2^\perp$ for Laplace transfer functions (Chap. 5) in the control literature. The transformation $\lambda = z^{-1}$ could have been employed to introduce consistency with the mathematics literature but this would unnecessarily complicate the notation.

$$\triangleq \|f\|_2^2 \triangleq \|f_1 - f_2 q_1\|_2^2 \qquad (9.2-11)$$

It will be assumed that in addition to being a stabilizing controller, q_0 satisfies (7.5–16). Not every stabilizing controller has this property, since v^* may have more poles at $z = 1$ than p^*. The final construction of the H_2^*-optimal q serves as proof of the existence of a q_0 with such properties.

Inspection of (9.2–11) shows that f_2 has no poles on or outside the UC except possibly for poles at $z = 1$ in the case where $v^*(z)$ has more than ℓ poles at $z = 1$. However, f_1 has no poles at $z = 1$ because q_0 satisfies (7.5–16). Hence the optimal q_1 *must* have the required number of zeros at $z = 1$ to produce an $f_2 q_1$ without any poles at $z = 1$ so that ϕ is finite. Also $f_2 q_1$ is strictly proper since q_1 is proper. Therefore $f_2 q_1$ is in H_2^*. Hence

$$\{f\}_+ = \{f_1\}_+ - f_2 q_1 \qquad (9.2-12)$$

$$\{f\}_- = \{f_1\}_- \qquad (9.2-13)$$

Then (9.2–8, 11, 12, 13) imply

$$\phi = \|\{f_1\}_-\|_2^2 + \|\{f_1\}_+ - f_2 q_1\|_2^2 \qquad (9.2-14)$$

Since $\{f_1\}_-$ is independent of q_1, the obvious solution to the minimization of (9.2–14) is

$$q_1 = f_2^{-1}\{f_1\}_+ \qquad (9.2-15)$$

However this solution is optimal only if q_1 is proper, stable and $f_2 q_1$ has no poles at $z = 1$. Careful inspection of (9.2–15), (9.2–11) shows that q_1 is proper and stable. Also $f_2 q_1 = \{f_1\}_+$ is in H_2^* and therefore it has no poles at $z = 1$. Hence (9.2–15) gives the optimal q_1.

Substitution of (9.2–15) into (9.1–3) yields the H_2^*-optimal controller $\tilde{q}_H(z)$:

$$\tilde{q}_H = q_0 + b_p^{*2}(1 - z^{-1})^{2\ell} f_2^{-1}\{f_1\}_+$$

$$= z b_p^*(p_M^* b_v^* v_M^*)^{-1}[(z b_p^*)^{-1} p_M^* b_v^* v_M^* q_0$$

$$+ \{(z b_p^* p_A^*)^{-1} b_v^* v_M^*\}_+ - \{(z b_p^* p_A^*)^{-1} p^* q_0 b_v^* v_M^*\}_+]$$

$$= z b_p^*(p_M^* b_v^* v_M^*)^{-1}[\{(z b_p^* p_A^*)^{-1} p^* b_v^* v_M^* q_0\}_{0-} + \{(z b_p^* p_A^*)^{-1} b_v^* v_M^*\}_+] \qquad (9.2-16)$$

where $\{\cdot\}_{0-}$ indicates that in the partial fraction expansion only the terms corresponding to poles on or outside the UC are retained. These are the poles of $(b_p^*)^{-1}b_v^*v_M^*$ on or outside the UC because $(p_A^*)^{-1}p^*q_0 = p_M q_0$ is stable since q_0 is a stabilizing controller. Also since q_0 satisfies (7.5–16), if π is an unstable (on or outside the UC) pole of $(b_p^*)^{-1}b_v^*v_M^*$ of multiplicity m, then $(1 - p^*q_0)$ has at least m zeros at $z = \pi$ — i.e.,

$$p^*(\pi)q_0(\pi) = 1 \qquad (9.2 - 17a)$$

$$\left.\frac{d^k}{dz^k}p^*(z)q_0(z)\right|_{z=\pi} = 0, \quad k = 1, \ldots, m - 1 \qquad (9.2 - 17b)$$

Thus (9.2–16) simplifies to

$$\tilde{q}_H = z b_p^*(p_M^* b_v^* v_M^*)^{-1} \left\{ (z b_p^* p_A^*)^{-1} b_v^* v_M^* \right\}_*$$

$\qquad\qquad\qquad\qquad\qquad\qquad\qquad\qquad\qquad\qquad\qquad\qquad \square$

In situations where future values of the setpoint, r are supplied to the controller to be followed by the system output after N_p time steps, $\tilde{q}_H(z)$ can be obtained from

$$\tilde{q}_H = z b_p^*(p_M^* b_M^* v_M^*)^{-1} \left\{ (z^{N_p+1} b_p^* p_A^*)^{-1} b_v^* v_M^* \right\}_* \qquad (9.2 - 18)$$

The proof follows that of Thm. 9.2-1 by changing the objective function to

$$\phi = \|(z^{-N_p} - \tilde{p}^*(z)\tilde{q}(z))r^*(z)\|_2^2 \qquad (9.2 - 19)$$

9.2.2   Design of the IMC Controller $\tilde{q}(z)$

As explained in Sec. 7.5.3 the H_2^*-optimal controller $\tilde{q}_H(z)$ may exhibit intersample rippling caused by poles of $\tilde{q}_H(z)$ close to (-1,0). As in Sec. 8.1.2, $\tilde{q}(z)$ is obtained as

$$\tilde{q}(z) = \tilde{q}_H(z)\tilde{q}_-(z)B(z) \qquad (9.2 - 20)$$

where $\tilde{q}_-(z)$ cancels all poles of $\tilde{q}_H(z)$ with negative real part and replaces them with poles at the origin. In this case, however, $B(z)$ is selected to preserve both the system type and the internal stability requirements described by Thm. 9.1-1 (ii). In this section we assume $\gamma(s) = 1$. Section 9.2.3 discusses the choice of $\gamma(s)$ further.

Similarly to Sec. 8.1.2 let κ_i, $i = 1, \ldots, \rho$ be the poles of $\tilde{q}_H(z)$ with negative real part. Then

$$\tilde{q}_-(z) = z^{-\rho} \prod_{j=1}^{\rho} \frac{z - \kappa_j}{1 - \kappa_j} \qquad (8.1-8)$$

Let π_i, $i = 1, \ldots, \xi$ be the unstable roots (including $z = 1$) of the least common denominator of $\tilde{p}^*(z), v^*(z)$ with multiplicity m_i. Recall that according to the assumption of Thm. 9.2-1, $v^*(z)$ has at least as many poles at $z = 1$ as $\tilde{p}^*(z)$ and each strictly unstable pole of $v^*(z)$ is also a pole of $\tilde{p}^*(z)$. The system type and the internal stability requirements can be unified as

$$\left. \frac{d^k}{dz^k}(1 - \tilde{q}_-(z)B(z)) \right|_{z=\pi_i} = 0, \; k = 0, \ldots, m_i - 1; \; i = 1, \ldots, \xi \qquad (9.2-21)$$

We can write

$$B(z) = \sum_{j=0}^{M-1} b_j z^{-j} \qquad (9.2-22)$$

where

$$M = \sum_{i=1}^{\xi} m_i \qquad (9.2-23)$$

and compute the coefficients b_j, $j = 0, \ldots, M - 1$ from (9.2–21). Note that since none of the π_i's is 0 or ∞, (9.2–21) is equivalent to

$$\left. \frac{d^k}{d\lambda^k}(1 - q_-(\lambda^{-1})B(\lambda^{-1})) \right|_{\lambda=\pi_i^{-1}} = 0, \; k = 0, \ldots, m_i - 1, \; i = 1, \ldots, \xi \quad (9.2-24)$$

Since both $\tilde{q}_-(\lambda^{-1})$ and $B(\lambda^{-1})$ are polynomials their derivatives with respect to λ can be computed easily. Equation (9.2–24) yields a system of M linear equations with M unknowns $(b_0, b_1, \ldots, b_{M-1})$. The resulting controller $\tilde{q}(z)$ combines the desirable properties of the H_2^*-optimal controller and deadbeat type controllers, as explained in Sec. 8.1.2.

9.2.3 Anti-aliasing Prefilter

If the designer decides to add a prefilter $\gamma(s)$ in the block structure (Fig. 7.1-1A), it should be such that the system type (asymptotic properties) and the internal stability requirements are satisfied.

Section 7.5.2 discussed in detail the design of $\gamma(s)$ so that the system type is preserved. When the only unstable poles of $\tilde{p}^*(z)$ are at $z = 1$ (i.e., of $\tilde{p}(s)$ at $s = 0$), the assumption of Thm. 9.2-1 that $v^*(z)$ has at least as many poles at

$z = 1$ as $\tilde{p}^*(z)$, ensures that the internal stability conditions are satisfied for any prefilter which preserves system type.

When $\tilde{p}^*(z)$ has unstable poles in addition to those at $z = 1$, it is not a simple manner to design γ such that condition (ii) of Thm. 9.1-1 is satisfied *after* $\tilde{q}(z)$ has been determined as outlined in the preceding two sections. The preferred approach is to design $\gamma(s)$ *first* according to Sec. 7.5.2. Then one computes $\tilde{p}^*_\gamma(z)$ and uses it instead of $\tilde{p}^*(z)$ in Thm. 9.2-1 in order to obtain $\tilde{q}_H(z)$ and subsequently $\tilde{q}(z)$. However, this means that the objective function which is minimized is not the one given by (9.2–9) but

$$\phi_\gamma = \|(1 - \tilde{p}^*_\gamma(z)\tilde{q}(z))v^*(z)\|_2^2 \qquad (9.2-25)$$

which does not correspond to the true physical problem. Usually (9.2–25) is a good approximation of (9.2–9).

9.2.4   Design for Common Input Forms

In this section we shall examine the H_2^*-optimal controller $\tilde{q}_H(z)$, given by (9.2–4), for specific systems and inputs.

(i) *MP System.*

When $p(s)$ is MP and also strictly proper (all physical systems are strictly proper), $p^*(z)$ will have a delay of one unit because of sampling. Hence $p_A^* = z^{-1}$, $p_M^* = zp^*$, and (9.2–4) yields

$$\tilde{q}_H = (p^*)^{-1}((b_p^*)^{-1}b_v^*v_M^*)^{-1}((b_p^*)^{-1}b_v^*v_M^* - \kappa)$$

$$= (p^*)^{-1}(1 - \kappa b_p^*(b_v^*v_M^*)^{-1}) \qquad (9.2-26)$$

where κ is the constant term in the partial fraction expansion of $(b_p^*)^{-1}b_v^*v_M^*$. Equivalently, since b_p^*, b_v^*, v_M^* are semi-proper, κ is the product of the constant terms of the PFE's of b_p^{*-1}, b_v^*, v_M^*. After some algebra we obtain

$$\kappa = v_0 \prod_{j=k'+1}^{k} \frac{1 - \pi_j}{1 - (\pi_j^H)^{-1}} \qquad (9.2-27)$$

where k', k, π_j are defined in Thm. 9.2-1 and v_0 is the first non-zero coefficient obtained by long division of $v^*(z)$ (equal to the constant term in the PFE of $v_M^*(z)$).

(ii) *Stable System.* $b_p^* = b_v^* = 1$

$$\tilde{q}_H(z) = z(p_M^*v_M^*)^{-1}\left\{z^{-1}p_A^{*-1}v_M^*\right\}_* \qquad (8.1-5)$$

This formula was stated in Thm. 8.1-1.

(iii) *Integrator.* $p(s) = \frac{1}{s} \Rightarrow p^*(z) = \frac{T}{z-1}$. For $b_p^* = b_v^* = 1$,

$$\tilde{q}_H(z) = z(p_M^* v_M^*)^{-1} \{z^{-1} p_A^{*-1} v_M^*\}_* \qquad (9.2-28)$$

The comments made in Sec. 5.2.2.iii for the continuous case, apply here as well.

(iv) *Type 1 design for system with one unstable pole.*
    Consider the MP system

$$p(s) = \tilde{p}(s) = \frac{b}{-s+b}, \quad b > 0 \qquad (9.2-29)$$

and assume that a step disturbance acts at the process input

$$v(s) = d(s) = \frac{b}{s(-s+b)} \qquad (9.2-30)$$

Then for a sampling time T we have

$$p^*(z) = \frac{1 - e^{bT}}{z - e^{bT}} \qquad (9.2-31)$$

$$v^*(z) = \frac{(1 - e^{bT})z}{(z-1)(z-e^{bT})} \qquad (9.2-32)$$

$$v_M^*(z) = zv^*(z) \qquad (9.2-33)$$

Note that $e^{bT} > 1$ since $b > 0$. The H_2^*-optimal controller can be obtained from (9.2–26). We have $b_p^* = b_v^*$ and so from (9.2–27)

$$\kappa = v_0 = 1 - e^{bT} \qquad (9.2-34)$$

Substitution of (9.2–31 through 34) into (9.2–26) yields

$$\tilde{q}_H(z) = \frac{(z - e^{bT})((1 + e^{bT})z - e^{bT})}{(1 - e^{bT})z^2} \qquad (9.2-35)$$

Since $\tilde{q}_H(z)$ has no poles with negative real parts,

$$\tilde{q}(z) = \tilde{q}_H(z) \qquad (9.2-36)$$

9.2.5   Integral Squared Error (ISE) for Step Inputs to Stable Systems

The H_2^*-optimal controller $\tilde{q}_H(z)$ minimizes the sum of squared errors (SSE) for a particular input. To correct intersample rippling, the IMC controller $\tilde{q}(z)$ is obtained through the modification discussed in Sec. 8.1.2.

The ISE can be computed for the closed-loop system with $\tilde{q}(z)$ from (7.5–1) which describes the continuous plant output. For the specific case of a step setpoint or disturbance input ($v = -r$ or d), we have $h_0(s)v^*(e^{sT}) = v(s) = s^{-1}$ and then (7.5–1), (7.5–2) yield

$$e(s) = (1 - p(s)\tilde{q}(e^{sT}))s^{-1} \qquad (9.2-37)$$

We have

$$ISE \triangleq \int_0^\infty e^2(t)dt = \frac{1}{2\pi}\int_{-\infty}^\infty |e(i\omega)|^2 d\omega = \|e\|_2^2 \qquad (9.2-38)$$

where $\|\cdot\|_2$ denotes the H_2-norm defined in Sec. 2.4.4.

For step inputs, we find from Table 8.1-1

$$\tilde{q}_H(z) = (p_M^*(z))^{-1} \qquad (9.2-39)$$

From (8.1–11) we have $B(z) = 1$ and therefore

$$\tilde{q}(z) = \tilde{q}_H(z)\tilde{q}_-(z) \qquad (9.2-40)$$

where $\tilde{q}_-(z)$ is defined in (8.1–8).

Hence we can write

$$ISE = \|(1 - p(s)(p_M^*(e^{sT}))^{-1}\tilde{q}_-(e^{sT}))s^{-1}\|_2^2 \qquad (9.2-41)$$

By following the steps used in the proof of Thm. 4.1-3 we can break (9.2–41) into two parts:

$$ISE = \|(1 - p_A(s))s^{-1}\|_2^2 + \|(1 - p_M(s)(p_M^*(e^{sT}))^{-1}\tilde{q}_-(e^{sT}))s^{-1}\|_2^2 \qquad (9.2-42)$$

where $p_A(s), p_M(s)$ are defined in (4.1–3 through 4.1–5).

Note that the first term in (9.2–45) is the minimum ISE for the continuous case. Hence the second term represents the additional ISE that is introduced because of the use of a discrete rather than a continuous controller (designed according to Secs. 8.1.1 and 8.1.2.)

9.3 The Discrete IMC Filter

The philosophy behind the IMC filter is the same as for stable systems (Sec. 8.2).
The filter structure is fixed and only a few parameters are adjusted to meet the
robustness objectives. The simplest filter form is

$$f_1(z) = \frac{(1-\alpha)z}{z-\alpha} \qquad (8.2-1)$$

9.3.1 Filter Form

The discrete IMC filter $f(z)$ has to satisfy the following requirements

(i) Asymptotic tracking of external system inputs (setpoints and/or disturbances) — i.e., $(1 - \tilde{p}^*\tilde{q}f)v^*$ has to be stable.

(ii) Internal stability — i.e., $\tilde{q}f$ and $(1 - \tilde{p}^*\tilde{q}f)\tilde{p}^*$ have to be stable.

Since $\tilde{q}(z)$ has been designed so that (i) and (ii) are satisfied for $f(z) = 1$, $f(z)$
should satisfy

$$\frac{d^k}{dz^k}(1 - f(z))\bigg|_{z=\pi_i} = 0, \; k = 0,\ldots,m_i - 1, \quad i = 1,\ldots,\xi \qquad (9.3-1)$$

where π_i, m_i were defined in Sec. 9.2.2. Note that (9.3-1) implies for $k = 0$:

$$f(z) = 1 \quad \text{at} \quad z = \pi_1,\ldots,\pi_\xi \qquad (9.3-2)$$

One can now select a filter of the form

$$f(z) = \phi(z)f_1(z) \qquad (9.3-3)$$

where

$$\phi(z) = \sum_{j=0}^{w} \beta_j z^{-j} \qquad (9.3-4)$$

and choose the coefficients $\beta_0,\ldots,\beta_w$ so that (9.3-1) is satisfied for some specified
α. The parameter α can be used as a tuning parameter.

Note that for $\xi = 1$, $\pi_1 = 1$, $m_1 = 1$, we only need $\phi(z) = 1$. For the
general case (9.3-1) can be transformed into a system of M linear equations
with $\beta_0,\ldots,\beta_w$ as unknowns, where M is given by (9.2-23). Lemma 8.2-1 can
help simplify the necessary algebra. One should select $w \geq M - 1$ so that the
system of linear equations has one or more solutions. When $w \geq M$ the system is

underdetermined and $\beta_0, \ldots, \beta_w$ can be obtained as the minimum norm solution. Note that for $M = 2$ one should select $w \geq 2$ in order to avoid the trivial solution $f(z) = 1$.

The case $\xi = 1$, $\pi_1 = 1$ was examined in detail in Sec. 8.2. Let us now examine the common case where $\xi > 1$, but $m_i = 1$ for $i = 2, \ldots, \xi$. Then (9.3–1) is equivalent to:

$$\left. \frac{d^k}{dz^k}(1 - f(z)) \right|_{z=\pi_1=1} = 0, \quad k = 0, \ldots, m_1 - 1 \tag{9.3 $-$ 5a}$$

$$f(\pi_i) = 1, \quad i = 2, \ldots, \xi \tag{9.3 $-$ 5b}$$

The following theorem holds:

Theorem 9.3-1. *For $\pi_1 = 1$, $\xi \geq 2$, $m_i = 1$ for $i = 2, \ldots, \xi$, the coefficients $\beta_0, \ldots, \beta_w$ must satisfy*

$$\beta_0 = 1 - \beta_1 - \ldots - \beta_w \tag{9.3 $-$ 6}$$

$$\begin{pmatrix} \Pi \\ N_w \end{pmatrix} \begin{pmatrix} \beta_1 \\ \vdots \\ \beta_w \end{pmatrix} = \begin{pmatrix} f_1(\pi_\xi)^{-1} - 1 \\ \vdots \\ f_1(\pi_2)^{-1} - 1 \\ -\alpha/(1-\alpha) \\ 0 \\ \vdots \\ 0 \end{pmatrix} \begin{matrix} \left.\vphantom{\begin{matrix}a\\b\\c\end{matrix}}\right\} \xi - 1 \\ \\ \left.\vphantom{\begin{matrix}a\\b\\c\end{matrix}}\right\} \triangleq \chi \\ \\ \left.\vphantom{\begin{matrix}a\\b\\c\end{matrix}}\right\} m_1 - 1 \end{matrix} \tag{9.3 $-$ 7}$$

where

$$\Pi = \begin{pmatrix} \pi_\xi^{-1} - 1 & \cdots & \pi_\xi^{-w} - 1 \\ \vdots & & \vdots \\ \pi_2^{-1} - 1 & \cdots & \pi_2^{-w} - 1 \end{pmatrix} \tag{9.3 $-$ 8}$$

and the elements ν_{ij} of the $(m_1 - 1) \times w$ matrix N_w are defined by (8.2–6)

$$\nu_{ij} = \begin{cases} 0 & \text{for } i > j \\ \frac{j!}{(j-i)!} & \text{for } i \leq j \end{cases} \tag{8.2 $-$ 6}$$

Proof. Follows directly from Thm. 8.2-1, (9.3–5) and the fact that $f_1(1) = 1$. $\square$

For $\xi = 1$, Thm. 9.3-1 reduces to Thm. 8.2-1. For $m_1 = 1$, the choice $w = \xi - 1$, reduces (9.3–7) to the Vandermonde form (5.3–4). In general one should select $w \geq M - 1 = \xi + m_1 - 2$ and obtain $\beta_1, \ldots, \beta_w$ as the minimum norm solution to (9.3–7):

$$\begin{pmatrix} \beta_1 \\ \vdots \\ \beta_w \end{pmatrix} = A^T (AA^T)^{-1} \chi \qquad (9.3-9)$$

where

$$A \triangleq \begin{pmatrix} \Pi \\ N_w \end{pmatrix} \qquad (9.3-10)$$

Note that from a numerical point of view it is preferable to compute the pseudo-inverse in (9.3–9) from a singular value decomposition of $\begin{pmatrix} \Pi \\ N_w \end{pmatrix}$.

9.3.2 Qualitative Interpretation of the Filter Function

The discussion in this section is in the spirit of that in Sec. 5.3.2 for continuous systems. With the filter $f(z)$ and the IMC controller $\tilde{q}(z)$ obtained in the first design step, the discrete complementary sensitivity function becomes for $p^* = \tilde{p}^*$:

$$\tilde{\eta}^*(z) = \tilde{p}^*(z)\tilde{q}(z)f(z) \qquad (9.3-11)$$

For open-loop stable systems, any stable filter that satisfies the system type requirements described by Thm. 7.5-1 is acceptable. For open-loop unstable systems however, the filter has to be unity at the unstable system poles, which limits the range of filter parameters α that can be chosen for reasonable performance, as we shall show next. For the effect of the unstable poles on $\tilde{\eta}^*$ to become negligible $f(z)$ has to approach $f_1(z)$ (8.2–1). Specifically, it follows from (9.3–3) that $\phi(z)$ has to approach unity. We will study the behavior of $\phi(z)$ for $w \to \infty$.

Consider the system studied in Sec. 9.2.4.iv. For internal stability and asymptotically error-free disturbance compensation we require

$$f(1) = f(e^{bT}) = 1 \qquad (9.3-12)$$

In the notation of Sec. 9.3.1, we have in this case $\xi = 2$, $\pi_1 = 1$, $\pi_2 = e^{bT}$, $m_1 = m_2 = 1$. Hence $f(z)$ is given by (9.3–3, 4, 6, 9) where

$$A = (e^{-bT} - 1 \quad \cdots \quad e^{-wbT} - 1) \qquad (9.3-13)$$

$$\chi = f_1(e^{bT})^{-1} - 1 = \frac{\alpha(1 - e^{-bT})}{1 - \alpha} \qquad (9.3-14)$$

From (9.3–9) it follows that

$$\sum_{j=1}^{w} \beta_j^2 = \chi^T (AA^T)^{-1} \chi = \frac{\alpha^2 (1 - e^{-bT})^2}{(1 - \alpha)^2 S_1} \qquad (9.3-15)$$

where

$$S_1 \triangleq \sum_{j=1}^{w} (e^{-jbT} - 1)^2$$

$$= \sum_{j=1}^{w} e^{-j2bT} - 2\sum_{j=1}^{w} e^{-jbT} + w$$

$$= \frac{1 - e^{-2bTw}}{e^{2bT} - 1} - 2\frac{1 - e^{-bTw}}{e^{bT} - 1} + w \qquad (9.3-16)$$

Since $|e^{-bT}| < 1$, it follows from (9.3–16) that $\lim_{w\to\infty} S_1 = \infty$ and $\lim_{w\to\infty} \sum_{j=1}^{w} \beta_j^2 = 0$. This fact, however, is not sufficient to produce an $f(z)$ that approximates the behavior of $f_1(z)$. For this to happen we need $\lim_{w\to\infty} \beta_0 = 1$. Let us compute this limit. From (9.3–9) we get

$$\beta_k = \frac{\alpha(1 - e^{-bT})(e^{-kbT} - 1)}{(1 - \alpha)S_1}, \quad k = 1, \ldots, w \qquad (9.3-17)$$

(9.3–6), (9.3–17) yield

$$\beta_0 = 1 - \frac{\alpha(1 - e^{-bT})S_2}{(1 - \alpha)S_1} \qquad (9.3-18)$$

where

$$S_2 \triangleq \sum_{j=1}^{w} (e^{-jbT} - 1) = \frac{1 - e^{-bTw}}{e^{bT} - 1} - w \qquad (9.3-19)$$

From (9.3–16), (9.3–19) it follows that $\lim_{w\to\infty} S_2/S_1 = -1$. Then (9.3–18) yields

$$\lim_{w\to\infty} \beta_0 = \frac{1 - \alpha e^{-bT}}{1 - \alpha} \qquad (9.3-20)$$

By writing $\alpha = e^{-T/\lambda}$ we get

$$\lim_{w\to\infty} \beta_0 = \frac{1 - e^{-T(1/\lambda+b)}}{1 - e^{-T/\lambda}} \qquad (9.3-21)$$

Hence in order for $\lim_{w\to\infty} \beta_0 \cong 1$ we need $1/\lambda \gg b$ or $\lambda b \ll 1$. In this case the behavior of $f(z)$ approaches that of $f_1(z)$ (compare to (5.3–25)) and if a λ in that range is sufficient for robustness, the unstable pole b produces no significant effect on the system behavior. If, however, one chooses a λ for which $\lambda b \gg 1$, then the $\lim_{w\to\infty} \beta_0$ is very far from 1 and as a result problems similar to those discussed in Sec. 5.3.2 for the continuous case appear.

This is illustrated in Fig. 9.3-1, where amplitude plots of f_1 and f are shown for different values of λ and w. We see that as w increases, f tends towards f_1. For $\lambda b \ll 1$, the approximation is very good, while for $\lambda b \gg 1$, the closer we get to f_1, the higher the peak in $|f|$ becomes.

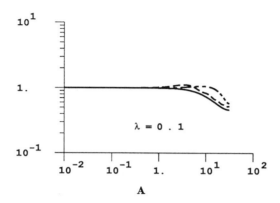

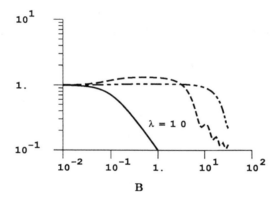

Figure 9.3-1. Effect of a RHP pole on the discrete IMC filter. $T = 0.1, b = 1$. Solid: f_1, Dash: $f_1\phi, w = 9$, Dot-Dash: $f_1\phi, w = 2$.

9.4   Robust Stability

For controllers designed via the IMC design procedure ($\tilde{\eta}^* = \tilde{p}^*\tilde{q}f$) Thm. 7.6-1 becomes Cor. 9.4-1.

Corollary 9.4-1 (Robust Stability). *Assume that all plants p in the family* Π^*

$$\Pi^* = \left\{ p : \left| \frac{p_\gamma^*(i\omega) - \tilde{p}_\gamma^*(i\omega)}{\tilde{p}^*(i\omega)} \right| < \bar{\ell}_m^*(\omega) \right\} \tag{9.4 - 1}$$

have the same number of RHP poles and that these poles do not become unobservable after sampling. Then the system is robustly stable if and only if the IMC filter satisfies

$$|f| < \frac{1}{|\tilde{p}^*\tilde{q}\bar{\ell}_m^*|}, \ \ 0 \le \omega \le \pi/T \tag{9.4 - 2}$$

where $\tilde{q}$ is a stabilizing controller for the nominal plant $\tilde{p}$.

For stable systems f is arbitrary. Therefore, there always exists a filter f which satisfies (9.4–2) regardless of the magnitude of the uncertainty $\bar{\ell}_m^*$. For unstable systems f is constrained to be unity at the poles of $\tilde{p}^*$ outside the UC. Thus, depending on $\bar{\ell}_m^*$, there might not exist any filter parameter α for which the constraint (9.4–2) is met. Indeed, there might not exist any filter — however complicated — which satisfies (9.4–2). A minimum amount of information is necessary or equivalently a maximum amount of uncertainty is allowed to stabilize an unstable system. The necessary information at $\omega = 0$ can be characterized easily.

Corollary 9.4-2. *Assume that a filter f is to be designed for a system or disturbance pole(s) at $s = 0$ — i.e., $f(1) = 1$. There exists an f such that the closed loop system is robustly stable for the family Π^* described by (9.4–1) only if $\bar{\ell}_m(0) < 1$.*

Note that contrary to Cor. 8.3-2, Cor. 9.4-2 is only necessary. For unstable systems, in general, the filter has to satisfy other constraints in addition to the one at $z = 1$.

9.5   Robust Performance

The results in Sec. 7.7.2 hold for unstable systems if it is assumed that all plants in the family Π^* have the same number of RHP poles and if the controller $\tilde{q}$ and filter f are stabilizing for the nominal plant $\tilde{p}^*$.

9.6 Summary of the IMC Design Procedure

The required information for the IMC design is the same as that for stable systems: process model, input type, performance specifications and uncertainty information. The input specification requires some care. If the physical disturbance enters at the plant input, the disturbance used in the design procedure has to include the unstable system poles for the resulting controller to yield offset free performance.

Design Procedure

Step 1: Nominal Performance
The stabilizing H_2-optimal controller $\tilde{q}$ is determined which minimizes

$$\|(1 - \tilde{p}^*\tilde{q})v^*\|_2$$

for the specified input v^*. The optimal controller $\tilde{q}$ can be found explicitly from (9.2–4).

$$\tilde{q}_H = zb_p^*(p_M^* b_v^* v_M^*)^{-1}\left\{(zb_p^* p_A^*)^{-1} b_v^* v_M^*\right\}_* \qquad (9.2-4)$$

If all the unstable plant/disturbance poles are at the origin, the entries in Table 8.1-1 can be used to find $\tilde{q}_H$ for typical inputs.

The controller $\tilde{q}_H$ is then modified as described in Sec. 9.2-2 to eliminate the problems of intersample rippling:

$$\tilde{q} = \tilde{q}_H \tilde{q}_{-B} \qquad (9.2-20)$$

Step 2: Robust Stability and Robust Performance
The controller $\tilde{q}$ is augmented by the IMC filter f

$$q = \tilde{q}f$$

In order for q to be stabilizing f has to be unity at all unstable system $\tilde{p}^*$ and input v^* poles $\pi_1, \ldots, \pi_k$. When the poles outside the UC are distinct, the one-parameter filter is determined through Thm. 9.3-1. The filter poles are a subset of the closed-loop poles. If they are made much slower than the mirror images of the poles of $\tilde{p}^*$ outside the UC undesirable performance and robustness properties result.

Robust Stability. Increase α (or λ in $\alpha = e^{-T/\lambda}$) until

$$|\tilde{p}^*\tilde{q}f\bar{\ell}_m^*| < 1, \qquad 0 \leq \omega \leq \pi/T \qquad (9.4-2)$$

is satisfied for $\alpha \geq \alpha^*$.

Robust Performance. Increase α, starting from α^*, until the following condition is met

$$|\hat{q}|\bar{\ell}_a + |1 - \tilde{p}\hat{q}|w \le 1, \qquad 0 \le \omega \le \pi/T \qquad (8.4-1)$$

where

$$\hat{q}(s) = \tilde{q}(e^{sT})f(e^{sT})h_0(s)\gamma(s)/T \qquad (7.5-10)$$

9.7   Application: Distillation Column Base Level Control

This example was discussed in detail in Sec. 5.7.1. It is briefly considered here again, because some interesting issues arise when a digital controller is designed for the process.

The process model is

$$\tilde{p}(s) = \frac{1}{s}(1 - 2e^{-s\theta}) \qquad (5.7-1)$$

Let us select a sampling time $T = \theta/N$, where N is an integer. Then the zero-order hold discrete equivalent is

$$\tilde{p}^*(z) = \mathcal{ZL}^{-1}\left\{h_0(s)\tilde{p}(s)\right\} = (1 - 2z^{-N})\mathcal{ZL}^{-1}\left\{h_0(s)\frac{1}{s}\right\} = (1 - 2z^{-N})\frac{T}{z-1}$$
$$(9.7-1)$$

In this very special case we find that if ζ is a finite zero of $\tilde{p}(s)$, then $e^{\zeta T}$ is a zero of $\tilde{p}^*(z)$ and therefore ζ is a zero of $\tilde{p}^*(e^{sT})$. This mapping does not hold for zeros in general, although it is always true for the poles of $\tilde{p}(s)$ and $\tilde{p}^*(z)$.

Because of this mapping the zeros of $\tilde{p}^*(z)$ that appear as poles of $\tilde{q}_H(z)$ are cancelled by zeros of $\tilde{p}(s)$ in (7.5-6). Therefore, any such zeros close to $(-1,0)$ do not produce intersample rippling even if the modification described in Sec. 9.2.2 is not made. However, this does not mean that the behavior of the control system will deteriorate if the suggested modification is introduced. Indeed, the steps proposed in Sec. 9.2 will result in a well-performing controller, regardless of whether $\tilde{q}_H(z)$ suffers from rippling problems or not.

Let us proceed to illustrate this point by simulating the response for the two controllers for a ramp disturbance $d(s) = s^{-2}$. For the simulations we choose $\theta = 5$ and $T = 1$, which implies that $N = 5$. It follows from (9.7–1) that the zeros of $\tilde{p}^*(z)$ are located at $2^{1/5}e^{i2k\pi/5}$, $k = 0, 1, 2, 3, 4$, where $2^{1/5}$ indicates the real fifth root of 2. Two of these zeros (the ones that correspond to $k = 2, 3$) have

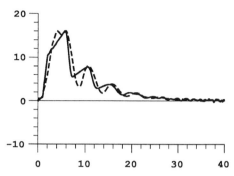

Figure 9.7-1. Distillation column base level control; response to $d(s) = s^{-2}$. Solid line: $\tilde{q}_H$. Dashed line: $\tilde{q}_H\tilde{q}_-B$.

negative real parts and will give rise to poles of $\tilde{q}_H(z)$ with negative real parts. The procedure of Sec. 9.2 yields

$$\tilde{q}_H(z) = \frac{z^3(17z - 16)(z - 1)}{(-2z^5 + 1)} \tag{9.7 - 2}$$

$$\tilde{q}_-(z) = \frac{z^2 + 1.8586z + 1.3195}{4.1781z^2} \tag{9.7 - 3}$$

$$B(z) = 2.0765 - 1.0765z^{-1} \tag{9.7 - 4}$$

Figure 9.7-1 shows the responses to $d(s) = s^{-2}$ for both $q = \tilde{q}_H$ and $q = \tilde{q}_H\tilde{q}_-B$. Clearly, $\tilde{q}_H$ produces no intersample rippling. One can also see that when the modification of Sec. 9.2.2 is made anyway, the response is essentially unaffected.

Finally, note that because the open-loop system is unstable, the controller has to be implemented in the classic feedback structure. Its expression can be obtained from (7.2–3) and its implementation presents no problem.

9.8   References

9.1. The parametrization of all stabilizing controllers presented in this section is the discrete equivalent of that presented in Sec. 5.1. It was obtained by extending that proposed by Zames and Francis (1983) to include systems with integrators.

9.2.1. The H^*-optimal controller can also be obtained with state-space methods (e.g., Kwakernaak and Sivan, 1972).

Part III

CONTINUOUS MULTI-INPUT MULTI-OUTPUT SYSTEMS

Chapter 10

FUNDAMENTALS OF MIMO FEEDBACK CONTROL

In the first section of this chapter some concepts from linear system theory are summarized. The singular value decomposition, which plays a key role in the rest of the book, is covered in detail. The control problem formulation introduced for SISO systems (model uncertainty description, input definition, performance objectives) is generalized to multivariable systems. The Nyquist stability criterion is extended to handle MIMO systems.

10.1 Definitions and Basic Principles

This introductory section is a self-contained summary. For the proofs and a deeper understanding the reader is referred to a basic text on linear systems.

10.1.1 Modeling

In the time domain a linear time invariant finite dimensional system can be described by the system of differential and algebraic equations

$$\dot{x} = Ax + Bu \qquad (10.1-1)$$

$$y = Cx + Du \qquad (10.1-2)$$

where $x \in \mathcal{R}^r, y \in \mathcal{R}^n$, and $u \in \mathcal{R}^m$ are the state, output and input vectors respectively and A, B, C, and D are constant matrices of appropriate dimensions. Taking the Laplace transform of (10.1–1) and (10.1–2) with zero initial conditions

$$sIx(s) = Ax(s) + Bu(s) \qquad (10.1-3)$$

or

$$x(s) = (sI - A)^{-1}Bu(s) \qquad (10.1-4)$$

$$y(s) = Cx(s) + Du(s) \qquad (10.1-5)$$

and substituting (10.1–4) into (10.1–5) we find

$$y(s) = (C(sI - A)^{-1}B + D)u(s) \qquad (10.1-6)$$

where

$$G(s) \triangleq C(sI - A)^{-1}B + D \qquad (10.1-7)$$

is referred to as the *system transfer matrix*. The elements $\{g_{ij}(s)\}$ of $G(s)$ are transfer functions expressing the relationship between specific inputs $u_j(s)$ and outputs $y_i(s)$. In this book, except for proofs and derivations we will use the transfer matrix rather than the state space description.

The matrix $G(s)$ will be assumed to be of full *normal rank* — i.e., rank $[G(s)] =$ min $\{m, n\}$ for every s in the set of complex numbers $\mathcal{C}$, except for a *finite* number of elements of $\mathcal{C}$.

In general, the elements $\{g_{ij}(s)\}$ will be allowed to include delays. The time domain realization of transfer matrices with delays is complex and will not be addressed here. In order to be physically realizable the transfer matrices have to be proper and causal.

Definition 10.1-1. *A system $G(s)$ is proper if all its elements $\{g_{ij}(s)\}$ are proper and strictly proper if all its elements are strictly proper. All systems which are not proper are improper.*

Definition 10.1-2. *A system $G(s)$ is causal if all its elements $\{g_{ij}(s)\}$ are causal. All systems which are not causal are noncausal.*

10.1.2   Poles

Definition 10.1-3. *The eigenvalues $\pi_i, i = 1, \ldots, n_p$, of the matrix A are called the poles of the system (10.1-1), (10.1-2). The pole polynomial $\pi(s)$ is defined as*

$$\pi(s) = \prod_{i=1}^{n_p}(s - \pi_i) \qquad (10.1-8)$$

Thus the poles are the roots of the *characteristic equation*

$$\pi(s) = 0 \qquad (10.1-9)$$

The poles determine the system's stability.

Theorem 10.1-1. *The system (10.1–1), (10.1–2) is stable if and only if all its poles $\{\pi_i\}$ are in the open left half plane.*

The following theorem allows us to determine the system poles directly from the transfer matrix $G(s)$ without performing a realization and constructing the matrix A first.

Theorem 10.1-2. *The pole polynomial $\pi(s)$ is the least common denominator of all non-identically-zero minors of all orders of $G(s)$.*

Example 10.1-1. Consider the matrix

$$G(s) = \frac{1}{(s+1)(s+2)(s-1)} \begin{pmatrix} (s-1)(s+2) & 0 & (s-1)^2 \\ -(s+1)(s+2) & (s-1)(s+1) & (s-1)(s+1) \end{pmatrix}$$

The minors of order 1 are the elements themselves. The minors of order 2 are

$$G_{1,2}^{1,2} = \frac{1}{(s+1)(s+2)} \qquad G_{1,3}^{1,2} = \frac{2}{(s+1)(s+2)}$$

$$G_{2,3}^{1,2} = \frac{-(s-1)}{(s+1)(s+2)^2}$$

(Here superscripts denote the rows and subscripts the columns used for computation of the minors). Considering the minors of all orders (i.e., orders 1 and 2) we find the least common denominator

$$\pi(s) = (s+1)(s+2)^2(s-1)$$

$\square$

10.1.3 Zeros

Recall that if ζ is a zero of the SISO system $g(s)$ then $g(\zeta) = 0$. Furthermore, we know that ζ is a zero of $g(s)$ if and only if ζ is a pole of $g^{-1}(s)$. The following definition consistently extends this concept of a zero to MIMO systems.

Definition 10.1-4. ζ *is a zero of $G(s)$ if the rank of $G(\zeta)$ is less than the normal rank of $G(s)$.*

In other words, since $G(s)$ is assumed to be of full normal rank, the transfer matrix $G(s)$ becomes rank deficient at the zero $s = \zeta$. The zero polynomial $\zeta(s)$ is defined as

$$\zeta(s) = \prod_{i=1}^{n_z} (s - \zeta_i) \qquad (10.1-10)$$

where n_z is the number of finite zeros of $G(s)$. Thus the zeros are the roots of

$$\zeta(s) = 0 \qquad (10.1-11)$$

The following theorem provides a method for calculating the zeros.

Theorem 10.1-3. *The zero polynomial $\zeta(s)$ is the greatest common divisor of the numerators of all order-r minors of $G(s)$, where r is the normal rank of $G(s)$,* *provided that these minors have all been adjusted in such a way as to have the pole polynomial $\pi(s)$ as their denominator.*

Example 10.1-2. Consider the system from Ex. 10.1-1. and adjust the denominators of all the minors of order 2 to be $\pi(s)$

$$G_{1,2}^{1,2} = \frac{(s-1)(s+2)}{\pi(s)} \qquad G_{1,3}^{1,2} = \frac{2(s-1)(s+2)}{\pi(s)}$$

$$G_{2,3}^{1,2} = \frac{-(s-1)^2}{\pi(s)}$$

and so

$$\zeta(s) = (s-1)$$

□

As Exs. 10.1-1. and 10.1-2. show, MIMO systems can have zeros and poles at the *same location*. Therefore it is generally not possible to find all the zeros of a square system from the condition $\det G(s) = 0$ because, when forming the determinant, zeros and poles at the same location cancel.

Definition 10.1-5. *A system $G(s)$ is nonminimum phase (NMP) if its transfer matrix contains zeros in the RHP or there exists a common time delay term that can be factored out of every matrix element.*

Note that the zero locations of a MIMO system are in no way related to the zero location of the individual SISO transfer functions constituting the MIMO system. Thus, it is possible for a MIMO system to be NMP even when all the SISO transfer functions are MP and vice versa.

Example 10.1-3. The system

$$G(s) = \frac{1}{s+1}\begin{pmatrix} s+3 & 2 \\ 3 & 1 \end{pmatrix}$$

has one finite zero at $s = +3$ though all the SISO transfer functions are MP. □

10.1.4   Vector and Matrix Norms

Let E be a *linear space* over the *field K* (typically K is the field of real $\mathcal{R}$ or complex numbers $\mathcal{C}$). We say that a real valued function $\| \cdot \|$ is a *norm on E* if and only if

$$\|x\| > 0 \quad \forall x \in E, x \neq 0 \tag{10.1 − 12a}$$

$$\|x\| = 0 \quad x = 0 \tag{10.1 - 12b}$$

$$\|\alpha x\| = |\alpha| \, \|x\| \quad \forall \alpha \in K, \quad \forall x \in E \tag{10.1 - 13}$$

$$\|x + y\| \le \|x\| + \|y\|, \quad \forall x, y \in E \tag{10.1 - 14}$$

A norm is a single number measuring the "size" of an element of E. Given a linear space E there may be many possible norms on E. Given the linear space E and a norm $\| \cdot \|$ on E, the pair $(E, \| \cdot \|)$ is called a *normed space*.

In this section let the linear space E be C^n. More precisely $x \in C^n$ means that $x = (x_1, x_2, \ldots x_n)$ with $x_i \in C, \forall i$. Three commonly used norms on C^n are given by

$$\|x\|_p \triangleq (|x_1|^p + |x_2|^p + \ldots + |x_n|^p)^{\frac{1}{p}} \quad p = 1, 2, \infty \tag{10.1 - 15}$$

where $\|x\|_\infty$ is interpreted as $\max_i |x_i|$. The norm $\|x\|_2$ is the usual Euclidean length of the vector x.

Let $E = C^{n \times n}$, the set of all $n \times n$ matrices with elements in C. E is a linear space. The following are norms on $C^{n \times n}$

$$\|A\|_1 = \max_j \sum_i |a_{ij}| \tag{10.1 - 16}$$

$$\|A\|_\infty = \max_i \sum_j |a_{ij}| \tag{10.1 - 17}$$

$$\|A\|_F = \left[\sum_i \sum_j |a_{ij}|^2 \right]^{\frac{1}{2}} \quad \text{Frobenius or Euclidean norm} \tag{10.1 - 18}$$

$$\|A\|_2 = \max_i \lambda_i^{\frac{1}{2}}(A^H A) \quad \text{Spectral norm} \tag{10.1 - 19}$$

where the superscript H is used to denote complex conjugate transpose. The eigenvalues $\lambda_i(A^H A)$ are (necessarily) real and nonnegative. Some useful relationships involving the spectral and Frobenius norms are

$$\|A\|_2 \le \|A\|_F \le \sqrt{n} \|A\|_2 \tag{10.1 - 20}$$

where $A \in C^{n \times n}$. These inequalities follow from the fact that $A^H A$ is positive semidefinite and

$$\max_i \lambda_i(A^H A) \le \|A\|_F^2 = \text{trace}[A^H A] \le n \max_i \lambda_i(A^H A) \tag{10.1 - 21}$$

Here we considered matrices as elements of a linear space. Next we shall consider matrices as representation of linear maps and shall relate the matrix norms to the vector norms of the domain and range spaces.

First let us adopt the following notation. From now on we shall use $|\cdot|$ for norms on $\mathcal{R}^n$ or $\mathcal{C}^n$ and $\|\cdot\|$ for norms on function spaces (see Sec. 10.1.6.) or for induced norms of linear operators.

Let $|\cdot|$ be a norm on E and let A be a linear map from E into E. Define the function

$$\|A\| \triangleq \sup_{x\neq 0} \frac{|Ax|}{|x|} \tag{10.1 - 22}$$

or equivalently

$$\|A\| \triangleq \sup_{|x|=1} |Ax| \tag{10.1 - 23}$$

The quantity $\|A\|$ is called the *induced norm of the linear map A* or the *operator norm induced by the norm $|\cdot|$*.

To interpret (10.1–23) geometrically, consider the set of all vectors of unit length — i.e., the unit sphere. Then $\|A\|$ is the least upper bound on the magnification of the elements of this set by the operator A.

It is easy to show from the definition (10.1–22) that any induced norm satisfies

$$|Ax| \leq \|A\| \cdot |x| \tag{10.1 - 24}$$

$$\|\alpha A\| = |\alpha| \cdot \|A\| \tag{10.1 - 25}$$

$$\|A + B\| \leq \|A\| + \|B\| \tag{10.1 - 26}$$

$$\|AB\| \leq \|A\| \cdot \|B\| \tag{10.1 - 27}$$

Any matrix norm $N(\cdot)$ which in addition to the axioms (10.1–12)-(10.1–14) satisfies

$$N(AB) \leq N(A)N(B) \tag{10.1 - 28}$$

is called *compatible*. (It is "compatible with itself.") An induced norm is an example of a compatible norm. It can be shown that for every compatible matrix norm $N(\cdot)$ there exists a vector norm $|\cdot|$ such that

$$|Ax| \leq N(A)|x| \tag{10.1 - 29}$$

We say that $N(\cdot)$ is an (operator) norm *compatible* with the (vector) norm $|\cdot|$.

It is left as an exercise for the reader to show that the norms $\|\cdot\|_1, \|\cdot\|_2$ and $\|\cdot\|_\infty$ are operator norms induced by the vector norms $|\cdot|_1, |\cdot|_2$ and $|\cdot|_\infty$ respectively. The Frobenius norm $\|\cdot\|_F$ is not an induced norm but it is compatible with $|\cdot|_2$. Also, if $\lambda(A)$ is an eigenvalue of A and x is a corresponding eigenvector, then for compatible matrix and vector norms

$$|Ax| = |\lambda(A)| \cdot |x| \le \|A\| \cdot |x| \tag{10.1-30}$$

or

$$|\lambda(A)| \le \|A\| \tag{10.1-31}$$

Let $\rho(A)$ be the *spectral radius* of A — i.e.,

$$\rho(A) = \max_i |\lambda_i(A)| \tag{10.1-32}$$

Because (10.1-31) holds for any eigenvalues of A

$$\rho(A) \le \|A\| \tag{10.1-33}$$

Thus the spectral radius forms a lower bound on any compatible matrix norm.

10.1.5 Singular Values and the Singular Value Decomposition

The *singular values* of a complex $n \times m$ matrix A, denoted $\sigma_i(A)$, are the k largest nonnegative square roots of the eigenvalues of $A^H A$ where $k = \min\{n, m\}$, that is

$$\sigma_i(A) = \sqrt{\lambda_i(A^H A)} \qquad i = 1, 2, \ldots, k \tag{10.1-34}$$

where we assume that the σ_i are ordered such that $\sigma_i \ge \sigma_{i+1}$. In the last section we asked the reader to show that the maximum singular value is the matrix norm induced by the vector norm $|\cdot|_2$ — i.e., the spectral norm. We can define the maximum ($\bar\sigma$) and minimum ($\underline\sigma$) singular values alternatively by

$$\bar\sigma(A) = \max_{x \ne 0} \frac{|Ax|_2}{|x|_2} = \|A\|_2 \tag{10.1-35}$$

$$\underline\sigma(A) = \left[\max_{x \ne 0} \frac{|A^{-1}x|_2}{|x|_2} \right]^{-1} = \|A^{-1}\|_2^{-1} \quad \text{if } A^{-1} \text{ exists} \tag{10.1-36}$$

$$= \min_{x \ne 0} \left[\frac{|A^{-1}x|_2}{|x|_2} \right]^{-1} = \min_{x \ne 0} \frac{|x|_2}{|A^{-1}x|_2} = \min_{x \ne 0} \frac{|Ax|_2}{|x|_2}$$

Thus $\bar{\sigma}$ and $\underline{\sigma}$ can be interpreted geometrically as the least upper bound and the greatest lower bound on the magnification of a vector by the operator A.

The smallest singular value $\underline{\sigma}(A)$ measures how near the matrix A is to being singular or rank deficient (a matrix is rank deficient if *both* its rows *and* columns are linearly dependent). To see this, consider finding a matrix L of minimum spectral norm that makes $A + L$ rank deficient. Since $A + L$ must be rank deficient there exists a nonzero vector x such that $|x|_2 = 1$ and $(A + L)x = 0$. Thus by (10.1–35) and (10.1–36)

$$\underline{\sigma}(A) \leq |Ax|_2 = |Lx|_2 \leq \|L\|_2 = \bar{\sigma}(L) \tag{10.1 – 37}$$

Therefore, L must have spectral norm of at least $\underline{\sigma}(A)$. Otherwise $A + L$ cannot be rank deficient. The property that

$$\underline{\sigma}(A) > \bar{\sigma}(L) \tag{10.1 – 38}$$

implies that $A + L$ is nonsingular (assuming square matrices) and will be a basic inequality used in the formulation of various robustness tests.

Definition 10.1-6. *A complex matrix A is Hermitian if $A^H = A$.*

Definition 10.1-7. *A complex matrix A is unitary if $A^H = A^{-1}$.*

A convenient way of representing a matrix that exposes its internal structure is known as the Singular Value Decomposition (SVD). For an $n \times m$ matrix A, the SVD of A is given by

$$A = U\Sigma V^H = \sum_{i=1}^{k} \sigma_i(A)u_i v_i^H \tag{10.1 – 39}$$

where U and V are unitary matrices with column vectors denoted by

$$U = (u_1, u_2, \ldots, u_n) \tag{10.1 – 40a}$$

$$V = (v_1, v_2, \ldots, v_m) \tag{10.1 – 40b}$$

and Σ contains a diagonal nonnegative definite matrix Σ_1 of singular values arranged in descending order as in

$$\Sigma = \begin{pmatrix} \Sigma_1 \\ 0 \end{pmatrix}; \quad n \geq m$$

or

$$\Sigma = (\Sigma_1 \quad 0); \quad n \leq m \tag{10.1 – 41}$$

and

$$\Sigma_1 = \text{diag}\,\{\sigma_1, \sigma_2, \ldots, \sigma_k\}; \qquad k = \min\{m, n\} \qquad (10.1-42)$$

where

$$\bar{\sigma} = \sigma_1 \geq \sigma_2 \geq \ldots \geq \sigma_k = \underline{\sigma}$$

It can be shown easily that the columns of V and U are unit eigenvectors of $A^H A$ and AA^H respectively. They are known as the *right* and *left singular vectors* of the matrix A. Trivially all unitary matrices have a spectral norm of unity. Thus by SVD an arbitrary matrix can be decomposed into a "rotation" (V^H) followed by scaling (Σ) followed by a "rotation" (U).

Example 10.1-4. The SVD of the matrix

$$A = \begin{pmatrix} 0.8712 & -1.3195 \\ 1.5783 & -0.0947 \end{pmatrix}$$

is

$$U = \frac{1}{\sqrt{2}}\begin{pmatrix} 1 & -1 \\ 1 & 1 \end{pmatrix}, \Sigma = \begin{pmatrix} 2 & 0 \\ 0 & 1 \end{pmatrix}, V = \frac{1}{2}\begin{pmatrix} \sqrt{3} & 1 \\ -1 & \sqrt{3} \end{pmatrix}$$

It is interpreted geometrically in Fig. 10.1-1. □

Let $\lambda(A)$ be the eigenvalue of minimum magnitude of A and $\underline{x}$ the associated eigenvector. Then from (10.1–36) we find

$$\underline{\sigma}(A) = \min_{x \neq 0} \frac{|Ax|_2}{|x|_2} \leq \frac{|A\underline{x}|_2}{|\underline{x}|_2} = |\lambda(A)| \qquad (10.1-43)$$

Combining (10.1–33) and (10.1–43) we conclude that $\underline{\sigma}$ and $\bar{\sigma}$ bound the magnitude of the eigenvalues:

$$\underline{\sigma}(A) \leq |\lambda_i(A)| \leq \bar{\sigma}(A) \qquad (10.1-44)$$

If A is Hermitian then the singular values and the eigenvalues coincide.

Define $u_1 = \bar{u}, u_n = \underline{u}, v_1 = \bar{v}, v_m = \underline{v}$. Then if follows that

$$A\bar{v} = \bar{\sigma}\ \bar{u} \qquad (10.1-45)$$

$$A\underline{v} = \underline{\sigma}\ \underline{u} \qquad (10.1-46)$$

From a systems point of view the vector $\bar{v}(\underline{v})$ corresponds to the *input* direction with the largest (smallest) amplification. Furthermore $\bar{u}(\underline{u})$ is the *output* direction in which the inputs are most (least) effective.

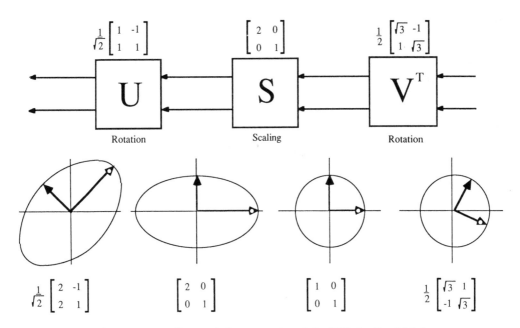

Figure 10.1-1. Geometric interpretation of the SVD for Ex. 10.1-4.

If A is square and nonsingular then

$$A^{-1} = V\Sigma^{-1}U^H \qquad (10.1 - 47)$$

is the SVD of A^{-1} but with the order of the singular values reversed. Let $\ell = n - j + 1$. Then it follows from (10.1–47) that

$$\sigma_j(A^{-1}) = 1/\sigma_\ell(A) \qquad (10.1 - 48)$$

$$u_j(A^{-1}) = v_\ell(A) \qquad (10.1 - 49a)$$

$$v_j(A^{-1}) = u_\ell(A) \qquad (10.1 - 49b)$$

and in particular

$$\bar{\sigma}(A^{-1}) = 1/\underline{\sigma}(A) \qquad (10.1 - 50)$$

$$\bar{u}(A^{-1}) = \underline{v}(A) \qquad (10.1 - 51a)$$

$$\underline{u}(A^{-1}) = \bar{v}(A) \qquad (10.1 - 51b)$$

When $G(i\omega)$ is a transfer matrix we can plot the singular values $\sigma_i(G(i\omega))(i = 1,\ldots,k)$ as a function of frequency. These curves generalize the SISO amplitude-ratio Bode plot to MIMO systems. In the MIMO case the amplification of the input vector sinusoid $ue^{i\omega t}$ depends on the *direction* of the complex vector u: the amplification is at least $\underline{\sigma}(G(i\omega))$ and at most $\bar{\sigma}(G(i\omega))$.

10.1.6 Norms on Function Spaces

In this section we will illustrate the extension of the concept of a *norm* to linear spaces whose elements are functions. Let us first consider the vector valued function $y(s)$ of dimension n. We define the set L_2^n to be the set of all vector functions with dimension n, which are square-integrable on the imaginary axis — i.e., for which the following quantity is finite:

$$\|y\|_2 = \left[\frac{1}{2\pi}\int_{-\infty}^{\infty} y(i\omega)^H y(i\omega)d\omega\right]^{\frac{1}{2}} \qquad (10.1 - 52)$$

Note that (10.1–52) defines the 2-norm of a function $y(s)$ through an *inner product*. For the special case when $y(s)$ has no poles in the closed RHP, Parseval's theorem yields an equivalent time domain expression for the 2-norm of $y(t)$ ($\|y\|_2$):

$$\|y\|_2 = \left[\int_0^\infty y(t)^T y(t) dt \right]^{\frac{1}{2}} \tag{10.1 - 53}$$

Assume now that the function $G(s)$ is *matrix valued* with dimension $m \times n$. Then the definition (10.1–52) becomes

$$\|G\|_2 = \left[\frac{1}{2\pi} \int_{-\infty}^{\infty} \text{trace} \left[G(i\omega)^H G(i\omega) \right] d\omega \right]^{\frac{1}{2}} \tag{10.1 - 54}$$

where $G(s)$ is in the set $L_2^{m \times n}$ of all matrix valued functions of dimension $m \times n$ for which (10.1–54) is finite. Equation (10.1–54) cannot be interpreted easily in a deterministic setting. We will discuss the implications later in our derivation of the Linear Quadratic Optimal Control problem.

Let us look next at the linear system

$$y(s) = G(s)u(s) \tag{10.1 - 55}$$

and pose the following problem: given a bound on $\|u\|_2$ what is the least upper bound on $\|y\|_2$? In other words, we are looking for the operator norm $\| \cdot \|_{i2}$ induced by $\| \cdot \|_2$.

Theorem 10.1-4. *Let $u \in L_2^n$ and $G \in L_2^{m \times n}$. Then $y \in L_2^m$ and the norm of the operator G induced by $\| \cdot \|_2$ is*

$$\|G\|_{i2} = \sup_{\omega} \bar{\sigma}(G(i\omega)) \triangleq \|G\|_\infty \tag{10.1 - 56}$$

where $\|G(i\omega)\|_\infty$ is the ∞-norm of the function G in the frequency domain.

Proof. We will sketch the proof for the SISO case ($y = gu$) and leave the rest as an exercise.

$$\|y\|_2^2 = \|g(i\omega)u(i\omega)\|_2^2 = \frac{1}{2\pi} \int_{-\infty}^{\infty} g(i\omega)^H g(i\omega) u(i\omega)^H u(i\omega) d\omega$$

$$\leq \sup_{\omega} |g(i\omega)|^2 \frac{1}{2\pi} \int_{-\infty}^{\infty} u(i\omega)^H u(i\omega) d\omega$$

Thus

$$\|y\|_2^2 \leq \|g\|_\infty^2 \|u\|_2^2$$

This proves that $\|g\|_\infty$ is an upper bound on the induced norm. To prove that it is in fact the least upper bound and thus equal to the induced norm we have to show that the bound can be reached for a specific u. The specific u is a "sinusoid" modified to be square integrable and occurring at the frequency where $|g(i\omega)|$ is maximum. $\square$

To appreciate the difference between the *2-norm of an operator* and the *operator norm induced by the 2-norm* (∞-norm) we refer the reader back to our discussion of the SISO H_2- and H_∞-optimal control problems (Chap. 2). We found that for H_2-optimal control the performance for a *specific input signal* is optimized which implies the minimization of the weighted 2-norm of the sensitivity operator. For H_∞-optimal control the performance for a *set of 2-norm bounded signals* is optimized. This implies minimization of the norm of the sensitivity operator *induced* by the 2-norm, which we showed to be equal to the ∞-norm of the sensitivity function (Thm. 10.1-4).

10.2 Classic Feedback

10.2.1 Definitions

The block diagram of a typical classic feedback loop is shown in Fig. 10.2-1A. Here C denotes the controller and P the plant transfer function. The transfer function P_d describes the effect of the disturbance d' on the process output y. P_m symbolizes the measurement device transfer function. The measured variable y_m is corrupted by measurement noise n. The controller determines the process input (manipulated variable) u on the basis of the error e. The objective of the feedback loop is to keep y close to the reference (setpoint) r.

Commonly we will use the simplified block diagram in Fig. 10.2-1B. Here d denotes the effect of the disturbance on the output. Exact knowledge of the output y is assumed ($P_m = 1, n = 0$).

10.2.2 Multivariable Nyquist Criterion

Consider the closed loop system in Fig. 10.2-1B when P is square (dim u = dim y). Let the *open* loop transfer matrix $P(s)C(s)$ be described in state space by

$$\dot{x} = A_o x + B_o(-e) \qquad (10.2-1)$$

$$y = C_o x + D_o(-e) \qquad (10.2-2)$$

where

$$e = y - r \qquad (10.2-3)$$

Combining (10.2–1) – (10.2–3) we obtain

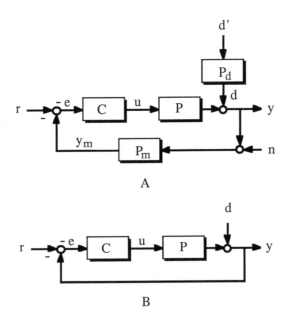

Figure 10.2-1. General (A) and simplified (B) block diagram of feedback control system.

$$\dot{x} = A_c x + B_c r \qquad (10.2 - 4)$$

$$y = C_c x + D_c r \qquad (10.2 - 5)$$

where

$$A_c = A_o - B_o (I + D_o)^{-1} C_o \qquad (10.2 - 6a)$$

$$B_c = B_o (I + D_o)^{-1} \qquad (10.2 - 6b)$$

$$C_c = (I + D_o)^{-1} C_o \qquad (10.2 - 6c)$$

$$D_c = (I + D_o)^{-1} D_o \qquad (10.2 - 6d)$$

We define the *open-loop characteristic polynomial (OLCP)*

$$\phi_{OL} = \det(sI - A_o) \qquad (10.2 - 7)$$

and the *closed-loop characteristic polynomial (CLCP)*

$$\phi_{CL} = \det(sI - A_c) \qquad (10.2 - 8)$$

Stability is determined by the zeros of the CLCP. We wish to express the CLCP in terms of $P(s)C(s)$. We define the *return difference operator $F(s)$*

$$F(s) = I + P(s)C(s) \qquad (10.2 - 9)$$

and state the following lemma.

Lemma 10.2-1 (Schur's formulae for partitioned determinants). *Let the square matrix G be partitioned as*

$$G = \begin{pmatrix} G_{11} & G_{12} \\ G_{21} & G_{22} \end{pmatrix}$$

Then the determinant can be expressed as

$$\det G = \det G_{11} \cdot \det(G_{22} - G_{21} G_{11}^{-1} G_{12}) \quad \text{if} \quad \det G_{11} \neq 0 \qquad (10.2 - 10)$$

or

$$\det G = \det G_{22} \cdot \det(G_{11} - G_{12} G_{22}^{-1} G_{21}) \quad \text{if} \quad \det G_{22} \neq 0 \qquad (10.2 - 11)$$

The determinant of $F(s)$ can be expressed as

$$\det F(s) = \det(I + C_o(sI - A_o)^{-1}B_o + D_o)$$

or by using (10.2–10)

$$\det F(s) = \det \begin{pmatrix} sI - A_o & B_o \\ -C_o & I + D_o \end{pmatrix} \div \det(sI - A_o)$$

$$= \det \begin{pmatrix} I_r & -B_o(I + D_o)^{-1} \\ 0 & I_n \end{pmatrix} \cdot \det \begin{pmatrix} sI - A_o & B_o \\ -C_o & I + D_o \end{pmatrix} \div \det(sI - A_o)$$

because the first term is unity. Combining the first two matrices we find

$$\det F(s) = \det \begin{pmatrix} sI - A_o + B_o(I + D_o)^{-1}C_o & 0 \\ -C_o & I + D_o \end{pmatrix} \div \det(sI - A_o)$$

or

$$\det F(s) = \frac{\det(sI - A_o + B(I + D_o)^{-1}C_o)}{\det(sI - A_o)} \cdot \det(I + D_o)$$

and because $\lim_{s \to \infty} F(s) = I + D_o$

$$\det F(s) = \frac{\phi_{CL}}{\phi_{OL}} \det F(\infty) \qquad (10.2 - 12)$$

and finally

$$\phi_{CL} = \frac{\det F(s)}{\det F(\infty)} \phi_{OL} \qquad (10.2 - 13)$$

If the open-loop system is stable then all RHP zeros of ϕ_{CL} have to be RHP zeros of $\det F(s) = \det(I + P(s)C(s))$ and we can determine stability directly from $\det F(s) = 0$. If the open-loop system is unstable we can generally not do so, because by forming the determinant unstable poles and zeros might cancel as discussed in Sec. 10.1.3. Multiplication by ϕ_{OL} brings back any unstable zeros which are cancelled when $\det F(s)$ is computed. Just as in the SISO case we can apply the principle of the argument to (10.2–12) and derive the multivariable Nyquist Stability Criterion.

Theorem 10.2-1 (Nyquist Stability Criterion). *Let the map of the Nyquist D contour under $\det F(s) = \det(I + P(s)C(s))$ encircle the origin n_F times in the clockwise direction. Let the number of open-loop unstable poles of PC be n_{PC}. Then the closed-loop system is stable if and only if*

$$n_F = -n_{PC}$$

Recall that for SISO systems we generally count the encirclements of (-1,0) by $p(s)c(s)$. This number is equal to the number of encirclements of the origin (0,0) by $1 + p(s)c(s)$.

10.2.3 Internal Stability

The concept of internal stability introduced in Sec. 2.3. applies to MIMO systems as well. For internal stability all elements in the 2×2 block transfer matrix in (10.2–14) have to be stable

$$\begin{pmatrix} y \\ u \end{pmatrix} = \begin{pmatrix} PC(I+PC)^{-1} & (I+PC)^{-1}P \\ C(I+PC)^{-1} & -C(I+PC)^{-1}P \end{pmatrix} \begin{pmatrix} r \\ u' \end{pmatrix} \qquad (10.2-14)$$

The Nyquist criterion developed in the last section is another test for internal stability. We will use both tests. Depending on the application one or the other allows us to make conclusions more easily.

10.2.4 Small Gain Theorem

Consider again the closed-loop system in Fig. 10.2-1B when P is square (dim $u = $ dim y) and the controller has been included in P so that we can set $C = I$.

Theorem 10.2-2 (Small Gain Theorem). *Assume that $P(s)$ is stable. Let $\rho(P(i\omega))$ be the spectral radius of $P(i\omega)$. Then the closed-loop system is stable if $\rho(P(i\omega)) < 1$, $\forall \omega$ or if $\|P(i\omega)\| < 1$, $\forall \omega$ where $\|\cdot\|$ denotes any compatible matrix norm.*

Proof. (By contradiction) Assume $\rho(P) < 1$, $\forall \omega$ and that the closed-loop system is unstable. We will employ the Nyquist stability criterion (Thm. 10.2-1). Instability implies that the image of $\det(I + P)$ encircles the origin as s traverses the Nyquist D contour. Because the image is closed there exists an $\epsilon \in [0, 1]$ and a frequency ω' such that

$$det(I + \epsilon P(i\omega')) = 0$$

(i.e., that the image goes through the origin).

$$\Leftrightarrow \prod_i \lambda_i (I + \epsilon P(i\omega')) = 0$$

$$\Leftrightarrow 1 + \epsilon \lambda_i (P(i\omega')) = 0 \quad \text{for some i}$$

$$\Leftrightarrow \lambda_i(P(i\omega')) = -\frac{1}{\epsilon} \quad \text{for some i}$$

$$\Rightarrow |\lambda_i(P(i\omega'))| \geq 1 \quad \text{for some i}$$

which is a contradiction because we assumed $\rho(P) < 1, \quad \forall \omega.$ □

Theorem 10.2-2 states that for an open-loop stable system, a sufficient condition for stability is to keep the "loop gain" $\rho(P)$ or $\|P\|$ less than unity. It is the multivariable extension of the Bode stability criterion for SISO systems which requires $|p(i\omega)| < 1, \quad \forall \omega$ for closed-loop stability. The Small Gain Theorem provides only a sufficient condition for stability and is therefore potentially conservative. It is useful because it does not require detailed information about the system.

10.3 Formulation of Control Problem

As discussed in Ch. 2, the following essentials have to be specified for any design procedure to yield a control algorithm that works satisfactorily in a real environment.

- process model

- model uncertainty bounds

- type of inputs (i.e., setpoints and disturbances)

- performance objectives

For a MIMO system it is much more difficult to make meaningful specifications than for a SISO system. For SISO systems any design procedure addresses the trade-off between performance and robustness and/or control action. For MIMO systems there is also a performance trade-off among the different outputs as well as a control action trade-off among the different inputs. These trade-offs are affected in a complex manner by the specifications.

10.3.1 Process Model

Because all our *analysis* procedures (e.g., for robust stability and performance) are frequency domain oriented, any linear time invariant model can be handled with equal ease. Models of high order and/or with time delays do not cause any problems. On the other hand, in particular for continuous MIMO systems, controller *synthesis* procedures become extremely complex when time delays are

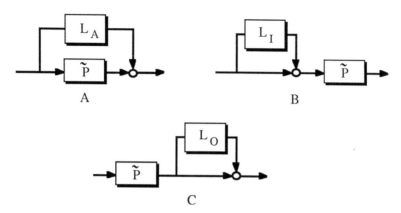

Figure 10.3-1. Additive (A), multiplicative input (B), and multiplicative output (C) uncertainty.

accounted for exactly. This added complexity is not justified by the final results: in practice the performance obtained with controllers based on Padé approximations is generally indistinguishable from that obtained with controllers based on irrational models. Thus while it is very convenient to describe many chemical processes with models involving time delays, it is unnecessarily cumbersome to account for these time delays explicitly during controller *synthesis*.

10.3.2 Model Uncertainty Description

Just as in the SISO case we will describe model uncertainty in the following manner: We will assume that the dynamic behavior of a plant is described not by a single linear time invariant model but by a *family* Π of linear time invariant models. While there are many different ways of parametrizing this family we concluded for SISO systems that a "Nyquist band" consisting of a union of disks of specified radius at each frequency was entirely adequate for most process control applications. This magnitude bounded additive (or multiplicative) uncertainty together with the H_∞-performance specification also allowed us to derive a very simple and exact condition for robust performance.

We can postulate similar uncertainty structures for MIMO systems (Fig. 10.3-1). For MIMO systems we have to distinguish multiplicative uncertainty at the plant input (L_I) and the plant output (L_O). The uncertainties L_I and L_O can be loosely interpreted as actuator and sensor uncertainty respectively. The plant P is related to the model $\tilde{P}$ and the uncertainty in the following manner.

$$P = \tilde{P} + L_A \qquad L_A = P - \tilde{P} \qquad\qquad (10.3-1)$$

$$P = \tilde{P}(I + L_I) \qquad L_I = \tilde{P}^{-1}(P - \tilde{P}) \qquad\qquad (10.3-2)$$

$$P = (I + L_O)\tilde{P} \qquad L_O = (P - \tilde{P})\tilde{P}^{-1} \qquad\qquad (10.3-3)$$

We can state a frequency dependent magnitude bound on these uncertainties in terms of a matrix norm. In principle, any matrix norm defined in Sec. 10.1.4. could be used. As we will see later, however, we can derive *necessary and sufficient* conditions for robust stability and performance only if we use the spectral norm. Thus we can state the following uncertainty bounds:

$$\bar{\sigma}(L_A) \leq \bar{\ell}_A(\omega) \qquad\qquad (10.3-4)$$

$$\bar{\sigma}(L_I) \leq \bar{\ell}_I(\omega) \qquad\qquad (10.3-5)$$

$$\bar{\sigma}(L_O) \leq \bar{\ell}_O(\omega) \qquad\qquad (10.3-6)$$

Note that contrary to the SISO case the three bounds are not equivalent and going from one uncertainty description to another generally increases the size of the family Π. For example, let us assume that we want to derive $\bar{\ell}_O$ from $\bar{\ell}_I$. From (10.3–2) and (10.3–3) we find

$$L_O = \tilde{P}L_I\tilde{P}^{-1} \qquad\qquad (10.3-7)$$

$$\bar{\sigma}(L_O) = \bar{\sigma}(\tilde{P}L_I\tilde{P}^{-1}) \leq \bar{\sigma}(\tilde{P})\bar{\sigma}(\tilde{P}^{-1})\bar{\sigma}(L_I) \leq \kappa(\tilde{P})\bar{\ell}_I$$

where the first inequality follows from (10.1–27) and

$$\kappa(\tilde{P}) \triangleq \bar{\sigma}(\tilde{P})\bar{\sigma}(\tilde{P}^{-1}) = \frac{\bar{\sigma}(\tilde{P})}{\underline{\sigma}(\tilde{P})} \qquad\qquad (10.3-8)$$

is the *condition number* of $\tilde{P}$. Thus

$$\bar{\ell}_O \leq \kappa(\tilde{P})\bar{\ell}_I \qquad\qquad (10.3-9)$$

Note that if one wanted to derive $\bar{\ell}_I$ from $\bar{\ell}_O$, one could write (10.3–7) as

$$L_I = \tilde{P}^{-1}L_O\tilde{P} \qquad\qquad (10.3-10)$$

and obtain the following upper bound for $\bar{\ell}_I$ by the same procedure:

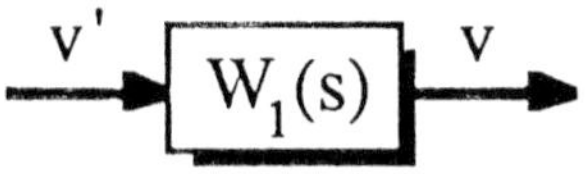

Figure 10.3-2. Input weight W_1 transforming normalized input v' to physical input v.

$$\bar{\ell}_I \leq \kappa(\tilde{P})\bar{\ell}_0 \qquad\qquad (10.3 - 11)$$

If the plant is unitary then $\kappa(\tilde{P}) = 1$ and $\bar{\ell}_O = \bar{\ell}_I$. If the plant is ill-conditioned — i.e., $\kappa(\tilde{P})$ is large — then the bounds formed by the RHS of (10.3–9) or (10.3–11) can be very conservative. Qualitatively, as we pass the uncertainty from the input through the ill-conditioned plant to the output it is strongly stretched in certain directions and the same happens when going from output to input uncertainty. Because it is not "round" anymore, the singular value bound describes it only in a conservative manner.

This example makes clear that for MIMO systems it is important to model uncertainty where it occurs and not necessarily where it is convenient mathematically. From this point of view all three types of uncertainty descriptions above are quite conservative because they "spread" the uncertainty (that might be caused by a single parameter or a single transfer matrix element) over the whole transfer matrix before defining a magnitude bound. In Chap. 11 we will introduce less conservative uncertainty descriptions which are closer to physical reality.

10.3.3 Input Specifications

As in the SISO case we will distinguish specific inputs and input sets. For notational convenience the inputs will be *normalized* (Fig. 10.3-2). It will be assumed that an input v entering the control loop is generated by passing the normalized input v' through a transfer matrix block $W_1(s)$, sometimes referred to as an *input weight* . The two types of *normalized* inputs of interest are

Specific input (vector of impulses):

$$v'(s) = \text{constant} \qquad\qquad (10.3 - 12)$$

Set of bounded inputs (all inputs with 2-norm bounded by unity):

$$\mathcal{V}' = \left\{ v' : \|v'\|_2^2 = \int_{-\infty}^{\infty} v'^H v' d\omega \leq 1 \right\} \qquad\qquad (10.3 - 13)$$

From these definitions follow the actual inputs v

Specific input:

$$v = W_1(s)v' \qquad (10.3-14)$$

Set of bounded inputs:

$$\mathcal{V} = \left\{ v : \|W_1^{-1}v\|_2^2 = \frac{1}{2\pi} \int_{-\infty}^{\infty} (W_1^{-1}v)^H (W_1^{-1}v)d\omega \leq 1 \right\} \qquad (10.3-15)$$

In general, for MIMO systems the controller which optimizes performance for a specific input [(10.3–12) or (10.3–14)], is not unique. The reason is that for *one* specific input vector there are many different controller transfer *matrices* which give rise to the same *one* manipulated variable vector and thus to the same optimal output. As we will show later, we can often find a unique controller by requiring it to be optimal in some sense for n different input vectors v'.

The set (10.3–15) can be interpreted as follows: If the spectrum of v is narrow and concentrated near ω^* (i.e., the input looks almost like $\mathrm{Re}\left\{\bar{v}'e^{i\omega^* t}\right\}$, where $\bar{v}'$ is a constant complex vector), then the power of v is limited by $(W_1^{-1}(i\omega^*)\bar{v}')^H (W_1^{-1}(i\omega^*)\bar{v}') \leq 1$. We expect W_1 to be large at low frequencies and small at high frequencies. Treating sets of inputs is attractive because at the design stage it is rarely possible to predict exactly what type of setpoint changes and disturbances are going to occur during actual operation. In principle, it is possible that if the input assumed for the design is not exactly equal to the input encountered in practice the performance could deteriorate significantly.

10.3.4   Control Objectives

In order for the controller to work well on the real plant the following objectives have to be met:

- Nominal stability

- Nominal performance

- Robust stability

- Robust performance

Nominal stability was treated in Sec. 10.2. The other objectives are going to be discussed next.

10.4 Nominal Performance

10.4.1 Sensitivity and Complementary Sensitivity Function

The most important relationships between the inputs and outputs in Fig. 10.2-1A $(P_d = P_m = I)$ are

$$e = (I + PC)^{-1}(d - r); \quad \text{for } n = 0 \tag{10.4 - 1}$$

$$y = PC(I + PC)^{-1}(r - n); \quad \text{for } d = 0 \tag{10.4 - 2}$$

We define the sensitivity function

$$E(s) \triangleq (I + PC)^{-1} \tag{10.4 - 3}$$

the complementary sensitivity function

$$H(s) \triangleq PC(I + PC)^{-1} \tag{10.4 - 4}$$

and the "generic" external input

$$v \triangleq d - r \tag{10.4 - 5}$$

For good performance it is desirable to make the sensitivity function as "small" as possible. This is only feasible over a finite frequency range because for strictly proper systems

$$\lim_{s \to +\infty} PC = 0 \tag{10.4 - 6}$$

and therefore

$$\lim_{s \to +\infty} E(s) = \lim_{s \to +\infty} (I + PC)^{-1} = I \tag{10.4 - 7}$$

Note that

$$E(s) + H(s) = I \tag{10.4 - 8}$$

which explains the name *complementary sensitivity function*. Ideally for performance $H(s)$ should be unity but because of (10.4–7) and (10.4–8) this can be achieved only over a finite frequency range.

The trade-offs between good reference following and disturbance suppression ($E \approx 0$) on one hand and suppression of measurement noise on the other ($H \approx 0$) are apparent from (10.4–8). For SISO systems, in addition to measurement noise, multiplicative uncertainty imposes a bound on the complementary sensitivity. We will see later that for MIMO systems, depending on the type of uncertainty, the imposed bound usually takes a much more complex form.

10.4.2 Asymptotic Properties of Closed-Loop Response (System Type)

In analogy to SISO systems, we wish to characterize the asymptotic closed loop response for disturbances/setpoints of the polynomial type (s^{-k}). For MIMO systems the situation is potentially more complicated because we could classify the behavior in each one of the channels separately. We will not do so here but extend the SISO definitions from Sec. 2.4.3. directly.

Definition 10.4-1. *Let $G(s)$ be the $n \times n$ open-loop transfer matrix and let m be the largest integer for which*

$$rank \left[\lim_{s \to 0} s^m G(s) \right] = n$$

Then the system $G(s)$ is said to be of Type m. (Note that $G(s)$ has at least $n \times m$ poles at the origin.)

Theorem 10.4-1. *Let the open-loop system $G(s)$ be of Type m. Then the sensitivity operator $E(s) = (I + G(s))^{-1}$ satisfies*

   Type m:

$$\lim_{s \to 0} s^{-k} E(s) = 0 \quad 1 \leq k < m \qquad (10.4 - 9)$$

Assume that the closed loop system is stable. Then as $t \to \infty$ the closed loop system perfectly tracks setpoint changes (perfectly rejects disturbances) of the form $\Sigma_{k=0}^{m} a_k s^{-k}$ where a_k are real constant vectors.

Proof. Follows directly from the Final Value Theorem.

10.4.3 Linear Quadratic (H_2-) Optimal Control

In analogy to the SISO case we could minimize the 2-norm of the error vector

$$\|e\|_2^2 = \frac{1}{2\pi} \int_{-\infty}^{\infty} e(i\omega)^H e(i\omega) d\omega \qquad (10.4 - 10)$$

for a particular input v. For MIMO systems, however, some modifications are required. First of all, some error components are usually more important than others. Also, we might be primarily interested in rejecting errors in a certain frequency range (for example, for low frequencies). This suggests the introduction of a frequency dependent (output) weight W_2 into the objective function (10.4-10)

$$\|e'\|_2^2 = \frac{1}{2\pi} \int_{-\infty}^{\infty} (W_2 e)^H (W_2 e) \, d\omega \qquad (10.4-11)$$

Furthermore, as we explained above, the controller, which solves this (weighted or unweighted) problem is not unique. Therefore we define an alternate problem: "Excite the system in *separate* experiments with n different linearly independent inputs v_i. Find the controller which minimizes the sum of squares of the 2-norms of the errors generated by the n experiments." From (10.4-1) we find that for one experiment the error is $e_i = E v_i$. Let us define $W_1 = (v_1, v_2, \dots, v_n)$. Then the columns of $E W_1$ are the errors from the n experiments. Consider now premultiplication by the output weight W_2 to generate $W_2 E W_1$, the matrix whose columns are the weighted errors e_i' from the n experiments. Then the controller C which minimizes the sum of squares of the weighted error 2-norms is defined implicitly by

$$\min_C \|W_2 E W_1\|_2^2 = \min_C \frac{1}{2\pi} \int_{-\infty}^{\infty} \text{trace} \left[(W_2 E W_1)^H (W_2 E W_1) \right] d\omega \qquad (10.4-12)$$

The H_2-optimal control problem (10.4-12) can be interpreted as the minimization of the average magnitude or, in mathematical terms, the *minimization of the 2-norm of the sensitivity operator E with input weight W_1 and output weight W_2*. It should be compared with the equivalent formula for the SISO case in Sec. 2.4.4. The weighted sensitivity is illustrated by the block diagram in Fig. 10.4-1: The n normalized error vectors e_i' are generated by the n normalized input vectors v_i'. In v_i' only the ith component is unity and all the other components are zero.

The definition of the MIMO H_2-objective must appear somewhat artificial. Though it is reasonable, it is certainty not something a control engineer would naturally formulate. The main motivation for this objective function is that powerful methods are available to minimize the weighted 2-norm of the sensitivity operator as defined by (10.4-12). Also (10.4-12) has a nice stochastic interpretation which will not be discussed in this book. Finally, more meaningful deterministic interpretations of (10.4-12) are available for special cases and are derived in Chap. 12.

The objective function (10.4-12) can be generalized to include, for example, a penalty term for excessive variations of the manipulated variables. This is well known and not discussed here.

The correct choice of weights for a particular practical problem is not trivial. Just as in the SISO case the weights should be regarded as tuning parameters which are chosen by the designer to achieve the best compromise between the conflicting objectives. The weight selection is guided by the expected system

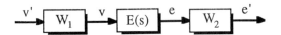

Figure 10.4-1. Sensitivity operator E with input weight W_1 and output weight W_2.

inputs and the relative importance of the outputs. If step setpoint changes for the different outputs are of primary importance then $W_1 = s^{-1}I$ is a reasonable weight. If regulation is more important then $W_1 = P_d$ should yield good performance for a vector d' of impulses. The error weight W_2 should reflect the relative importance of the errors as well as the relevant frequency range. A typical weight which penalizes low-frequency errors (i.e., offset) heavily would be $W_2 = s^{-1}\bar{W}_2$ where $\bar{W}_2$ is a constant diagonal matrix. Note, however, that there is no need to include the factor s^{-1} in *both* W_1 and W_2 if simply no offset to step-like inputs is desired.

10.4.4 H_∞-Optimal Control

The inputs v are assumed to belong to a set of norm-bounded functions with a frequency-dependent weight as discussed in Sec. 10.3.3.

$$\mathcal{V} = \left\{v : \|W_1^{-1}v\|_2^2 \le 1\right\} \qquad (10.4-13)$$

Equivalently we can define the set of normalized inputs $v' = W_1^{-1}v$

$$\mathcal{V}' = \left\{v' : \|v'\|_2^2 \le 1\right\} \qquad (10.4-14)$$

(We refer to Fig. 10.4-1 for an interpretation of the weights.) This input class is much more general than what we considered for the H_2-problem where $v'(t)$ was assumed to consist of impulses.

Each input $v \in \mathcal{V}$ gives rise to an error e. This error is processed through the output weight W_2 (Fig. 10.4-1) which reflects the relative importance of the individual error components and also the frequency range over which the error is to be made small. The controller is to be designed to minimize the worst *normalized* (i.e., weighted) error e' which can result from any input $v \in \mathcal{V}$.

$$\min_C \max_{v \in \mathcal{V}} \|e'\|_2 = \min_C \max_{v' \in \mathcal{V}'} \|W_2 E W_1 v'\|_2 \qquad (10.4-15)$$

From Thm. 10.1-4 we find for $\mathcal{V}'$ defined by (10.4-14)

$$\max_{v' \in \mathcal{V}'} \|W_2 E W_1 v'\|_2 = \sup_\omega \bar\sigma(W_2 E W_1(i\omega)) = \|W_2 E W_1\|_\infty \qquad (10.4-16)$$

With (10.4-16) the H_∞-optimal control problem becomes

$$\min_C \|W_2 E W_1\|_\infty = \min_C \sup_\omega \bar\sigma(W_2 E W_1(i\omega)) \qquad (10.4-17)$$

Thus, the H_∞-optimal controller minimizes the maximum magnitude or, in mathematical terms, *minimizes the ∞-norm* of the sensitivity function E with input weight W_1 and output weight W_2. According to this frequency domain interpretation the H_2-optimal controller minimizes the *average* value and the H_∞-optimal controller the *peak value* of the weighted sensitivity function.

Let us assume for simplicity that W_1 and W_2 are scalar and let the optimum value of the objective function (10-4-17) be k. Then for the optimal controller the sensitivity function satisfies:

$$\|E\|_\infty = \sup_\omega \bar\sigma(E(i\omega)) < k|W_1 W_2|^{-1} \qquad (10.4-18)$$

Inequality (10.4-18) implies that the maximum singular value of the sensitivity function lies below the bound $k|W_1 W_2|^{-1}$. Typically this bound is selected by the designer to be low at low frequencies and to increase with frequency. Often the designer wishes to specify a minimum bandwidth and to limit the magnitude of the sensitivity operator in order to avoid excessive disturbance amplification. In the H_∞ formulation this can be done explicitly by specifying $W_1 W_2$ accordingly. The H_2 optimal control objective is to minimize the *average* weighted sensitivity and large disturbance amplification is (in principle) possible at certain frequencies. In both cases (H_2 and H_∞), however, the weights are basically tuning parameters selected to reflect input types, relative importance of outputs and desired sensitivity function shapes.

The H_∞ performance requirement is usually written as

$$\|W_2 E W_1\|_\infty < 1 \qquad (10.4-19)$$

where it has been assumed that W_1 and W_2 have been scaled such that a unity bound on the RHS makes sense. For example, in the case (10.4-18), when the weights are scalars either W_1 or W_2 are specified to include k.

Note that for high frequencies $\bar\sigma(PC)$ is small and therefore

$$\bar\sigma(E) = \bar\sigma((I + PC)^{-1}) \cong 1 \qquad \omega \text{ large} \qquad (10.4-20)$$

Thus tight performance specifications are only meaningful in the low frequency range where PC is "large." Then the performance specification (10.4-18) with $k = 1$ reduces to

$$\underline\sigma(I + PC) \cong \underline\sigma(PC) > |W_1 W_2| \qquad \omega \text{ small} \qquad (10.4-21)$$

In analogy to the SISO case (Sec. 2.4) the smallest loop gain measured by $\underline{\sigma}(PC)$ has to be shaped to fall above the performance weight $|W_1 W_2|$.

10.5 Summary

The singular values provide a practical framework for extending the concept of gain to MIMO systems. In particular the gain of the system G depends on the direction of the input vector u but it is bounded by the smallest and largest singular value.

$$\underline{\sigma}(G(i\omega))|u(i\omega)| \le |G(i\omega)u(i\omega)| \le \bar{\sigma}(G(i\omega))|u(i\omega)| \qquad (10.1-35,36)$$

The operator norm of G induced by the 2-norm ($\|\cdot\|_2$) is the ∞-norm of the transfer function matrix G:

$$\|G\|_{i2} = \sup_\omega \bar{\sigma}(G(i\omega)) \triangleq \|G\|_\infty \qquad (10.1-56)$$

If signal magnitude is measured by the 2-norm, then, by definition, a signal passing through the system G is amplified at most by $\|G\|_\infty$.

The closed-loop characteristic polynomial (ϕ_{CL}) can be obtained from the open loop characteristic polynomial (ϕ_{OL}) by

$$\phi_{CL} = \frac{\det F(s)}{\det F(\infty)} \phi_{OL} \qquad (10.2-13)$$

where $F(s)$ is the return difference operator

$$F(s) = I + PC(s) \qquad (10.2-9)$$

Closed loop stability can be determined by checking the encirclements of the origin by the map of the Nyquist D contour under $\det F(s)$ (Thm. 10.2-1).

An open-loop stable system is closed-loop stable if the loop gain ($\rho(PC(i\omega))$ or $\|PC(i\omega)\|$) is less than unity for all frequencies ω (Thm. 10.2-2).

For MIMO systems we have to distinguish uncertainty at the process input

$$P = \tilde{P}(I + L_I), \quad L_I = \tilde{P}^{-1}(P - \tilde{P}), \quad \bar{\sigma}(L_I) \le \bar{\ell}_I(\omega) \qquad (10.2-2,5)$$

and the process output

$$P = (I + L_O)\tilde{P}, \quad L_O = (P - \tilde{P})\tilde{P}^{-1}, \quad \bar{\sigma}(L_O) \le \bar{\ell}_O(\omega) \qquad (10.3-3,6)$$

In general, it is not possible to convert from one uncertainty description to the other without introducing conservatism.

The controller design techniques aim to make a measure of the sensitivity operator E

$$E = (I + PC)^{-1} \tag{10.4 - 3}$$

small. The H_2-optimal controller minimizes the 2-norm ("average") of the sensitivity operator with input weight W_1 and output weight W_2

$$\min_C \|W_2 E W_1\|_2^2 = \min_C \frac{1}{2\pi} \int_{-\infty}^{\infty} \text{trace}\left[(W_2 E W_1)^H (W_2 E W_1)\right] d\omega \tag{10.4 - 12}$$

The H_∞-optimal controller minimizes the ∞-norm ("peak") of the weighted sensitivity function

$$\min_C \|W_2 E W_1\|_\infty = \min_C \sup_\omega \bar{\sigma}(W_2 E W_1(i\omega)) \tag{10.4 - 17}$$

10.6 References

10.1. Most of these concepts are covered, for example, by Kwakernaak & Sivan (1972).

10.1.2 and 10.1.3. The definitions, theorems, and examples were taken from Postlethwaite & MacFarlane (1979). Alternate definitions of zeros are available and are summarized by Holt & Morari (1985a). Numerically it is most reliable to find the poles by computing the eigenvalues of A and the zeros by solving a generalized eigenvalue problem (Laub & Moore, 1978).

10.1.4–10.1.6. This material is covered comprehensively by Desoer & Vidyasagar (1975) who also prove Thm. 10.1-4. For a limited discussion in the matrix context the reader is referred to Gantmacher (1959) and Bellman (1970). A very good discussion on matrix and vector norms can be found in Stewart (1973), where the term "consistent" instead of "compatible" is used. The physical interpretation of SVD was adopted from Bruns & Smith (1982).

10.2.2. Lemma 10.2-1 can be found in Gantmacher (1959). The derivation of the closed-loop characteristic polynomial was adopted from Postlethwaite & MacFarlane (1979).

10.2.4. Desoer & Vidyasagar (1975) present the Small Gain Theorem in a general context.

10.3.2. The different types of uncertainty descriptions were used by Doyle & Stein (1981).

10.4.2. More general definitions of System Type were proposed by Sandell & Athans (1973) and Wolfe & Meditch (1977).

Chapter 11

ROBUST STABILITY AND PERFORMANCE

Many approaches can be taken to describe the uncertainty associated with a MIMO model for a physical system. It must be emphasized that however attractive an uncertainty description may seem from a practical point of view it is only useful if it permits the derivation of "tight" conditions for robust stability and robust performance. Two types of descriptions will be discussed in this chapter: "unstructured" and "structured" uncertainty. Both lead to "tight" (necessary and sufficient) robustness conditions. The necessity is only meaningful, however, if the assumed uncertainty is an accurate description of the true uncertainty. Otherwise the mathematically tight robustness conditions can be very conservative from a practical point of view (see the discussion of Thm. 2.5-1).

Unstructured Uncertainty. The uncertainty is expressed in terms of a specific *single* perturbation of the type introduced in Sec. 10.3.2. Similar to the SISO case, the conditions for robust stability can then be expressed as bounds on transfer matrices which are directly related to performance [e.g., $\bar{\sigma}(\tilde{H})$ or $\bar{\sigma}(\tilde{E})$]. Though the bounds derived using unstructured uncertainty are necessary and sufficient, they are generally *conservative* from a practical point of view since the actual uncertainty can rarely be lumped into a single norm-bounded perturbation without including many more possible plants than actually needed.

Structured Uncertainty. The individual sources of uncertainty are identified and represented directly – there is no need to lump them together. This generally leads to an uncertainty description with multiple perturbations (Δ_i's). By assuming norm bounds on these uncertainties (e.g., $\bar{\sigma}(\Delta_i) \leq 1$), it is possible to derive necessary and sufficient and, from a practical point of view, *non-conservative* conditions for robustness using the structured singular value μ. One disadvantage of this procedure is that the resulting conditions are not in terms of a simple bound on $\bar{\sigma}(\tilde{H})$ or $\bar{\sigma}(\tilde{E})$, but involve $\mu(M)$ where M may be a complicated function of $\tilde{E}$ and $\tilde{H}$.

To alleviate this problem, we will outline a general technique for deriving *sufficient* robustness conditions which can be expressed in terms of bounds on any arbitrary transfer matrix of interest. Though these conditions are mathematically conservative, they are appealing from an engineering point of view because they allow the designer to see how particular forms of uncertainty restrict, for example, the sensitivity operator.

11.1 Robust Stability for Unstructured Uncertainty

In this section, the uncertainty which may occur in different parts of the system is lumped into one single perturbation L. We refer to this uncertainty as "unstructured." More precisely, "unstructured" uncertainty means that *several* sources of uncertainty are described with a *single* perturbation which is a full matrix with the *same dimensions* as the plant P.

11.1.1 Uncertainty Description

Let $P \in \Pi$ be any member of the set of possible plants Π, and let $\tilde{P} \in \Pi$ denote the nominal model of the plant. To describe unstructured uncertainty the following four single perturbations are commonly used: additive (L_A), multiplicative output (L_O), multiplicative input (L_I), and inverse multiplicative output (L_E) perturbations (Fig. 11.1-1). Some of these were introduced in Sec. 10.3.2.

$$P = \tilde{P} + L_A \quad \text{or} \quad L_A = P - \tilde{P} \qquad (11.1-1)$$

$$P = (I + L_O)\tilde{P} \quad \text{or} \quad L_O = (P - \tilde{P})\tilde{P}^{-1} \qquad (11.1-2)$$

$$P = \tilde{P}(I + L_I) \quad \text{or} \quad L_I = \tilde{P}^{-1}(P - \tilde{P}) \qquad (11.1-3)$$

$$P = (I - L_E)^{-1}\tilde{P} \quad \text{or} \quad L_E = (P - \tilde{P})P^{-1} \qquad (11.1-4)$$

The conditions for robust stability are different depending on which single perturbation is chosen to describe the uncertainty.

In each of the cases above the magnitude of the perturbation L may be measured in terms of a bound on $\bar{\sigma}(L)$

$$\bar{\sigma}(L) \le \bar{\ell}(\omega) \qquad \forall \omega \qquad (11.1-5)$$

where

$$\bar{\ell}(\omega) = \max_{P \in \Pi} \bar{\sigma}(L)$$

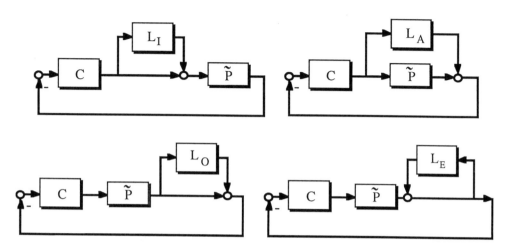

Figure 11.1-1. Four common uncertainty descriptions involving single perturbations: multiplicative input uncertainty (L_I); additive uncertainty (L_A); multiplicative output uncertainty (L_O); inverse multiplicative output uncertainty (L_E).

The bound $\bar{\ell}(\omega)$ can also be interpreted as a scalar *weight* on a normalized perturbation $\Delta(s)$

$$L(s) = \bar{\ell}(s)\Delta(s), \quad \bar{\sigma}(\Delta(i\omega)) \leq 1 \quad \forall\omega \qquad (11.1-6)$$

Generally the magnitude bound $\bar{\ell}(\omega)$ will *not* constitute a tight description of the "real" uncertainty. This means that the set of plants satisfying (11.1-6) will be larger than the original set Π.

We will also assume that the set of uncertain plants is "connected." This implies that all plants in the set are obtained by continuously deforming the model in the frequency domain — just like the Nyquist bands were generated for SISO systems in Sec. 2.2.2.

11.1.2 General Robust Stability Theorem

When L is of the form (11.1–6) each one of the block diagrams in Fig. 11.1-1 can be put into the form shown in Fig. 11.1-2 where the perturbation Δ satisfies $\bar{\sigma}(\Delta) \leq 1$. (We will demonstrate this in detail in Secs. 11.1.3 through 11.1.5.) If the nominal system is stable then M is stable and Δ is a perturbation which can destabilize the system. The following theorem establishes conditions on M

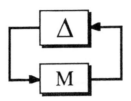

Figure 11.1-2. General $M - \Delta$ structure for robustness analysis.

so that it cannot be destabilized by Δ.

Theorem 11.1-1. *Assume that M is stable and that the perturbation Δ is of such a kind that the perturbed closed-loop system is stable if and only if the map of the Nyquist D contour under $\det(I - M\Delta)$ does not encircle the origin. Then the closed-loop system in Fig. 11.1-2 is stable for all perturbations Δ ($\bar{\sigma}(\Delta) \le 1$) if and only if one of the following three equivalent conditions is satisfied:*

$$\det(I - M\Delta(i\omega)) \ne 0 \qquad \forall \omega, \forall \Delta \ni \bar{\sigma}(\Delta) \le 1 \qquad (11.1-7)$$

$$\Leftrightarrow \rho(M\Delta(i\omega)) < 1 \qquad \forall \omega, \forall \Delta \ni \bar{\sigma}(\Delta) \le 1 \qquad (11.1-8)$$

$$\Leftrightarrow \bar{\sigma}(M(i\omega)) < 1 \qquad \forall \omega \qquad (11.1-9a)$$

$$\Leftrightarrow \|M\|_\infty < 1 \qquad (11.1-9b)$$

Proof. Assume there exists a perturbation Δ' such that $\bar{\sigma}(\Delta') \le 1$ and the image of $\det(I - M\Delta'(s))$ encircles the origin as s traverses the Nyquist contour. Because the Nyquist contour and its map are closed, there exists an $\epsilon \in [0, 1]$ and an ω' such that $\det(I - M\epsilon\Delta'(i\omega')) = 0$. Since $\bar{\sigma}(\epsilon\Delta') = \epsilon\bar{\sigma}(\Delta') \le 1$, $\epsilon\Delta'$ is just another perturbation from the set. Thus the closed-loop system is stable for all perturbations in the set if and only if (11.1–7) is satisfied.

Assume now there exists a perturbation Δ' and a frequency ω' such that $\rho(M\Delta'(i\omega')) < 1$ but that

$$\det(I - M\Delta'(i\omega')) = 0$$

$$\Leftrightarrow \prod_i \lambda_i(I - M\Delta'(i\omega')) = 0$$

$$\Leftrightarrow 1 - \lambda_i(M\Delta'(i\omega')) = 0 \quad \text{for some i}$$

$$\Rightarrow \rho(M\Delta'(i\omega')) \geq 1$$

which is a contradiction. Therefore (11.1–8) is sufficient for robust stability. Because of (10.1–33), (11.1–9) is also sufficient.

To prove necessity of (11.1–8) assume there is a Δ' for which $\bar{\sigma}(\Delta') \leq 1$ and $\rho(M\Delta') = 1$. Then $|\lambda_i(M\Delta')| = 1$ for some i. Δ' can always be chosen such that $\lambda_i(M\Delta') = +1$ and therefore $\det(I - M\Delta') = 0$. To prove necessity of (11.1–9) let $\bar{\sigma}(M) = 1$. Define $D = \text{diag}\{1, 0, \ldots, 0\}$ and $\Delta' = VDU^H$, where U and V are the matrices of the left and right singular vectors of $M(M = U\Sigma V^H)$. Clearly $\bar{\sigma}(\Delta') = 1$ and $\det(I - M\Delta') = \det(I - U\Sigma V^H V D U^H) = \det(I - U\Sigma D U^H) = \det(I - \Sigma D) = 0$. □

Theorem 11.1-1 states that if $\bar{\sigma}(M) < 1$, there is no perturbation Δ ($\bar{\sigma}(\Delta) \leq 1$) which makes $\det(I - M\Delta(s))$ encircle the origin as s traverses the Nyquist D contour. Note that we *assumed* that the absence of encirclements is necessary and sufficient for robust stability. This is the case, for example, when all perturbations Δ are stable or when all members P of the set Π of possible plants have the same number of RHP poles. We will generally assume one or the other. Using more complicated arguments it can be shown that the number of RHP poles may change as long as they appear and disappear by crossing the imaginary axis and not by moving away from or toward RHP zeros. This is also what we meant by a "connected" set of uncertain plants in Sec. 11.1.1. The connectedness condition is very difficult to check however.

In principle we could use a different norm to bound the uncertainty Δ. Assume $\|\Delta\| \leq 1$ where $\|\cdot\|$ is any compatible matrix norm. If Δ is stable, then it follows directly from the Small Gain Theorem (Thm. 10.2-2) that the closed-loop system in Fig. 11.1-2 is stable for all perturbations Δ ($\|\Delta\| \leq 1$) if $\|M\| < 1$. The Small Gain Theorem is only sufficient, however, and therefore potentially conservative. Thus even when $\|M\| = 1$, there is generally no Δ ($\|\Delta\| \leq 1$) which leads to instability. Our objective is to make all tests for robust stability and performance "tight" — i.e., necessary and sufficient. Therefore magnitude bounds on the uncertainty will always be given in terms of the spectral norm.

Next we will use Thm. 11.1-1 to derive conditions for robust stability for the different uncertainty descriptions (11.1–2)-(11.1–4). The derivation for the additive uncertainty is left as an exercise.

11.1.3 Multiplicative Output Uncertainty

Let

$$P = (I + L_O)\tilde{P} \text{ or } L_O = (P - \tilde{P})\tilde{P}^{-1} \qquad (11.1-2)$$

By comparing Fig. 11.1-1 and 11.1-2 we find

$$M = -\tilde{P}C(I + \tilde{P}C)^{-1}\bar{\ell}_O \qquad (11.1-10)$$

Corollary 11.1-1. *Under the assumption of Thm. 11.1-1 the closed-loop system is stable for all perturbations L_O ($\bar{\sigma}(L_O) \leq \bar{\ell}_O$) if and only if*

$$\bar{\sigma}(\tilde{P}C(I+\tilde{P}C)^{-1})\bar{\ell}_O = \bar{\sigma}(\tilde{H})\bar{\ell}_O < 1, \quad \forall \omega \quad \Leftrightarrow \quad \|\tilde{H}\bar{\ell}_O\|_\infty < 1 \qquad (11.1-11)$$

This result is a direct extension of the SISO result expressed through Thm. 2.5-1.

The robust stability condition (11.1-11) can always be satisfied for open loop stable systems since $\tilde{H} = 0$ (no feedback) is always possible. However, good disturbance rejection and good command following require $\tilde{H} \cong I$ (i.e., $\bar{\sigma}(\tilde{H}) \cong 1$). Condition (11.1-11) says that the system has to be "detuned" ($\bar{\sigma}(\tilde{H}) < 1$) at frequencies where $\ell_O(\omega) \geq 1$.

Note that for high frequencies $\tilde{P}C$ is "small"

$$\bar{\sigma}(\tilde{P}C(I+\tilde{P}C)^{-1}) = \underline{\sigma}^{-1}(I + (\tilde{P}C)^{-1}) \cong \bar{\sigma}(\tilde{P}C)$$

and therefore (11.1-11) becomes

$$\bar{\sigma}(\tilde{P}C) < \bar{\ell}_O^{-1} \qquad \omega \text{ large}$$

The design implication is that the controller gain for high frequencies is limited by uncertainty. In analogy to the SISO case (Sec. 2.5) the loop gain $\bar{\sigma}(\tilde{P}C)$ has to be "shaped" to fall below the uncertainty bound $\bar{\ell}_O^{-1}$.

11.1.4 Multiplicative Input Uncertainty

Let

$$P = \tilde{P}(I + L_I) \text{ or } L_I = \tilde{P}^{-1}(P - \tilde{P}) \qquad (11.1-3)$$

By comparing Figs. 11.1-1 and 11.1-2 we find

$$M = -(I + C\tilde{P})^{-1}C\tilde{P}\bar{\ell}_I \qquad (11.1-12)$$

Corollary 11.1-2. *Under the assumption of Thm. 11.1-1 the closed loop system is stable for all perturbations* L_I $(\bar{\sigma}(L_I) \leq \bar{\ell}_I)$ *if and only if*

$$\bar{\sigma}(\tilde{H}_I)\bar{\ell}_I < 1, \quad \forall \omega \quad \Leftrightarrow \quad \|\tilde{H}_I \bar{\ell}_I\|_\infty < 1 \qquad (11.1-13)$$

where

$$\tilde{H}_I = (I + C\tilde{P})^{-1} C\tilde{P} \qquad (11.1-14)$$

$\tilde{H}_I$ is the nominal closed-loop transfer function as seen from the *input* of the plant. It is desirable to have this transfer function close to I in order to reject disturbances affecting the inputs to the plant. However, since performance is usually measured at the output of the plant it may be of interest to use (11.1–13) in order to derive a bound in terms of $\tilde{H}$. To derive this bound $\tilde{P}$ is assumed to be square and the inequality

$$\bar{\sigma}(\tilde{H}_I) = \bar{\sigma}(\tilde{P}^{-1}\tilde{H}\tilde{P}) \leq \bar{\sigma}(\tilde{P}^{-1})\bar{\sigma}(\tilde{H})\sigma(\tilde{P}) = \kappa(\tilde{P})\bar{\sigma}(\tilde{H})$$

is used; the bound for robust stability is:

$$\bar{\sigma}(\tilde{H})\bar{\ell}_I(\omega) < \frac{1}{\kappa(\tilde{P})} \qquad \forall \omega \qquad (11.1-15)$$

Condition (11.1–15) has been used to introduce the condition number $\kappa(\tilde{P})$ as a stability sensitivity measure with respect to input uncertainty, but this is misleading. The condition number enters the stability condition (11.1–15) mainly as the result of the conservative step introduced by going from an input (11.1–13) to an output uncertainty description (11.1–15). For $\kappa(\tilde{P})$ large, (11.1–15) may be arbitrarily conservative even though the uncertainty is tightly described in terms of a norm-bounded input uncertainty such that (11.1–13) is both necessary and sufficient.

Note that for high frequencies $C\tilde{P}$ is "small"

$$\bar{\sigma}(C(I + \tilde{P}C)^{-1}\tilde{P}) = \underline{\sigma}^{-1}(I + (C\tilde{P})^{-1}) \cong \bar{\sigma}(C\tilde{P})$$

and therefore (11.1–13) becomes

$$\bar{\sigma}(C\tilde{P}) < \bar{\ell}_I^{-1} \qquad \omega \text{ large}$$

The design implication is that the controller gain for high frequencies is limited by uncertainty. The loop gain $\bar{\sigma}(C\tilde{P})$, which is generally *not* equal to $\bar{\sigma}(\tilde{P}C)$ (see Sec. 11.1.3), has to be "shaped" to fall below the uncertainty bound $\bar{\ell}_I^{-1}$.

11.1.5   Inverse Multiplicative Output Uncertainty

Let

$$P = (I - L_E)^{-1}\tilde{P} \text{ or } L_E = (P - \tilde{P})P^{-1} \qquad (11.1-4)$$

By comparing Figs. 11.1-1 and 11.1-2 we find

$$M = (I + \tilde{P}C)^{-1}\bar{\ell}_E \qquad (11.1-16)$$

Corollary 11.1-3. *Under the assumption of Thm. 11.1-1 the closed loop system is stable for all perturbations L_E ($\bar{\sigma}(L_E) \leq \bar{\ell}_E$) if and only if*

$$\bar{\sigma}((I + \tilde{P}C)^{-1})\bar{\ell}_E = \bar{\sigma}(\tilde{E})\bar{\ell}_E < 1, \quad \forall \omega \quad \Leftrightarrow \quad \|\tilde{E}\bar{\ell}_E\|_\infty < 1 \qquad (11.1-17)$$

For minimum phase systems the nominal sensitivity function $\tilde{E}$ may be arbitrarily small ("perfect control") and (11.1–17) can always be satisfied. Therefore, condition (11.1–17) seems to imply that for minimum phase systems arbitrarily good nominal performance ($\tilde{E}$ small) is possible regardless of how large the uncertainty is. This is not quite true. The pitfall is that any real system has to be strictly proper, and $E = I$ as well as $\tilde{E} = I$ must be required as $\omega \to \infty$. Consequently, to satisfy (11.1–17) it is necessary that $\bar{\sigma}(L_E) = \bar{\sigma}((P - \tilde{P})P^{-1}) \leq 1$ as $\omega \to \infty$ for all possible P. This condition is usually violated in practice, because the relative order of the actual plant is higher than that of the model.

Corollaries 11.1-1 and 11.1-3 prescribe two fundamentally different ways of handling uncertainty: to guarantee robust stability Cor. 11.1-1 requires the system to be detuned (low gain), while Cor. 11.1-3 requires that the control be tightened (high gain). In practice, it is desirable to combine the two approaches: By tightening the control at low frequencies, better performance is obtained. Eventually, at higher frequencies, the system has to be detuned to guarantee robust stability. In fact, it can be shown that it is possible to combine Cor. 11.1-1 and 11.1-3 over different frequency ranges.

11.1.6   Example: Input Uncertainty for Distillation Column

Consider the distillation column described in the Appendix where the overhead composition is to be controlled at $y_D = 0.99$ and the bottom composition at $x_B = 0.01$ using the distillate D and boilup V as manipulated inputs. By linearizing the nonlinear model at steady state and by assuming that the dynamics may be approximated by a first-order system with time constant $\tau = 75$ min, the following linear model is derived

$$\tilde{P} = \frac{1}{\tau s + 1} \begin{pmatrix} -0.878 & 0.014 \\ -1.082 & -0.014 \end{pmatrix} \qquad (11.1-18)$$

A simple decentralized control system with two PI controllers is chosen

$$C(s) = \frac{1 + \tau s}{s} \begin{pmatrix} -0.15 & 0 \\ 0 & -7.5 \end{pmatrix} \qquad (11.1-19)$$

This controller can be shown to give acceptable nominal performance. Assume there is relative uncertainty of magnitude $w_I(s)$ on *each* manipulated variable:

$$w_I(s) = 0.2\frac{5s + 1}{0.5s + 1} \qquad (11.1-20)$$

This implies a relative uncertainty of up to 20% in the low frequency range which increases at high frequencies, reaching a value of 1 at $\omega \cong 1$ min^{-1}. This increase with frequency allows for a time delay of about one minute, and may represent the effect of the flow dynamics which were neglected when developing the model. This relative uncertainty can be written in terms of two *scalar* multiplicative perturbations Δ_D and Δ_V.

$$D = (1 + w_I(s)\Delta_D)D_c, \qquad |\Delta_D| \le 1 \qquad \forall \omega \qquad (11.1-21a)$$

$$V = (1 + w_I(s)\Delta_V)V_c, \qquad |\Delta_V| \le 1 \qquad \forall \omega \qquad (11.1-21b)$$

Here D and V are the actual inputs, while D_c and V_c are the desired values of the flow rates as computed by the controller. Equations (11.1–21) can be approximated by an "unstructured" single perturbation $L_I = w_I\Delta_I$, where Δ_I is a "full" 2×2 matrix

$$\begin{pmatrix} D \\ V \end{pmatrix} = (I + w_I(s)\Delta_I)\begin{pmatrix} D_c \\ V_c \end{pmatrix}, \qquad \bar{\sigma}(\Delta_I) \le 1 \qquad \forall \omega \qquad (11.1-22)$$

with $\bar{\ell}_I(\omega) = |w_I(i\omega)|$. Inequality (11.1–13) indicates that robust stability is guaranteed if $\bar{\sigma}(\tilde{H}_I) < 1/\bar{\ell}_I(\omega)$ $\forall \omega$. From Fig. 11.1-3 it is seen that this condition is violated over a wide frequency range. By other means it can be shown, however, that the system is robustly stable. The reason for the conservativeness of condition (11.1–13) in this instance is that the use of unstructured uncertainty (11.1–22) includes plants not included in the "true" uncertainty description (11.1–21). These problems may be avoided by using the structured singular value $\mu(\tilde{H}_I)$ as discussed in Sec. 11.2.

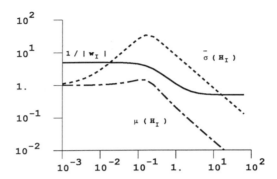

Figure 11.1-3. Robust stability for diagonal input uncertainty is guaranteed since $\mu(H_I) \leq 1/|w_I|$, $\forall \omega$. The use of unstructured uncertainty and $\bar{\sigma}(H_I)$ is conservative. (Reprinted with permission from *Chem. Eng. Sci.*, **42**, 1769 (1987), Pergamon Press, plc.)

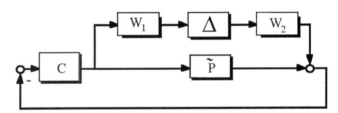

Figure 11.1-4. System with weighted additive uncertainty. Rearranging this system to fit Fig. 11.1-2 gives $M = W_1 C (I + \tilde{P}C)^{-1} W_2$.

11.1.7 Integral Control and Robust Stability

Because of the importance of integral action in the context of process control we will derive specifically conditions under which controllers with integral action can be designed in the presence of uncertainty. We will keep the uncertainty description as general as possible. We define Π_A as the set of plants which is generated by a single weighted additive norm perturbation (Fig. 11.1-4)

$$\Pi_A = \left\{ P : P = \tilde{P} + L_A \right\}, \quad L_A = W_2 \Delta W_1, \quad \bar{\sigma}(\Delta) \leq 1 \quad \forall \omega \qquad (11.1-23)$$

Π_A includes additive uncertainty (11.1-1) ($W_1 = \bar{\ell}_A$, $W_2 = I$), multiplicative output uncertainty (11.1-2) ($W_1 = \tilde{P}\bar{\ell}_O$, $W_2 = I$) and multiplicative input uncertainty (11.1-3) ($W_1 = I$, $W_2 = \tilde{P}\ell_I$) as special cases.

Robust stability for a system under "perfect control" ($\tilde{H} = I$, $\forall \omega$) will be studied first. Though "perfect control" cannot be realized in practice it is a useful conceptual tool.

Theorem 11.1-2. (Perfect Control). *Let the set of plants be given by* Π_A. *Under the assumption of Thm. 11.1-1 robust stability may be achieved for the system under "perfect control"* ($\tilde{H} = I$) *if and only if*

$$\bar{\sigma}(W_1 \tilde{P}^{-1} W_2) < 1 \qquad \forall \omega \qquad (11.1-24)$$

Proof. Rearranging the block diagram in Fig. 11.1-4 into the form in Fig. 11.1-2 yields

$$M = -W_1 C(I + \tilde{P}C)^{-1} W_2 = -W_1 \tilde{P}^{-1} \tilde{H} W_2 \qquad (11.1-25)$$

□

(11.1–24) follows from (11.1–9) for $\tilde{H} = I$.

Corollary 11.1-4. *For specific choices of weighting matrices (11.1-24) is equivalent to the following:*

Additive Uncertainty:

$$\bar{\ell}_A < \underline{\sigma}(\tilde{P}) \qquad (11.1-26)$$

Multiplicative Uncertainty:

$$\bar{\ell}_O < 1 \text{ or } \bar{\ell}_I < 1 \qquad (11.1-27)$$

Arbitrary Weights:

$$\det P \neq 0 \quad \forall \omega, \quad \forall P \in \Pi_A \qquad (11.1-28)$$

Proof. (11.1–26) and (11.1–27) follow from (11.1–24) by substitution of the appropriate weights. (11.1–28) is a direct consequence of (11.1–7), (11.1–9) and (11.1–24).

□

This corollary implies that robust "perfect control" is possible if and only if none of the plants P in the set Π_A has zeros on the imaginary axis (i.e., $\det P \neq 0$). The necessity of this condition is obvious since perfect control ($E = \tilde{E} = 0$) is impossible for plants with RHP zeros. Also, because of the particular norm bounded uncertainty description, RHP zeros can only arise from LHP zeros crossing the imaginary axis. Therefore, checking for zeros on the imaginary axis is sufficient to guarantee that there are no plants with RHP zeros in the set Π_A.

A general condition for robust stability for the set Π_A follows from (11.1–9) with M defined by (11.1–25)

$$\bar{\sigma}(W_1 \tilde{P}^{-1} \tilde{H} W_2) < 1 \qquad (11.1-29)$$

For stable plants (11.1–29) can be satisfied simply by setting $\tilde{H} = 0$ (open loop). Integral action implies "perfect control" at steady state and imposes the performance requirement $\tilde{H}(0) = I$. If (11.1–29) is satisfied for $\omega = 0$ with $\tilde{H}(0) = I$ then a controller with sufficient roll-off ($\tilde{H}$ small enough) can always be found such that the system is robustly stable. Thus we have the following theorem.

Theorem 11.1-3 (Integral Control). *Assume all plants $P \in \Pi_A$ are stable. Then robust stability may be achieved for a system under integral control if and only if*

$$\bar{\sigma}(W_1 \tilde{P}^{-1} W_2) < 1 \qquad \text{for } \omega = 0 \qquad (11.1-30)$$

or for specific choices of weighting matrices

Additive Uncertainty:

$$\bar{\ell}_A(0) < \underline{\sigma}(\tilde{P}(0)) \qquad (11.1-31)$$

Multiplicative Uncertainty:

$$\bar{\ell}_O(0) < 1 \text{ or } \bar{\ell}_I(0) < 1 \qquad (11.1-32)$$

Arbitrary Weights:

$$\det P(0) \neq 0, \qquad \forall P \in \Pi_A \qquad (11.1-33)$$

Proof. Follows directly from Thm. 11.1-2 and Cor. 11.1-4 for $\omega = 0$.   □

Theorem 11.1-3 is the MIMO extension of Cor. 4.3-2. For MIMO systems the requirement that the multiplicative error must not exceed 100% is equivalent to the requirement that the gain matrix must remain nonsingular.

11.2 Robust Stability for Structured Uncertainty

11.2.1 Uncertainty Description

In this section, we will describe the uncertainty in a "structured" manner by identifying the sources and locations of uncertainty in the system. Usually, this leads to an uncertainty description with multiple perturbations (Δ_i). These perturbations may correspond to uncertainty in the model parameters, uncertainty with respect to the manipulated variables (input or actuator uncertainty) and the outputs (measurement uncertainty), etc. By using such a mechanistic approach, we can norm-bound each perturbation (e.g., $\bar{\sigma}(\Delta_i) \leq 1$) without introducing much conservativeness and get a "tight" description of the uncertainty set.

However, we should not necessarily describe the uncertainty as rigorously as possible. Rather, we should take an "engineering approach" and describe the uncertainty only as rigorously as necessary. This means, for example, that some sources of uncertainty (occurring at different places in the system) should be lumped into an "unstructured" multiplicative perturbation, if this does not add much conservativeness. This leads to a *practical uncertainty description*: some sources of uncertainty are described in a "structured" manner (e.g., parametric uncertainty), while the rest (usually uncertain high-frequency dynamics) is lumped into a single "unstructured" perturbation. This will be illustrated through an example later.

Consider the uncertainty as perturbations on the nominal system. Each perturbation Δ_i is assumed to be a *norm-bounded* transfer matrix

$$\bar{\sigma}(\Delta_i) \leq 1 \qquad \forall \omega \qquad (11.2-1)$$

Weighting matrices are used to normalize the uncertainty such that the bound is unity at all frequencies; that is, the actual perturbation L_i is

$$L_i = W_2 \Delta_i W_1 \qquad (11.2-2)$$

If Δ_i represents a real parameter variation we may restrict Δ_i to be real, but in general Δ_i may be any rational transfer matrix satisfying (11.2-1). Just like in Sec. 11.1 the choice of the singular value $\bar{\sigma}$ as the norm for bounding Δ_i is not arbitrary, but is needed to obtain the necessity in the theorems which follow.

The perturbations (uncertainties) which may occur at different places in the feedback system can be collected and placed into one large block diagonal perturbation matrix

$$\Delta = \text{diag}\{\Delta_1, \ldots, \Delta_m\} \qquad (11.2-3)$$

which satisfies

$$\bar{\sigma}(\Delta) \leq 1 \qquad \forall \omega \qquad\qquad (11.2-4)$$

The blocks Δ_i in (11.2–3) can have any size and may also be repeated. For example, repetition can be needed in order to handle correlations between the uncertainties in different elements. The nominal closed loop system with no uncertainty ($\Delta = 0$) is assumed to be stable. The perturbations (uncertainty) give rise to stability problems because of the additional feedback paths created by the uncertainty. This is shown explicitly by writing the uncertainty as perturbations on the nominal system in the form shown in Fig. 11.1-2. M is the nominal closed-loop system "as seen from" the various uncertainties, and is stable since the nominal system is assumed stable. More precisely, M is the *interconnection matrix*, the nominal transfer function from the output of the perturbations Δ_i to their inputs. Constructing M is conceptually straightforward, but may be tedious for specific problems. Many practical problems can be cast into the $M - \Delta$ form shown in Fig. 11.1-2 as we will demonstrate through examples. Indeed such a transformation is always possible when the plant is a *linear fractional transformation* of the Δ_i's.

In analogy to the well known scalar case P is a linear fractional transformation of Δ when it is of the form

$$P = N_{11} + N_{12}\Delta(I - N_{22}\Delta)^{-1}N_{21}$$

$$= N_{11} + N_{12}(I - \Delta N_{22})^{-1}\Delta N_{21}$$

where the N_{ij}'s are matrices of appropriate dimension which do not involve Δ and Δ is block diagonal.

11.2.2   Structured Singular Value

Let X_ν be the set of all complex perturbations with a specific block diagonal structure and spectral norm less than ν:

$$X_\nu = \{\Delta = \mathrm{diag}\{\Delta_1, \Delta_2, \ldots \Delta_m\}|\bar{\sigma}(\Delta_i) \leq \nu\} \qquad (11.2-5)$$

By following the steps in the proof of Thm. 11.1-1 but with $\Delta \in X_\nu$, it can easily be shown that robust stability is guaranteed if and only if

$$\det(I - M\Delta) \neq 0 \qquad \forall \Delta \in X_\nu \qquad\qquad (11.2-6)$$

$$\Leftrightarrow \rho(M\Delta) < 1 \qquad \forall \Delta \in X_\nu \qquad\qquad (11.2-7)$$

or

$$\Leftarrow \nu < \bar{\sigma}^{-1}(M) \tag{11.2 - 8}$$

Note that (11.2–8) is only a sufficient condition for (11.2–6). When we proved necessity of the similar condition (11.1–9) we made use of the fact that the perturbation set includes *all* $\Delta(\bar{\sigma}(\Delta) \le 1)$. Here, however, we restrict the set of permissible Δ's to X_ν. In general (11.2–8) can be arbitrarily conservative. Therefore as an alternative to $\bar{\sigma}$ let us define the structured singular value which takes into account the structure of the perturbation Δ.

Definition 11.2-1. *The function $\mu(M)$, called the Structured Singular Value (SSV) is defined such that $\mu^{-1}(M)$ is equal to the smallest $\bar{\sigma}(\Delta)$ needed to make $(I - M\Delta)$ singular — i.e.*

$$\mu^{-1}(M) = \min_\nu \{\nu | \det(I - M\Delta) = 0 \text{ for some } \Delta \in X_\nu\} \tag{11.2 - 9}$$

If no Δ exists such that $\det(I - M\Delta) = 0$, then $\mu(M) = 0$.

Condition (11.2–6) and Def. 11.2-1 yield the following theorem for robust stability.

Theorem 11.2-1. *Assume that the nominal system M is stable and that the perturbation Δ is of such a kind that the perturbed closed-loop system is stable if and only if the map of the Nyquist D contour under $\det(I - M\Delta)$ does not encircle the origin. Then the closed-loop system in Fig. 11.1-2 is stable for all perturbations $\Delta \in X_{\nu=1}$ if and only if*

$$\mu(M(i\omega)) < 1 \qquad \forall \omega \tag{11.2 - 10}$$

Theorem 11.2-1 may be interpreted as a "generalized small gain theorem" which also takes into account the *structure* of Δ. The SSV is defined to obtain the tightest possible bound on M such that (11.2–6) is satisfied. It is important to note that $\mu(M)$ depends *both* on the matrix M *and* on the *structure* of the perturbation Δ. $\mu(M)$ is a generalization of the spectral radius $\rho(M)$ and maximum singular value $\bar{\sigma}(M)$: let the perturbations be of the form

$$X_1 = \{\Delta | \Delta = \delta I, |\delta| \le 1\}$$

Then it is easy to show that $\mu(M) = \rho(M)$. If the perturbations are unstructured (Δ is a full matrix) then $\mu(M) = \bar{\sigma}(M)$ as we know from Thm. 11.1-1.

The definition of μ may be extended by restricting Δ to a smaller set — e.g., real Δ_i's or several identical Δ_i's ("repeated Δ's"). A detailed discussion of these issues is beyond the scope of this book.

Definition 11.2-1 is not in itself useful for computing μ since the optimization problem implied by it does not appear to be easily solvable. Fortunately, several properties of μ can be proven which make it a powerful tool for applications.

Properties of μ

1. $\mu(\alpha M) = |\alpha|\mu(M)$, where α is a scalar.

2. From the discussion above we conclude

$$\rho(M) \leq \mu(M) \leq \bar{\sigma}(M) \qquad\qquad (11.2-11)$$

3. Let $\mathcal{U}$ be the set of all unitary matrices with the same block diagonal structure as Δ. If $U \in \mathcal{U}$ and $\Delta \in X$, then $U\Delta \in X$ and $\mu(MU) = \mu(M)$. Therefore from (11.2–11)

$$\rho(MU) \leq \mu(M) \qquad \forall U \in \mathcal{U} \qquad\qquad (11.2-12)$$

Indeed it can be shown that

$$\max_{U \in \mathcal{U}} \rho(MU) = \mu(M) \qquad\qquad (11.2-13)$$

This optimization problem is not convex.

4. Let $\mathcal{D}$ be the set of real positive diagonal matrices $D = \text{diag}\,\{d_i I_i\}$ where the size of each block (i.e., the size of I_i) is equal to the size of the blocks Δ_i. If $D \in \mathcal{D}$ and $\Delta \in X$, then $D\Delta D^{-1} \in X$ and $\mu(DMD^{-1}) = \mu(M)$. Therefore from (11.2–11)

$$\mu(M) \leq \bar{\sigma}(DMD^{-1}) \qquad \forall D \in \mathcal{D} \qquad\qquad (11.2-14)$$

which suggests to determine an upper bound of $\mu(M)$ from

$$\mu(M) \leq \inf_{D \in \mathcal{D}} \bar{\sigma}(DMD^{-1}) \qquad\qquad (11.2-15)$$

It can be shown that the optimization problem is convex and that equality is reached in (11.2-15) for three or fewer blocks. Numerical evidence suggests that the bound (11.2-15) is tight for four or more blocks.

Extensive numerical experimentation has shown that the minimization of $\|DMD^{-1}\|_F$ yields very good approximations for the optimal D which minimizes $\bar{\sigma}(DMD^{-1})$. This is theoretically justified from the property (10.1–20):

$$\frac{1}{\sqrt{n}}\|A\|_F \le \bar{\sigma}(A) \le \|A\|_F$$

where n is the dimension of A. Clearly a significant reduction in $\|DMD^{-1}\|_F$ will result in a significant reduction of $\bar{\sigma}(DMD^{-1})$. Hence the D that minimizes $\|DMD^{-1}\|_F$ is usually a very good approximation of the D that minimizes $\bar{\sigma}(DMD^{-1})$.

Minimization of $\|DMD^{-1}\|_F$. Let m be the number of blocks in Δ. One of the scalars can be kept constant ($d_m = 1$) without loss of generality. Obtain the optimal $d_1, \ldots, d_{m-1}$ as follows: for a specific j ($1 \le j \le m-1$) partition

$$D = \text{diag} \left\{ D_j^a, d_j I_j, D_j^c \right\}$$

$$M = \begin{pmatrix} M_j^{a_1} & M_j^{a_2} & M_j^{a_3} \\ M_j^{b_1} & M_j^{b_2} & M_j^{b_3} \\ M_j^{c_1} & M_j^{c_2} & M_j^{c_3} \end{pmatrix}$$

Then

$$DMD^{-1} = \begin{pmatrix} D_j^a M_j^{a_1}(D_j^a)^{-1} & D_j^a M_j^{a_2} d_j^{-1} & D_j^a M_j^{a_3}(D_j^c)^{-1} \\ d_j M_j^{b_1}(D_j^a)^{-1} & M_j^{b_2} & d_j M_j^{b_3}(D_j^c)^{-1} \\ D_j^c M_j^{c_1}(D_j^a)^{-1} & D_j^c M_j^{c_2} d_j^{-1} & D_j^c M_j^{c_3}(D_j^c)^{-1} \end{pmatrix}$$

$$\|DMD^{-1}\|_F^2 = \|D_j^a M_j^{a_1}(D_j^a)^{-1}\|_F^2 + \|D_j^a M_j^{a_3}(D_j^c)^{-1}\|_F^2 + \|M_j^{b_2}\|_F^2$$

$$+ \|D_j^c M_j^{c_1}(D_j^a)^{-1}\|_F^2 + \|D_j^c M_j^{c_3}(D_j^c)^{-1}\|_F^2$$

$$+ d_j^2 \left[\|M_j^{b_1}(D_j^a)^{-1}\|_F^2 + \|M_j^{b_3}(D_j^c)^{-1}\|_F^2 \right]$$

$$+ d_j^{-2} \left[\|D_j^a M_j^{a_2}\|_F^2 + \|D_j^c M_j^{c_2}\|_F^2 \right]$$

$$\triangleq \alpha_j^4 + d_j^2 \beta_j^4 + d_j^{-2} \gamma_j^4$$

where $\alpha_j, \beta_j, \gamma_j$ are positive real numbers, independent of d_j.

The optimal D is determined iteratively. Start with some initial guesses for $d_1, \ldots, d_{m-1}$, e.g., $D = I$ or D equal to the optimal D for the previous ω that was considered and set $k = 0$.

For iteration k:

$$j := 1 + \mod(k, (m - 1))$$

$$k := k + 1$$

$$d_j := \gamma_j / \beta_j$$

This procedure converges rapidly.

Example 11.2-1. We continue the distillation column example of Sec. 11.1.6. The input uncertainty is expressed through (11.1-21) or equivalently (11.1–22) where the perturbation matrix Δ_I is *diagonal*. The interconnection matrix $M = w_I(s)\tilde{H}_I$ and from Thm. 11.2-1 the system is robustly stable if and only if

$$\mu(\tilde{H}_I) < |w_I(i\omega)|^{-1} = \bar{\ell}_I^{-1}(\omega) \qquad \forall \omega \qquad (11.2 - 16)$$

where $\mu(\tilde{H}_I)$ is computed with respect to the diagonal matrix Δ_I. From Fig. 11.1-3 we see that (11.2–16) is satisfied and robust stability is guaranteed with the controller (11.1–19). □

11.2.3 Simultaneous Multiplicative Input and Output Uncertainty

Consider the system in Fig. 11.2-1 with both multiplicative input and output uncertainty. The possible plants are given by

$$P = (I + L_O)\tilde{P}(I + L_I) \qquad (11.2 - 17a)$$

$$L_I = W_{2I}\Delta_I W_{1I}, \qquad \bar{\sigma}(\Delta_I) \leq 1 \qquad \forall \omega \qquad (11.2 - 17b)$$

$$L_O = W_{2O}\Delta_O W_{1O}, \qquad \bar{\sigma}(\Delta_O) \leq 1 \qquad \forall \omega \qquad (11.2 - 17c)$$

The reader should verify that the plant is a linear fractional transformation of the uncertainty $\Delta = \mathrm{diag}\{\Delta_O, \Delta_I\}$:

$$P = N_{11} + N_{12}\Delta(I - N_{22}\Delta)^{-1}N_{21}$$

where

$$N_{11} = \tilde{P}$$
$$N_{12} = (W_{1O} \quad \tilde{P}W_{1I})$$
$$N_{21} = \begin{pmatrix} W_{2O}\tilde{P} \\ W_{2I} \end{pmatrix}$$
$$N_{22} = \begin{pmatrix} 0 & W_{2O}\tilde{P}W_{1I} \\ 0 & 0 \end{pmatrix}$$

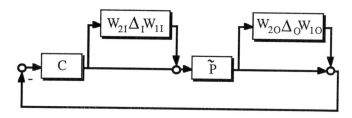

Figure 11.2-1. System with weighted multiplicative input and output uncertainty.

The perturbation block Δ_I represents the multiplicative input uncertainty. If its source is uncertainty in the manipulated variables, then

$$\Delta_I : \text{ diagonal}, \quad W_{1I} = \text{diag}\{w_{Ii}\}, \quad W_{2I} = I \qquad (11.2-18)$$

where w_{Ii} represents the relative uncertainty of each manipulated input.

The block Δ_O represents the multiplicative output uncertainty. If its source is uncertainty or neglected deadtimes involved in one or more of the measurements, then

$$\Delta_O : \text{ diagonal}, \quad W_{1O} = \text{diag}\{w_{Oi}\}, \quad W_{2O} = I \qquad (11.2-19)$$

where w_{Oi} represents the relative uncertainty for each measurement. These sources of input and output uncertainty are present in any plant. Δ_I and Δ_O are restricted to be *diagonal* matrices, since there is little reason to assume that the actuators or measurements influence each other. However, some of the unmodelled dynamics of the plant $\tilde{P}$ itself, which has cross terms, may be approximated by lumping them into Δ_I and Δ_O, thus making either one of them a full matrix.

To examine the constraints on the nominal system imposed by the robust stability requirement for this uncertainty description, let $\Delta = \text{diag}\{\Delta_I, \Delta_O\}$ and rearrange the system in Fig. 11.2-1 into the form in Fig. 11.1-2. The interconnection matrix M becomes:

$$M = \begin{bmatrix} -W_{1I}C\tilde{P}(I+C\tilde{P})^{-1}W_{2I} & -W_{1I}C(I+\tilde{P}C)^{-1}W_{2O} \\ W_{1O}\tilde{P}(I+C\tilde{P})^{-1}W_{2I} & -W_{1O}\tilde{P}C(I+\tilde{P}C)^{-1}W_{2O} \end{bmatrix}$$

$$= \begin{bmatrix} W_{1I} & 0 \\ 0 & W_{1O} \end{bmatrix} \begin{bmatrix} -\tilde{P}^{-1}\tilde{H}\tilde{P} & -\tilde{P}^{-1}\tilde{H} \\ \tilde{E}\tilde{P} & -\tilde{H} \end{bmatrix} \begin{bmatrix} W_{2I} & 0 \\ 0 & W_{2O} \end{bmatrix} \qquad (11.2-20)$$

and robust stability is guaranteed for all Δ such that $\bar{\sigma}(\Delta) < 1$ if and only if $\mu(M) \leq 1, \forall \omega$. μ is computed with respect to the structure of Δ which in turn depends on the structure assumed for Δ_I and Δ_O. Note that (11.1-11) and (11.1-13) follow as special cases when the weights are assumed to be scalar, Δ_I and Δ_O are full matrices and either $\Delta_I = 0$ or $\Delta_O = 0$.

11.2.4 Batch Reactor: Simultaneous Parametric and Unstructured Uncertainty

Consider a perfectly mixed batch reactor where an exothermic reaction is taking place. The reaction temperature T is controlled by the temperature T_c of the fluid in the cooling jacket (the fluid in the cooling jacket may be boiling, and T_c may be adjusted by changing the pressure). A heat balance for the batch reactor gives

$$C_p \frac{dT}{dt} = (-\Delta H_r)r - UA(T - T_c) \qquad (11.2-21)$$

where

T	reactor temperature (K)
T_c	coolant temperature (K)
r	reaction rate (function of T) (mol/s)
ΔH_r	heat of reaction (negative constant) (J/mol)
C_p	total heat capacity of fluid in reactor (J/K)
UA	overall heat transfer coefficient (J/sK)

Linearizing the reaction rate at the operating point T^0

$$r = r^0 + k_T(T - T^0)$$

results in a linear transfer function from T_c to T

$$T(s) = \frac{UA/C_p}{s+a} T_c(s) \qquad (11.2-22)$$

where

$$a = \frac{UA - (-\Delta H_r)k_T}{C_p} \qquad (11.2-23)$$

Two sources of uncertainty will be considered for the linear model (11.2–22): the effect of nonlinearity expressed as uncertainty in the pole location a and neglected high-frequency dynamics.

Pole Uncertainty (Δ_E). Most of the terms in (11.2-23) are nearly constant, except $k_T = \partial r/\partial T$ which is a strong function of temperature. From (11.2-23) we see that the reactor may be open-loop stable ($a > 0$) at low temperatures where k_T is small, and unstable at high temperatures where the reactor is more temperature sensitive. To describe the effect of temperature on a, let

$$|a - \tilde{a}| \le r_a \tilde{a}$$

where $\tilde{a}$ is the nominal pole location and r_a the relative "uncertainty" of the real constant a. If $r_a > 1$ the plant may be stable or unstable. Equivalently, the possible a's may be written in terms of a norm-bounded perturbation Δ_E

$$a = \tilde{a}(1 + r_a \Delta_E), \qquad |\Delta_E| \le 1, \qquad \Delta_E \quad \text{real} \qquad (11.2-24)$$

This uncertainty may be expressed as an inverse multiplicative perturbation $(I + w_E \Delta_E)^{-1}$ on the plant

$$\frac{1}{s+a} = \frac{1}{s+\tilde{a}} \cdot \frac{1}{1 + w_E(s)\Delta_E}, \qquad w_E(s) = \frac{r_a}{1 + s/\tilde{a}} \qquad (11.2-25)$$

Neglected Dynamics (Δ_O). Uncertainty in the high frequency dynamics cannot be modelled in a "structured" manner using parametric uncertainty. It is most conveniently expressed as multiplicative uncertainty, for example output multiplicative uncertainty ($I + w_O \Delta_O$). Physically, this uncertainty may account for neglected (and unknown) dynamics for changing the cooling temperature T_c (if T_c is manipulated indirectly with pressure), neglected actuator dynamics (the valve used to control pressure) and neglected dynamics introduced by the heat capacity of the walls.

The following considerations can assist in arriving at a choice for w_O. Naturally $|w_O|$ should be small at low frequencies and increase with frequency. One could view the neglected dynamics as an unknown delay with upper bound $\bar{\delta}$. This would lead to a multiplicative uncertainty of the form (4.6–4) which in turn can be approximated by $w_O = 2\bar{\delta}s(\bar{\delta}s + 2)^{-1}$ (see Fig. 4.4-2).

A block diagram representation of the uncertainty is depicted in Fig. 11.2-2. Note that in general both blocks (Δ_E and Δ_O) are needed: We *cannot* lump the pole uncertainty (Δ_E) into the output uncertainty (Δ_O) if the pole is allowed to cross the imaginary axis. This would result in $|w_O(j\omega)| \to \infty$ at $\omega = 0$. Similarly we *cannot* lump the output uncertainty into the pole uncertainty. The reason is that the inverse multiplicative uncertainty description (Δ_E) cannot be used to model neglected or uncertain RHP zeros (this would require an unstable perturbation Δ_E). It is therefore not suited for handling neglected high-frequency dynamics which most certainly include RHP zeros.

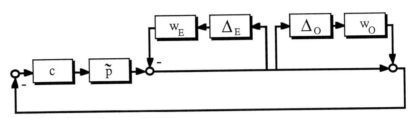

Figure 11.2-2. Reactor control loop with parametric uncertainty represented as inverse multiplicative output uncertainty Δ_E and unstructured multiplicative output uncertainty Δ_O.

Combining the two scalar perturbations into one block perturbation $\Delta = \text{diag}\,\{\Delta_E, \Delta_O\}$ and rearranging Fig. 11.2-2 to match Fig. 11.1-2 yields the following interconnection matrix:

$$M = \begin{bmatrix} w_E \tilde{\epsilon} & -w_O \tilde{\eta} \\ w_E \tilde{\epsilon} & -w_O \tilde{\eta} \end{bmatrix} \qquad (11.2-26)$$

If, in addition to the real Δ_E's, all *complex* Δ_E's with $|\Delta_E| < 1$ are considered possible, then robust stability is guaranteed if and only if $\mu(M) < 1$ or using Def. 11.2-1 if and only if

$$|w_E \tilde{\epsilon}| + |w_O \tilde{\eta}| < 1 \qquad (11.2-27)$$

Because of the identity $\tilde{\eta} + \tilde{\epsilon} = 1$, this bound is *impossible* to satisfy if $|w_E|$ and $|w_O|$ are both "large" (that is, close to one or larger) over the same frequency range. For $r_a > 1$ the pole may cross the imaginary axis, and $|w_E| > 1$ for $\omega < \omega^* = \tilde{a}\sqrt{r_a^2 - 1}$ and $|w_E| < 1$ for $\omega > \omega^*$. In that situation, robust stability is guaranteed only if the unstructured relative uncertainty given in terms of $|w_O(j\omega)|$ reaches one at a frequency *higher* than ω^*.

If pole uncertainty were the only source of uncertainty ($w_O = 0$), the robust stability bound would be $|\tilde{\epsilon}| < |w_E|^{-1}$. Since the plant is minimum phase, this bound could always be satisfied by increasing the gain and making $\tilde{\epsilon}$ small, regardless of the size of r_a.

In summary, the pole location uncertainty is handled by "tightening" the control at low frequencies. Indeed, $\tilde{\epsilon}$ small ("tight" control) is needed in order to stabilize an unstable plant. To realize robust stability in face of uncertain high-frequency dynamics, however, it is necessary to detune the system and make $\tilde{\eta}$ small ($\tilde{\epsilon} \cong 1$) at frequencies where $w_O(\omega)$ is larger than one. Thus we cannot stabilize an unstable plant if there are RHP-zeros or model uncertainty in the same frequency range as the location of the unstable pole.

The reactor example in this section is meant primarily to illustrate the modelling of uncertainty and the implications of different types of uncertainty on controller tuning. The approximate analysis performed here does not guarantee in any way stability of the nonlinear system (11.2–21).

11.2.5 Independent Uncertainty in the Transfer Matrix Elements

In many cases the uncertainty is most easily described in terms of uncertainties of the individual transfer matrix elements. This kind of uncertainty description may arise from an experimental identification of the system. In general, it is *not* a good representation of the actual sources of uncertainty, but it is included here because it has been proposed in the literature on several occasions.

Let us assume that each element p_{ij} in the plant P is independent, but confined to a disk with radius $a_{ij}(\omega)$ centered at $\tilde{p}_{ij}$ in the Nyquist plane

$$|p_{ij} - \tilde{p}_{ij}| \le a_{ij} \qquad (11.2-28)$$

or equivalently

$$|p_{ij} - \tilde{p}_{ij}| \le r_{ij}|\tilde{p}_{ij}| \qquad (11.2-29)$$

where a_{ij} and r_{ij} are the additive and multiplicative (relative) uncertainty respectively. The main limitations of these uncertainty descriptions is that correlations between the elements cannot be handled, which is potentially *very* conservative. Defining the scalar complex perturbation Δ_{ij} (11.2–28) becomes

$$p_{ij} - \tilde{p}_{ij} = \Delta_{ij}a_{ij}, \qquad |\Delta_{ij}| \le 1 \qquad (11.2-30)$$

or equivalently, in matrix form

$$P - \tilde{P} = \begin{pmatrix} \Delta_{11}a_{11} & \Delta_{12}a_{12} & \cdots \\ \Delta_{21}a_{21} & \cdots & \cdots \\ \cdots & \cdots & \Delta_{nn}a_{nn} \end{pmatrix} \qquad (11.2-31)$$

Introducing weighting matrices W_1 and W_2 it is possible to rewrite (11.2-31) in terms of the "large" diagonal perturbation matrix Δ_e

$$P - \tilde{P} = W_{e2}\Delta_e W_{e1} \qquad (11.2-32)$$

where $W_{e2} \in \mathcal{R}^{n \times n^2}, W_{e1} \in \mathcal{R}^{n^2 \times n}$ and $\Delta_e \in \mathcal{C}^{n^2 \times n^2}$ are defined as

$$W_{e2} = (I \quad I \dots I), \quad W_{e1} = \begin{pmatrix} a_1 & & & \\ & a_2 & & \\ & & \ddots & \\ & & & a_n \end{pmatrix}, \quad a_i = \begin{pmatrix} a_{1i} \\ a_{2i} \\ \vdots \\ a_{ni} \end{pmatrix} \qquad (11.2-33)$$

$$\Delta_e = \text{diag}\left\{\Delta_{11}, \Delta_{21}, \dots, \Delta_{nn}\right\}, \qquad |\Delta_{ij}| \leq 1$$

A block diagram representation of (11.2–32) is given in Fig. 11.1-4 with $W_2 = W_{e2}$ and $W_1 = W_{e1}$. The interconnection matrix M (Fig. 11.1-2) is $M = -W_{e1}C(I + \tilde{P}C)^{-1}W_{e2} = -W_{e1}\tilde{P}^{-1}\tilde{H}W_{e2}$. Thus we have robust stability for the uncertainty (11.2–30) if and only if

$$\mu(W_{e1}\tilde{P}^{-1}\tilde{H}W_{e2}) < 1 \qquad \forall \omega \qquad (11.2-34)$$

For the special case $\tilde{H} = \tilde{\eta}I$, (11.2-34) provides an explicit bound on the complementary sensitivity

$$\bar{\sigma}(\tilde{H}) = |\tilde{\eta}| < \mu^{-1}(W_{e1}\tilde{P}^{-1}W_{e2}) \qquad \forall \omega \qquad (11.2-35)$$

11.2.6 Condition Number and Relative Gain Array as Sensitivity Measures

We would like to learn for what *class of systems* independent element uncertainty as discussed in the preceding section, imposes severe constraints on the complementary sensitivity [(11.2–34) and (11.2–35)]. We can then determine for each system *a priori* if a detailed analysis of independent element uncertainty and its effect on robust stability is justified. The proofs of all results in this section are omitted because they are straightforward but tedious.

Theorem 11.2-2. (Condition number criterion.) *Assume the nominal response is decoupled, $\tilde{H} = \text{diag}\{\tilde{\eta}_i\}$. Under the assumption of Thm. 11.2-1 robust stability is guaranteed for element uncertainty (11.2-29) if*

$$|\tilde{\eta}_i| < \frac{1}{r_{max}\sqrt{n}\kappa^*(\tilde{P})} \qquad \forall \omega, \quad \forall i \qquad (11.2-36)$$

Here we have used the following definitions:

Maximum relative uncertainty:

$$r_{max} = \max_{ij} r_{ij} \qquad (11.2-37)$$

Minimized condition number:

$$\kappa^*(G) = \min_{D_1, D_2} \kappa(D_1 G D_2) \qquad (11.2-38)$$

where D_1 and D_1 are real diagonal matrices. Note that because r_{ij} and r_{max} are independent of the scaling of the system inputs and outputs and because κ^* is obtained by minimizing over all scaling matrices, the inequality (11.2–36) is *scaling invariant*. It indicates that for systems with high condition number κ^* only small errors r_{ij} are allowed. Otherwise robust stability cannot be guaranteed. Because (11.2–36) is only sufficient, a comparison with the necessary and sufficient condition (11.2–35) yields

$$\mu(W_{e1} \tilde{P}^{-1} W_{e2}) \leq r_{max} \sqrt{n} \kappa^*(\tilde{P}) \qquad (11.2-39)$$

Numerical experience suggests that this inequality is quite tight. An exact condition is available for 2×2 systems at steady state, as we will show next.

The uncertainty description (11.2–30) assumes that Δ_{ij} are *complex* scalars. This may be reasonable at non-zero frequencies, but does not make any physical sense at steady state ($\omega = 0$) where $\tilde{P}, P$ and Δ_{ij} must be *real*. Conditions (11.2–35) and (11.2–36) may therefore be conservative at $\omega = 0$ where complex perturbations cannot occur. If all perturbations are real and all bounds are equal ($r_{ij} = r \quad \forall i, j$) we find for 2×2 systems:

$$\mu_{real}(W_{e1} \tilde{P}^{-1} W_{e2}) = r \kappa^*(\tilde{P}) \quad (\omega = 0) \qquad (11.2-40)$$

We know from Thm. 11.1-3 that for robust integral control for stable systems it is necessary and sufficient to bound the steady state uncertainty. Thus we have the following theorem.

Theorem 11.2-3 (2×2 systems). *Assume that the uncertainties of the elements in $\tilde{P}(0)$ are independent and real and have equal relative magnitude bounds r. Then for open loop stable systems, robust stability and integral control may be achieved if and only if*

$$\kappa^*(\tilde{P}(0)) < r^{-1} \qquad (11.2-41)$$

If the magnitude bounds on the relative uncertainties are not equal, and r is replaced by r_{max}, then (11.2–41) provides a sufficient condition for robust stability and integral control. A comparison with (11.1–33) indicates that (11.2–41) implies nonsingularity at steady state.

Theorems 11.2-2 and 11.2-3 give clear interpretations of the minimized condition number as a sensitivity measure: $\kappa^*(\tilde{P}(0))$ and $\kappa^*(\tilde{P}(j\omega))$ are good measures of sensitivity only if the plant uncertainties are given in terms of *independent* (uncorrelated) norm-bounded elements with *equal relative error bounds*. For other uncertainty structures the minimized condition number may be misleading, and bounds on the uncertainties such as (11.2–41) may be arbitrarily conservative. This will be illustrated by a subsequent example.

Conditions (11.2–36) and (11.2–41) provide some insight into the effects of plant ill-conditioning but from a numerical point of view they are hardly more convenient than the general condition (11.2–35) because they involve nonconvex optimization problems. Fortunately, accurate bounds on κ^* can be obtained from the Relative Gain Array.

Relative Gain Array (RGA). The RGA Λ of a matrix M is defined as

$$\Lambda(M) = M \times (M^{-1})^T \qquad (11.2-42)$$

where $\times$ denotes the element-by-element (Schur) product. If M is a transfer matrix then $\Lambda(M)$ is a function of frequency. It can be easily shown that $\Lambda(M)$ has the following properties: all rows and columns of Λ sum to one

$$\sum_i \lambda_{ij} = \sum_j \lambda_{ij} = 1 \qquad (11.2-43)$$

and Λ is independent of scaling

$$\Lambda(D_1 M D_2) = \Lambda(M) \qquad (11.2-44)$$

where D_1 and D_2 are arbitrary nonsingular diagonal matrices. Also a permutation of rows (columns) of M leads to the same permutation of rows (columns) of $\Lambda(M)$. When the argument M is omitted in $\Lambda(M)$ we generally mean the RGA of the plant, i.e., $\Lambda = \Lambda(P)$. When we speak of a "system M with a large RGA" we mean that some norm of $\Lambda(M)$ is large.

The following inequalities show that plants with large elements in the RGA are *always* ill-conditioned

$$\kappa(P) \geq \kappa^*(P) \geq \|\Lambda\|_m - 1/\kappa^*(P) \geq \|\Lambda\|_m - 1 \qquad (11.2-45)$$

where

$$\|\Lambda\|_m = 2 \cdot \max \{\|\Lambda\|_1, \|\Lambda\|_\infty\} \qquad (11.2-46)$$

Vice versa, a large value of $\kappa^*(P)$ implies large elements in the RGA. At least for 2×2 systems we have

$$\kappa^*(P) \leq \|\Lambda\|_m \qquad (11.2-47)$$

and it is conjectured that a similar inequality holds for larger systems. We can combine (11.2–45) and (11.2–47) to show the close relationship between κ^* and Λ at least for 2×2 systems

$$\|\Lambda\|_m - \frac{1}{\kappa^*(P)} \leq \kappa^*(P) \leq \|\Lambda\|_m \qquad (11.2-48)$$

Example 11.2-2. Let us examine again the distillation column introduced in Sec. 11.1.6 but with reflux L and boilup V as manipulated inputs. The steady-state gain matrix is

$$\tilde{P}(0) = \begin{pmatrix} 0.878 & -0.864 \\ 1.082 & -1.096 \end{pmatrix} \qquad (11.2-49)$$

and

$$\lambda_{11} = 35.07, \quad \|\Lambda\|_m = 138.275, \quad \kappa^*(\tilde{P}) = 138.268, \quad \kappa(\tilde{P}) = 141.7$$

From the high condition number $\kappa^*(\tilde{P})$, one might conclude that the plant may become singular for very small perturbations. This would be true if the uncertainty had the form of independent element errors, but not necessarily otherwise. To illustrate this point consider conditions for using integral control $(\tilde{H}(0) = I)$ under two different assumptions about the uncertainty.

Case 1. The elements are assumed independent and norm bounded with equal relative error r. Theorem 11.1-3 and (11.2–35) imply that robust stability with integral control may be achieved if and only if $\mu(W_{e1}\tilde{P}^{-1}W_{e2}) < 1$ for $\omega = 0$, where μ is computed with respect to the *real* perturbation matrix Δ_e.

$$W_{e2} = \begin{pmatrix} 1 & 0 & 1 & 0 \\ 0 & 1 & 0 & 1 \end{pmatrix}, \quad W_{e1} = r \begin{pmatrix} 0.878 & 0 \\ 1.082 & 0 \\ 0 & 0.864 \\ 0 & 1.096 \end{pmatrix}$$

$$W_{e1}\tilde{P}^{-1}W_{e2} = r \begin{pmatrix} 35.07 & -27.65 & 35.07 & -27.65 \\ 34.07 & -27.65 & 34.07 & -27.65 \\ 43.22 & -34.07 & 43.22 & -34.07 \\ 43.22 & -35.07 & 43.22 & -35.07 \end{pmatrix}$$

This gives $\mu_{real}(W_{e1}\tilde{P}^{-1}W_{e2}) = 138.268r$ which is equal to $r\kappa^*(P)$ as expected from (11.2–40). From Thm. 11.2-3 robust stability with integral action is possible if and only if $r < \kappa^*(P)^{-1} = 0.0072$. In practice, the variation in each element (mainly due to nonlinearities) is much larger than 0.7%, and integral control does not seem to be possible for this distillation column according to this analysis.

Case 2. A more realistic uncertainty description for this high purity distillation column is the following additive uncertainty

$$P - \tilde{P} = \begin{pmatrix} d & -d \\ -d & d \end{pmatrix}$$

which may be written as in Fig. 11.1-4 in terms of one real scalar Δ-block with

$$P - \tilde{P} = W_2 \Delta W_1, \quad W_2 = |d| \begin{pmatrix} 1 \\ -1 \end{pmatrix}, \quad W_1 = (1 \quad -1), \quad |\Delta| \le 1$$

This structure of the uncertainty arises from the material balance constraints which cannot be violated. Using Thm. 11.1-3, robust stability and integral control ($\tilde{H}(0) = I$) are possible if and only if $\bar{\sigma}(W_1 \tilde{P}^{-1} W_2) < 1$ for $\omega = 0$. Here $W_1 \tilde{P}^{-1} W_2 = 0 \cdot |d|$ and therefore robust stability and integral control are possible for *any value of d* and the elements may even change sign without causing stability problems. Thus, despite the high condition number, the system is not at all sensitive to this physically-motivated model error. □

11.3   Robust Performance

11.3.1 H_∞-Performance Objective

We require that the performance objective defined in Sec. 10.4.4 be satisfied for all plants P in the uncertainty set Π

$$\|W_2 E(P) W_1\|_\infty = \sup_\omega \bar{\sigma}(W_2 E(P) W_1) < 1 \quad \forall P \in \Pi \qquad (11.3-1)$$

Note that W_1 and W_2 are the *performance weights* and are entirely unrelated to the *uncertainty weights* in Sec. 11.2 for which the same symbols were used. In order to be able to evaluate (11.3–1) we assume the uncertainty to be norm bounded and of the form introduced in Sec. 11.2.1. Thus after appropriate scaling it can be expressed as a block diagonal matrix

$$\Delta_u = \text{diag}\{\Delta_1, \ldots, \Delta_m\} \qquad (11.3-2)$$

which satisfies

$$\bar{\sigma}(\Delta_u) \le 1 \quad \forall \omega \qquad (11.3-3)$$

(Here the subscript u stands for "uncertainty.") In a procedure similar to the one for constructing M in Sec. 11.2.1 we can construct the matrix G shown in Fig. 11.3-1A. The input vector consists of the outputs from the uncertainty block Δ_u and the normalized inputs v'. The output vector is formed by the inputs to the

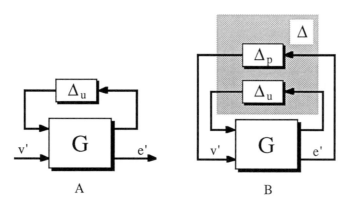

Figure 11.3-1. Block diagram structure for checking robust performance. Full perturbation matrix Δ_p in (B) leads to robust performance test via SSV.

uncertainty block Δ_u and the normalized outputs e'. If we partition G into four blocks consistent with the dimensions of the two input and two output vectors we can identify G_{11} as the matrix M shown in Fig. 11.1-2 and G_{22} as the weighted *nominal* sensitivity function $W_2 E(\tilde{P}) W_1$.

The robust performance objective (11.3–1) can now be expressed in terms of G.

$$\|F(G, \Delta_u)\|_\infty = \sup_\omega \bar{\sigma}(F(G, \Delta_u)) < 1 \qquad (11.3-4)$$

where the transfer matrix from v' to e'

$$e' = F(G, \Delta_u) v' \qquad (11.3-5)$$

is described by the *Linear Fractional Transformation* (LFT)

$$F(G, \Delta_u) = G_{22} + G_{21} \Delta_u (I - G_{11} \Delta_u)^{-1} G_{12} \qquad (11.3-6)$$

Comparing condition (11.1–9) for robust stability and the formally identical condition (11.3–4) for robust performance we conclude: *the system $F(G, \Delta_u)$ satisfies the robust performance condition (11.3–4)* **if and only if** *it is robustly stable for the norm bounded matrix perturbation Δ_p ($\bar{\sigma}(\Delta_p) \leq 1$).* (Here the subscript p stands for "performance.") We have expressed this equivalence between robust performance and robust stability in Fig. 11.3-1B: conditions (11.3–1) and (11.3–4) are satisfied if and only if the system G is robustly stable with respect to the block diagonal perturbation

$$\Delta = \text{diag}\left\{\Delta_u, \Delta_p\right\}, \qquad \bar{\sigma}(\Delta) \le 1 \qquad (11.3-7)$$

Δ_p is generally a full matrix of appropriate dimensions. A necessary and sufficient condition for robust stability in the presence of norm bounded block diagonal perturbations can be expressed in terms of the structured singular value μ (Thm. 11.2-1).

Theorem 11.3-1. *The nominally stable system G (Fig. 11.3-1) subjected to the block diagonal uncertainty Δ_u ($\bar{\sigma}(\Delta_u) \le 1$) satisfies the robust performance condition $\|F(G, \Delta_u)\|_\infty < 1$ if and only if*

$$\mu_\Delta(G) < 1 \qquad \forall \omega \qquad (11.3-8)$$

where μ is computed with respect to the block diagonal perturbation $\Delta = \text{diag}\left\{\Delta_u, \Delta_p\right\}$.

Theorem 11.3-1 is probably the main reason for measuring performance in terms of the ∞-norm and bounding uncertainty in the same manner. It is then possible to express robust performance in terms of robust stability and to test for either one in a *nonconservative* manner by calculating μ. Indeed, if the uncertainty is modeled *exactly* by Δ_u — i.e., if all plants in this norm-bounded set do actually occur in practice, the conditions for robust stability and performance are necessary and sufficient.

Some care is necessary to interpret the robust performance test correctly when $\mu(\omega) = \beta(\omega) > 1$. It means that if each one of the uncertainty blocks is *reduced* by a factor β^{-1} then the *relaxed* performance specification $\bar{\sigma}(W_2 E W_1) \le \beta$ can be met. $\mu > 1$ does *not* give any explicit information on how much the performance violates the specification (11.3–1) for the uncertainty Δ_u.

Because $\Delta_1 = \text{diag}\left\{\Delta_u, 0\right\}$ and $\Delta_2 = \text{diag}\left\{0, \Delta_p\right\}$ are special cases of Δ ($\bar{\sigma}(\Delta) \le 1$) we find

$$\mu_\Delta(G) \ge \max\left\{\mu_{\Delta_u}(G_{11}), \mu_{\Delta_p}(G_{22}) = \bar{\sigma}(G_{22})\right\} \qquad (11.3-9)$$

Inequality (11.3–9) implies that for robust performance ($\mu_\Delta(G) < 1$) it is necessary that the system is robustly stable ($\mu_{\Delta_u}(G_{11}) < 1$) and satisfies the performance specifications in the absence of uncertainty ($\bar{\sigma}(G_{22}) < 1$), which is not very surprising. This suggests that *robust performance* might not always be a very important issue: If both the nominal performance and the robust stability condition are satisfied with some margin then the robust performance condition should also be satisfied. The next two sections will shed some light on this issue.

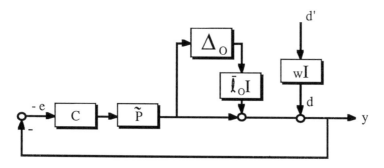

Figure 11.3-2. System with multiplicative output uncertainty and performance weight w.

11.3.2 Multiplicative Output Uncertainty

Consider the robust performance problem for the system depicted in Fig. 11.3-2. The set of plants is described by

$$\Pi = \left\{ P = (I + \bar{\ell}_O \Delta_O)\tilde{P}, \ \bar{\sigma}(\Delta_O) \leq 1 \right\} \qquad (11.3 - 10)$$

The H_∞ performance specification places a bound on the sensitivity operator

$$\bar{\sigma}(Ew) < 1 \qquad \forall \omega, \forall P \in \Pi \qquad (11.3 - 11)$$

where w is a scalar weight. Defining $v' = d' = w^{-1}d$ and $e' = e$ we can put the block diagram in Fig. 11.3-2 into the form shown in Fig. 11.3-1 with

$$G = \begin{pmatrix} -\tilde{H}\bar{\ell}_O & -\tilde{H}w \\ \tilde{E}\bar{\ell}_O & \tilde{E}w \end{pmatrix} \qquad (11.3 - 12)$$

According to Thm. 11.3-1, the robust performance condition (11.3–11) is met if and only if $\mu(G) < 1$ where μ is evaluated with respect to the block diagonal matrix $\Delta = \text{diag}\{\Delta_u, \Delta_p\}$ and Δ_u and Δ_p are full. Alternatively, we can start from (11.3–11) and derive a sufficient condition. Straightforward algebra yields for the multiplicative output uncertainty (11.1–2), (11.1–6)

$$E = \tilde{E}(I + \bar{\ell}_O \Delta_O \tilde{H})^{-1} \qquad (11.3 - 13)$$

We substitute (11.3–13) into (11.3–11)

$$\bar{\sigma}(w\tilde{E}(I + \bar{\ell}_O \Delta_O \tilde{H})^{-1}) < 1$$

$$\Leftarrow \bar{\sigma}(w\tilde{E})\bar{\sigma}(I + \bar{\ell}_0\Delta_O\tilde{H})^{-1} < 1$$

$$\Leftrightarrow \bar{\sigma}(w\tilde{E}) < \underline{\sigma}(I + \bar{\ell}_0\Delta_O\tilde{H})$$

$$\Leftrightarrow \bar{\sigma}(w\tilde{E}) < 1 - \bar{\sigma}(\bar{\ell}_0\Delta_O\tilde{H})$$

$$\Leftrightarrow \bar{\sigma}(w\tilde{E}) + \bar{\sigma}(\bar{\ell}_0\tilde{H}) < 1 \qquad (11.3-14)$$

For the specific case of a multiplicative output uncertainty (11.3–14) is a sufficient condition for robust performance and therefore an upper bound on μ.

$$\mu(G) \le \bar{\sigma}(w\tilde{E}) + \bar{\sigma}(\bar{\ell}_0\tilde{H}) \qquad (11.3-15)$$

A comparison with Thm. 2.6-1 or an examination of the steps leading to (11.3–14) reveals that (11.3–15) is an *equality* for SISO systems. Because the "robust stability term" ($\bar{\sigma}(\bar{\ell}_0\tilde{H})$) and the "nominal performance term" ($\bar{\sigma}(w\tilde{E})$) appear *additively*, robust performance can be easily achieved by satisfying both robust stability and nominal performance with some margin ($\bar{\sigma}(\bar{\ell}_0\tilde{H}) \le \alpha$, $\bar{\sigma}(w\tilde{E}) \le 1 - \alpha$, $\alpha < 1$). This is exact for SISO systems but can be somewhat conservative for MIMO systems. Thus, "robust performance" is not very critical for SISO systems or MIMO systems with multiplicative output uncertainty: an examination of robust stability and nominal performance suffices as an approximate check for robust performance.

11.3.3 Multiplicative Input Uncertainty

Next we study the robust performance problem for the system shown in Fig. 11.3-3. The set of plants is described by

$$\Pi = \left\{ P = \tilde{P}(I + \bar{\ell}_I\Delta_I), \ \bar{\sigma}(\Delta_I) \le 1 \right\} \qquad (11.3-16)$$

The performance specification is again given by (11.3–11). The interconnection matrix G (Fig. 11.3-1A) derived from the block diagram in Fig. 11.3-3 is

$$G = \begin{pmatrix} -\tilde{P}^{-1}\tilde{H}\tilde{P}\bar{\ell}_I & -\tilde{P}^{-1}\tilde{H}w \\ \tilde{E}\tilde{P}\bar{\ell}_I & \tilde{E}w \end{pmatrix} \qquad (11.3-17)$$

and robust performance is guaranteed according to Thm. 11.3-1 if and only if $\mu(G) < 1$ for G defined by (11.3–17) and μ evaluated with respect to the block diagonal matrix $\Delta = \text{diag}\{\Delta_u, \Delta_p\}$ where Δ_u and Δ_p are full. Similarly as in the last section we can start from the requirement (11.3–11). Comparing Fig. 11.3-2 and 11.3-3 we find

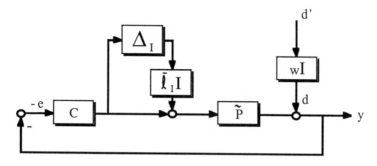

Figure 11.3-3. System with multiplicative input uncertainty and performance weight w.

$$\bar{\ell}_O \Delta_O = \bar{\ell}_I \tilde{P} \Delta_I \tilde{P}^{-1} \qquad (11.3-18)$$

We substitute (11.3–18) into (11.3–13) to obtain the condition for robust performance

$$\bar{\sigma}(w\tilde{E}(I + \bar{\ell}_I \tilde{P} \Delta_I \tilde{P}^{-1} \tilde{H})^{-1}) < 1 \qquad (11.3-19)$$

$$\Leftarrow \bar{\sigma}(w\tilde{E}) < \underline{\sigma}(I + \bar{\ell}_I \tilde{P} \Delta_I \tilde{P}^{-1} \tilde{H})$$

$$\Leftarrow \bar{\sigma}(w\tilde{E}) < 1 - \bar{\sigma}(\bar{\ell}_I \tilde{P} \Delta_I \tilde{P}^{-1} \tilde{H})$$

$$\Leftarrow \bar{\sigma}(w\tilde{E}) + \bar{\sigma}(\tilde{P})\bar{\sigma}(\tilde{P}^{-1})\bar{\sigma}(\bar{\ell}_I \tilde{H}) < 1$$

$$\Leftrightarrow \bar{\sigma}(w\tilde{E}) + \kappa(\tilde{P})\bar{\sigma}(\bar{\ell}_I \tilde{H}) < 1 \qquad (11.3-20)$$

We can rewrite (11.3–19) as

$$\bar{\sigma}(w(I + \bar{\ell}_I \tilde{H} C^{-1} \Delta_I C)^{-1} \tilde{E}) < 1 \qquad (11.3-21)$$

and follow the same steps as above to find

$$\bar{\sigma}(w\tilde{E}) + \kappa(C)\bar{\sigma}(\bar{\ell}_I \tilde{H}) < 1 \qquad (11.3-22)$$

We leave it to the reader to show that conditions similar to (11.3–20) and (11.3–22) but involving the sensitivity and complementary sensitivity (11.1–14) at the plant input can be derived, which are sufficient for robust performance.

$$\kappa(\tilde{P})\bar{\sigma}(w\tilde{E}_I) + \bar{\sigma}(\bar{\ell}_I\tilde{H}_I) < 1 \qquad (11.3-23)$$

$$\kappa(C)\bar{\sigma}(w\tilde{E}_I) + \bar{\sigma}(\bar{\ell}_I\tilde{H}_I) < 1 \qquad (11.3-24)$$

We will concentrate the following discussion on (11.3–20) and (11.3–22). Conditions (11.3–23) and (11.3–24) can be interpreted similarly.

Note first that even when robust stability and nominal performance are satisfied with a reasonable margin ($\bar{\sigma}(\bar{\ell}_I\tilde{H}) < 1$ and $\bar{\sigma}(w\tilde{E}) < 1$) the robust performance condition can be violated by an *arbitrarily* large amount if either the controller C or the plant $\tilde{P}$ is ill-conditioned. On the other hand if either $\kappa(\tilde{P})$ or $\kappa(C)$ is small, the input uncertainty can be treated more or less like output uncertainty and it is not necessary to pay special attention to robust performance. It should be emphasized, however, that both (11.3–20) and (11.3–22) are only *sufficient*. If the plant is ill-conditioned, any controller designed for good nominal performance will also be ill-conditioned because it tends to invert the plant. Under these circumstances (11.3–20) and (11.3–22) can be *arbitrarily* conservative compared to the exact condition involving μ.

Nevertheless, (11.3–20) and (11.3–22) give rough guidelines for controller design to avoid robust performance problems. For a well-conditioned plant a simple decoupling (inverse-based) controller is also well conditioned and should give good robust performance. For a badly conditioned plant decoupling should be avoided and for robust performance much attention has to be paid to the modelling of the uncertainty and the control system design.

11.3.4 H_2-Performance Objective

We wish to evaluate a bound on the 2-norm of the weighted sensitivity

$$\|W_2EW_1\|_2^2 = \frac{1}{2\pi}\int_{-\infty}^{\infty} \text{trace}[(W_2EW_1)^H(W_2EW_1)]d\omega \qquad (11.3-25)$$

for a family Π of plants. Assume that a bound $\beta_0(\omega)$ can be found such that

$$\sup_{P\in\Pi} \bar{\sigma}(W_2EW_1) = \beta_0(\omega) \qquad (11.3-26)$$

Then, because for $A \in \mathcal{C}^{n\times n}$, $\text{trace}(A^HA) \leq n\bar{\sigma}^2(A)$

$$\sup_{P\in\Pi} \|W_2EW_1\|_2^2 \leq \frac{n}{2\pi}\int_{-\infty}^{\infty} \beta_0^2(\omega)d\omega \qquad (11.3-27)$$

where n is the maximum rank of $W_2EW_1(\omega)$. For the type of norm-bounded uncertainty introduced in Sec. 11.2.1 the bound $\beta_0(\omega)$ can be found from μ as

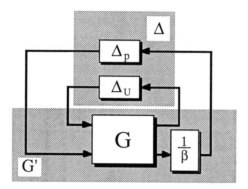

Figure 11.3-4. Robust performance block diagram with additional block $\beta^{-1}I$.

follows. Modify the block diagram in Fig. 11.3-1 by introducing an additional block $\beta^{-1}I$ ($\beta > 0$) as shown in Fig. 11.3-4. Define

$$G' = \begin{pmatrix} G_{11} & G_{12} \\ \beta^{-1}G_{21} & \beta^{-1}G_{22} \end{pmatrix} \qquad (11.3-28)$$

Then

$$\mu(G'(\beta)) = 1 \Leftrightarrow \beta = \beta_0 \qquad (11.3-29)$$

defines a function $\beta_0(\omega)$ such that

$$\sup_{\bar{\sigma}(\Delta_u) \le 1} \bar{\sigma}(F(G, \Delta_u)) = \beta_0(\omega) \qquad (11.3-30)$$

where F is the perturbed weighted sensitivity described by the LFT (11.3–6). Equation (11.3–26) follows directly from (11.3–30).

Inequality (11.3–27) provides only a bound for the H_2 objective, which can be conservative. Alternatively we can compute for *one* specific input v the worst ISE that can result from any plant in the set Π or equivalently for any $\Delta \in X$. This can be done *exactly* without conservatism as shown next.

The 2-norm of the weighted error for a specific input v ($\|W_2 Ev\|_2$) is given by (11.3–25) with

$$W_1 = \begin{pmatrix} v & 0 \end{pmatrix} \qquad (11.3-31)$$

where the 0-matrix is chosen to make W_1 square. Because W_2EW_1 is now of rank one and trace$(A^H A) = \bar{\sigma}^2(A)$ when A is of rank one, (11.3–27) becomes the equality

$$\sup_{P \in \Pi} \|W_2 E v\|_2^2 = \sup_{P \in \Pi} \|W_2 E W_1\|_2^2 = \frac{1}{2\pi} \int_{-\infty}^{\infty} \beta_0^2(\omega) d\omega \qquad (11.3-32)$$

For the specific case of an SISO system with multiplicative uncertainty, $W_2 = 1$, $W_1 = v$, and we can find G' from (11.3–12) or (11.3–17)

$$G' = \begin{pmatrix} -\tilde{\eta}\bar{\ell}_m & -\tilde{\eta}v \\ \beta^{-1}\tilde{\epsilon}\bar{\ell}_m & \beta^{-1}\tilde{\epsilon}v \end{pmatrix} \qquad (11.3-33)$$

Here we know from Sec. 11.3.2

$$\mu(G') = |\tilde{\eta}\bar{\ell}_m| + |\beta^{-1}\tilde{\epsilon}v| \qquad (11.3-34)$$

Setting $\mu(G') = 1$ and solving for β we find

$$\beta_0 = |\tilde{\epsilon}v|(1 - |\tilde{\eta}\bar{\ell}_m|)^{-1} \qquad (11.3-35)$$

Substituting this expression for β_0 in (11.3–32) we find the same result as in Section 2.6.1.

Definition (11.3–29) implies that at each frequency that value of β has to be found which makes the SSV unity. It is obvious that $\mu(G'(\beta))$ is a monotonic function of β: as β increases the destabilizing effect of the uncertainty decreases. More precisely, if the system in Fig. 11.3-6 is stable for β_1 and $\Delta_i, \bar{\sigma}(\Delta_i) \leq 1$, then it is also stable for any $\beta_2 > \beta_1$. Therefore $\mu(G'(\beta_2)) \leq \mu(G'(\beta_1))$. In the computations we usually employ the upper bound of μ rather than μ itself. The following theorem makes the iterations necessary to solve (11.3–29) in terms of the upper bound very simple.

Theorem 11.3-2. *Let*

$$M^x = \begin{pmatrix} M_{11} & M_{12} \\ xM_{21} & xM_{22} \end{pmatrix} \qquad (11.3-36)$$

where x is a positive scalar and let $D = \text{diag}\{D_1, D_2\}$. Then the upper bound of the SSV $\mu(M^x)$, $\inf_{D \in \mathcal{D}} \bar{\sigma}(DM^xD^{-1})$, (see (11.2–15)), is a non-decreasing function of x.

Proof. Let $0 < x_2 \leq x_1$. Then we can write $x_2 = \alpha x_1$ where $0 < \alpha \leq 1$. From (11.3–36) we have

$$DM^{x_2}D^{-1} = \begin{pmatrix} D_1 & 0 \\ 0 & D_2 \end{pmatrix} \begin{pmatrix} I & 0 \\ 0 & \alpha I \end{pmatrix} M^{x_1}D^{-1} = \begin{pmatrix} I & 0 \\ 0 & \alpha I \end{pmatrix} DM^{x_1}D^{-1}$$

$$\Rightarrow \bar{\sigma}(DM^{x_2}D^{-1}) \le \bar{\sigma}\begin{pmatrix} I & 0 \\ 0 & \alpha I \end{pmatrix}\bar{\sigma}(DM^{x_1}D^{-1})$$

$$\Leftrightarrow \bar{\sigma}(DM^{x_2}D^{-1}) \le \bar{\sigma}(DM^{x_1}D^{-1}) \qquad \forall D \in \mathcal{D}$$

$$\Rightarrow \inf_{D \in \mathcal{D}} \bar{\sigma}(DM^{x_2}D^{-1}) \le \inf_{D \in \mathcal{D}} \bar{\sigma}(DM^{x_1}D^{-1})$$

11.3.5   Application: High-Purity Distillation

Consider the distillation column described in the Appendix where the overhead composition is to be controlled at $y_D = 0.99$ and the bottom composition at $x_B = 0.01$ using the reflux L and boilup V as manipulated variables. After linearization the model is

$$\tilde{P} = \frac{1}{75s+1}\begin{pmatrix} 0.878 & -0.864 \\ 1.082 & -1.096 \end{pmatrix} \tag{11.3 – 37}$$

In a similar manner as in Sec. 11.1.6 we will assume a full block input uncertainty with weight

$$w_I(s) = 0.2\frac{5s+1}{0.5s+1} \tag{11.3 – 38}$$

The performance specification is simply

$$\bar{\sigma}(E) < |w_P|^{-1} \qquad \forall P \in \Pi, \ \forall \omega \tag{11.3 – 39}$$

where

$$w_P = 0.5\frac{10s+1}{10s} \tag{11.3 – 40}$$

The performance weight $w_P(s)$ implies that we require integral action ($w_P(0) = \infty$). It allows an amplification of disturbances at high frequencies by a factor of two at most ($\lim_{\omega\to\infty}|w_P(i\omega)|^{-1} = 2$). A particular sensitivity function which exactly matches the performance bound (11.3–40) at low frequencies and satisfies it easily at high frequencies is $E = \frac{20s}{20s+1}I$. This corresponds to a first order response with time constant 20 min.

For robust performance

Table 11.3-1. State space realization of "μ-optimal" controller, $C_\mu(s) = C(sI - A)^{-1}B + D$.

$$A = \begin{pmatrix} -1.002 \cdot 10^{-7} & 0 & 0 & 0 & 0 & 0 \\ 0 & -3.272 \cdot 10^{-6} & 0 & 0 & 0 & \\ 0 & 0 & -0.1510 & 0 & 0 & 0 \\ 0 & 0 & 0 & -9.032 & 0 & 0 \\ 0 & 0 & 0 & 0 & -538.2 & 0 \\ 0 & 0 & 0 & 0 & 0 & -586.8 \end{pmatrix}$$

$$B = \begin{pmatrix} -65.13 & -90.09 \\ 72.24 & 90.31 \\ 5.492 & -4.394 \\ -90.86 & -113.6 \\ 1867 & -1494 \\ 672.2 & 840.3 \end{pmatrix}$$

$$C = \begin{pmatrix} 0.6564 & 0.7171 & 4.949 & 5.033 & -1691 & -311.2 \\ 0.6555 & 0.5425 & 4.941 & -5.040 & -1689 & 311.6 \end{pmatrix}$$

$$D = \begin{pmatrix} 5866 & -3816 \\ 5002 & -4878 \end{pmatrix}$$

$$\mu(G) < 1 \qquad \forall \omega \qquad\qquad (11.3 - 41)$$

where G is defined by (11.3–17) with $\bar{\ell}_I = w_I$ and $w = w_P$. We will consider three different controllers: an inverse-based controller $C_1(s)$ (in this case equivalent to a steady state decoupler with PI controllers), a diagonal PI-controller $C_2(s)$ and a "μ-optimal" controller $C_\mu(s)$, found by approximate minimization of the LHS of (11.3–41).

$$C_1(s) = c_1(s)G_{LV}^{-1}(s) \qquad\qquad (11.3 - 42a)$$

$$c_1(s) = 0.7s^{-1} \qquad\qquad (11.3 - 42b)$$

$$C_2(s) = c_2(s) \begin{pmatrix} 1 & 0 \\ 0 & -1 \end{pmatrix} \qquad\qquad (11.3 - 43a)$$

$$c_2(s) = 2.4(75s + 1)s^{-1} \qquad\qquad (11.3 - 43b)$$

$$C_\mu(s): \quad \text{Table 11.3-1}$$

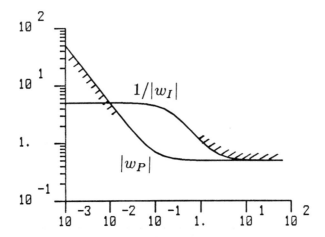

Figure 11.3-5. Performance and robustness bounds for multivariable loop shaping.

One way of designing controllers which meet the nominal performance and robust stability specifications is to use multivariable loop shaping. For nominal performance, $\underline{\sigma}(\tilde{P}C)$ must be above $|w_P|$ for low frequencies (Sec. 10.4.4). For robust stability with input uncertainty, $\bar{\sigma}(C\tilde{P})$ must lie below $1/|w_I|$ for high frequencies (Sec. 11.1.4) (Fig. 11.3-5).

For the inverse-based controller (11.3–42) we get $\bar{\sigma}(C_1\tilde{P}) = \underline{\sigma}(\tilde{P}C_1) = |c_1|$ and it is trivial to choose a $c_1(s)$ to satisfy these conditions. The choice $c_1(s) = 0.7s^{-1}$ yields a controller which has much better nominal performance than required, and which can allow about two times more uncertainty than assumed. This is also seen from Fig. 11.3-6 and 11.3-7 where the nominal performance and robust stability conditions (10.4–19) and (11.1–13) are displayed graphically.

For the diagonal controller (11.3–43) we find $\bar{\sigma}(C_2\tilde{P}) = 1.972|c_2|$ and $\underline{\sigma}(\tilde{P}C_2) = 0.0139|c_2|$, and the difference between these two singular values is so large that no choice of c_2 is able to satisfy both nominal performance and robust stability. This is shown in Figs. 11.3-6 and 11.3-7 for $c_2(s)$ defined by (11.3–43b).

The sufficient conditions for robust performance (11.3–20) and (11.3–22) suggest that the ill-conditioned controller C_1 ($\kappa(C_1) = 141.7$) for the ill-conditioned plant $\tilde{P}$ ($\kappa(\tilde{P}) = 141.7$) may give very poor robust performance even though both the nominal performance ($\bar{\sigma}(w_P\tilde{E}) < 1$) and robust stability conditions ($\bar{\sigma}(w_I\tilde{H}_I) < 1$) are individually satisfied. On the other hand, for a controller with a low condition number ($\kappa(C_2) = 1$) we expect to get robust performance for

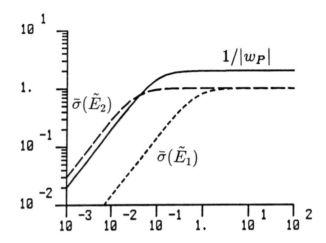

Figure 11.3-6. Nominal performance test for controllers C_1 and C_2.

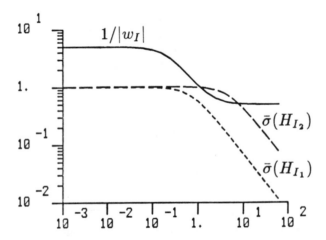

Figure 11.3-7. Robust stability test for controllers C_1 and C_2.

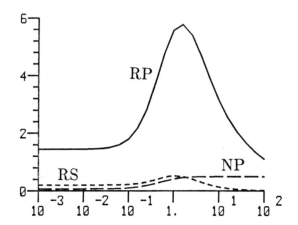

Figure 11.3-8. μ-plots for inverse-based controller $C_1(s)$.

"free" *provided* nominal performance and robust stability are satisfied. However, as we saw for C_2 in Fig. 11.3-6 it is rarely possible to achieve good nominal performance for an ill-conditioned plant with a scalar controller.

The exact test for robust performance (11.3–41) is plotted in Figs. 11.3-8 and 11.3-9 for C_1 and C_2. As expected, the inverse-based controller $C_1(s)$ is far from meeting the robust performance requirements (μ_{RP} is about 5.8), even though the controller was shown to achieve both nominal performance and robust stability. On the other hand, the performance of the diagonal controller $C_2(s)$ is much less affected by uncertainty ($\mu_{RP} = 1.71$).

The μ-synthesis method used to design the "μ-optimal" controller gives controllers of very high order, but by employing model reduction, we were able to find a "μ-optimal" controller with six states (Table 11.3-1). The robust performance test for this controller is shown in Fig. 11.3-10. (The μ-plot is not quite flat as it should be for the truly optimal case.) The peak value for μ is 1.06, which means that this controller almost satisfies the robust performance condition. This value for μ_{RP} is significantly lower than for the diagonal PI controller C_2.

The time responses in Fig. 11.3-11 confirm the predictions by μ. Note in particular the poor performance of C_1 in the presence of model uncertainty. The large value of $\mu(0)$ for the diagonal PI controller leads to a very sluggish approach to steady state when compared to the μ-optimal controller.

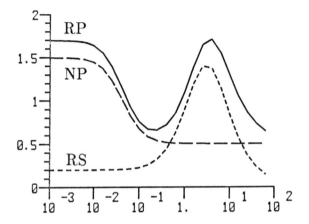

Figure 11.3-9. μ-plots for diagonal controller $C_2(s)$.

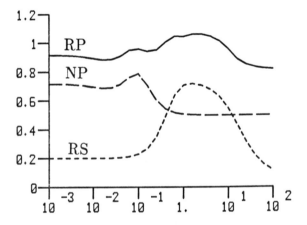

Figure 11.3-10. μ-plots for "μ-optimal" controller $C_\mu(s)$.

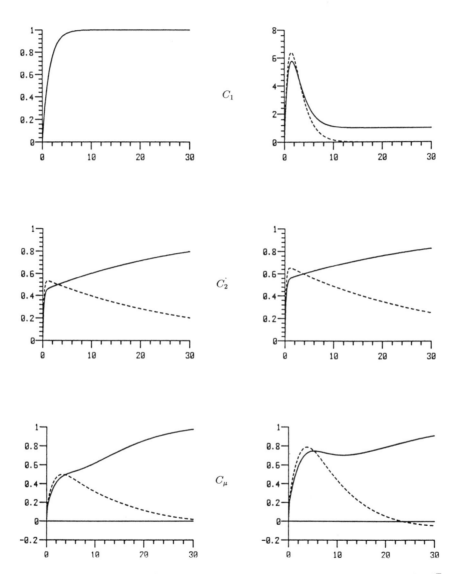

Figure 11.3-11. Time responses for the three controllers for a setpoint change $r = (1,0)^T$. Left column: no model error, right column $L_I = \text{diag}\{0.2, -0.2\}$.

11.4 Robustness Conditions in Terms of Specific Transfer Matrices

In Secs. 11.2 and 11.3 we derived necessary and sufficient conditions of the form

$$\mu_\Delta(M) < k(\omega), \quad \forall\omega \qquad (11.4-1)$$

for robust stability and performance. The implications of (11.4–1) may not be easy to understand for the engineer. A simple robustness bound of the form $\bar{\sigma}(T) < k'(\omega) \ \forall\omega$ may provide more insight, where T denotes a transfer matrix of engineering significance — e.g., the sensitivity $\tilde{E}$ — the complementary sensitivity $\tilde{H}$ or the loop gain $\tilde{P}C$. The goal of this section is to derive such bounds.

More specifically we will show first that in all cases of practical interest M can be related to the transfer matrix T of engineering significance by a *Linear Fractional Transformation* (LFT)

$$M = N_{11} + N_{12}T(I - N_{22}T)^{-1}N_{21} \qquad (11.4-2)$$

(Sometimes a superscript on N — e.g., N^T — will be used to denote the dependence of N on the particular choice of T.) Then we will derive from N a bound on $\bar{\sigma}(T)$ which guarantees that (11.4–1) is satisfied. Since one objective is to assist the engineer with the bound in the controller design, it is important that N be independent of C.

11.4.1 How to find the LFT

In many cases the LFT (11.4-2) can be found by inspection. In other cases the following three-step procedure may be used (Fig. 11.4-1).

1) Write M as a LFT of C:

$$M = G_{11} + G_{12}C(I - G_{22}C)^{-1}G_{21} \qquad (11.4-3)$$

The matrix G is easy to construct by inspection of the block diagram.

2) Write the controller C as a LFT of the transfer matrix of interest (T).

$$C = J_{11} + J_{12}T(I - J_{22}T)^{-1}J_{21} \qquad (11.4-4)$$

J is most easily found by solving the expression $T(C)$ for C.

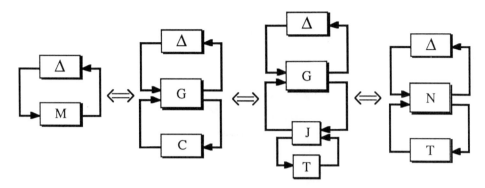

Figure 11.4-1. Equivalent representations of system M with perturbation Δ.

3) Given G and J, N follows easily because any interconnection of LFT's is again an LFT

$$N = \begin{pmatrix} N_{11} & N_{12} \\ N_{21} & N_{22} \end{pmatrix} = \begin{pmatrix} G_{11} + G_{12}J_{11}(I - G_{22}J_{11})^{-1}G_{21} & G_{12}(I - J_{11}G_{22})^{-1}J_{12} \\ J_{21}(I - G_{22}J_{11})^{-1}G_{21} & J_{22} + J_{21}G_{22}(I - J_{11}G_{22})^{-1}J_{12} \end{pmatrix}$$
$$(11.4-5)$$

For the special case $J_{11} = 0$ this reduces to

$$N = \begin{pmatrix} G_{11} & G_{12}J_{12} \\ J_{21}G_{21} & J_{22} + J_{21}G_{22}J_{12} \end{pmatrix} \qquad (11.4-6)$$

Without proof we remark that when T is a *closed-loop* transfer function then $N_{22} = 0$.

If N^H is known, then it is easy to derive N for other closed-loop transfer functions T. Note that $\tilde{H}$ is an LFT of $\tilde{E}$:

$$\tilde{H} = I - \tilde{E} \qquad (11.4-7)$$

Let

$$\tilde{H} = J_{11} + J_{12}\tilde{E}(I - J_{22}\tilde{E})^{-1}J_{21} \qquad (11.4-8)$$

then we find by comparing (11.4–7) and (11.4–8)

$$J = \begin{pmatrix} I & -I \\ I & 0 \end{pmatrix} \qquad (11.4-9)$$

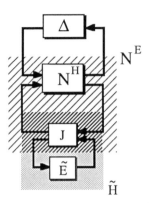

Figure 11.4-2. Illustration for finding N^E from N^H.

The transformations are interpreted in the block diagram in Fig. 11.4-2. The matrix N^E can be found from (11.4–5) with N^H instead of G, and J defined by (11.4–9)

$$N^E = \begin{pmatrix} N_{11}^H + N_{12}^H N_{21}^H & -N_{12}^H \\ N_{21}^H & 0 \end{pmatrix} \qquad (11.4-10)$$

Here we have set $N_{22}^H = 0$ as explained above.

Example 11.4-1. Consider the system with simultaneous multiplicative input and output uncertainty studied in Sec. 11.2.3. For $W_{1I} = W_{1O} = wI$ and $W_{2I} = W_{2O} = I$, (11.2–20) becomes

$$M = w \begin{pmatrix} -C\tilde{P}(I + C\tilde{P})^{-1} & -C(I + \tilde{P}C)^{-1} \\ \tilde{P}(I + C\tilde{P})^{-1} & -\tilde{P}C(I + \tilde{P}C)^{-1} \end{pmatrix} \qquad (11.4-11)$$

Recall that $\Delta = \text{diag}\{\Delta_I, \Delta_O\}$. Let us express M as a LFT of $\tilde{H}$ using the three-step procedure

(1) It is easier to find the matrix G directly from Fig. 11.2-1 than from (11.4–11):

$$G = \begin{pmatrix} 0 & 0 & -I \\ P & 0 & -P \\ P & I & -P \end{pmatrix} \qquad (11.4-12)$$

Note that G_{11} is the upper left 2×2 block of G corresponding to Δ.

(2) Solving

$$\tilde{H} = \hat{P}C(I + \hat{P}C)^{-1} \tag{11.4 - 13}$$

for C yields

$$C = \hat{P}^{-1}\tilde{H}(I - \tilde{H})^{-1} \tag{11.4 - 14}$$

Comparing (11.4–14) with (11.4–4) we find

$$J = \begin{pmatrix} 0 & \tilde{P}^{-1} \\ I & I \end{pmatrix} \tag{11.4 - 15}$$

Substituting (11.4–12) and (11.4–15) into (11.4–5) yields

$$N_{11}^H = \begin{pmatrix} 0 & 0 \\ \tilde{P} & 0 \end{pmatrix}, \quad N_{12}^H = \begin{pmatrix} -\tilde{P}^{-1} \\ -I \end{pmatrix}, \quad N_{21}^H = (\tilde{P} \quad I), \quad N_{22}^H = 0 \tag{11.4 - 16}$$

To find M as a LFT of $\tilde{E}$, use N^H and (11.4–10) to get:

$$N_{11}^E = \begin{pmatrix} -I & -\tilde{P}^{-1} \\ 0 & -I \end{pmatrix}, \quad N_{12}^E = \begin{pmatrix} \tilde{P}^{-1} \\ I \end{pmatrix}, \quad N_{21}^E = (\tilde{P} \quad I), \quad N_{22}^E = 0 \tag{11.4 - 17}$$

$\square$

11.4.2 New Properties of μ

The results in this section apply to any complex matrices although in our case these will be transfer matrices.

Theorem 11.4-1. *Let M be written as a LFT of T*

$$M = N_{11} + N_{12}T(I - N_{22}T)^{-1}N_{21} \tag{11.4 - 18}$$

and let k be a given constant. Assume $\mu_\Delta(N_{11}) < k$ and $\det(I - N_{22}T) \neq 0$. Then

$$\mu_\Delta(M) < k \tag{11.4 - 19}$$

if

$$\bar{\sigma}(T) \leq c_T \tag{11.4 - 20}$$

where c_T solves

$$\mu_{\tilde{\Delta}} \begin{pmatrix} N_{11} & N_{12} \\ kc_T N_{21} & kc_T N_{22} \end{pmatrix} = k \qquad (11.4-21)$$

and $\tilde{\Delta} = \text{diag}\{\Delta, T\}$.

Proof. Assume that T is defined such that $\bar{\sigma}(T) \le c_T$ $\forall T$. Then at each frequency holds

$$\mu_{\Delta}(M) < k \qquad \forall T$$

$$\Leftrightarrow \det(I + M\Delta) \ne 0 \qquad \forall \Delta \ni \bar{\sigma}(\Delta) \le 1/k, \quad \forall T$$

$$\Leftrightarrow \det \begin{pmatrix} I + N_{11}\Delta & -N_{12}T \\ N_{21}\Delta & I - N_{22}T \end{pmatrix} \ne 0 \quad \forall \Delta \ni \bar{\sigma}(\Delta) \le 1/k, \quad \forall T$$

The last step follows from (11.4–18) and Schur's formula (Lemma 10.2-1).

$$\Leftrightarrow \det \left[I + \begin{pmatrix} \frac{1}{k}N_{11} & c_T N_{12} \\ \frac{1}{k}N_{21} & c_T N_{22} \end{pmatrix} \begin{pmatrix} k\Delta & 0 \\ 0 & -\frac{1}{c_T}T \end{pmatrix} \right] \ne 0 \quad \forall \Delta \ni \bar{\sigma}(\Delta) \le 1/k, \quad \forall T$$

$$\Leftrightarrow \mu_{\tilde{\Delta}} \begin{pmatrix} \frac{1}{k}N_{11} & c_T N_{12} \\ \frac{1}{k}N_{21} & c_T N_{22} \end{pmatrix} < 1, \quad \forall T$$

$$\Leftrightarrow \mu_{\tilde{\Delta}} \begin{pmatrix} N_{11} & kc_T N_{12} \\ N_{21} & kc_T N_{22} \end{pmatrix} < k, \qquad \forall T$$

$$\Leftrightarrow \mu_{\tilde{\Delta}} \begin{pmatrix} N_{11} & N_{12} \\ kc_T N_{21} & kc_T N_{22} \end{pmatrix} < k, \quad \forall T$$

□

In general, c_T can be found numerically using the implicit expression (11.4–21). This search is straightforward because of Thm. 11.3-2. In the special case when $N_{11} = N_{22} = 0$, c_T can be computed explicitly.

Theorem 11.4-2. *Assume $N_{11} = N_{22} = 0$. Then c_T satisfying (11.4–21) is given by*

$$c_T = k\mu_{\tilde{\Delta}}^{-2} \begin{pmatrix} 0 & N_{12} \\ N_{21} & 0 \end{pmatrix} \qquad (11.4-22)$$

Proof. Let $\tilde{\Delta} = \text{diag}\{\Delta_1, \Delta_2\}$, $\bar{\sigma}(\Delta_i) \le 1$; Δ_i may have additional structure

$$\mu_{\tilde{\Delta}} \begin{pmatrix} 0 & N_{12} \\ kc_T N_{21} & 0 \end{pmatrix} < k$$

$$\Leftrightarrow \det \left[I + \frac{1}{k} \begin{pmatrix} 0 & N_{12} \\ kc_T N_{21} & 0 \end{pmatrix} \begin{pmatrix} \Delta_1 & 0 \\ 0 & \Delta_2 \end{pmatrix} \right] \neq 0$$

$$\Leftrightarrow \det \begin{pmatrix} I & \frac{1}{k} N_{12} \Delta_2 \\ c_T N_{21} \Delta_1 & I \end{pmatrix} \neq 0$$

$$\Leftrightarrow \det(I - \frac{c_T}{k} N_{12} \Delta_2 N_{21} \Delta_1) \neq 0$$

$$\Leftrightarrow \det \begin{pmatrix} I & \sqrt{\frac{c_T}{k}} N_{12} \Delta_2 \\ \sqrt{\frac{c_T}{k}} N_{21} \Delta_1 & I \end{pmatrix} \neq 0$$

$$\Leftrightarrow \mu_{\tilde{\Delta}} \begin{pmatrix} 0 & N_{12} \\ N_{21} & 0 \end{pmatrix} = \sqrt{\frac{k}{c_T}}$$

□

Let us first discuss the assumptions made in Thm. 11.4-1. In general, the bound (11.4–19) results from a robust stability or robust performance condition and T is a particular transfer matrix of interest (e.g., $\tilde{H}$ or $\tilde{E}$). In this case M is (internally) stable and $\det(I - N_{22}T) \neq 0$ $\forall \omega$ as required by the assumption. Furthermore since $\mu_\Delta(M) \geq \mu_\Delta(N_{11})$, the condition $\mu_\Delta(N_{11}) < k$ is necessary for a solution of (11.4–21) to exist. If $\mu(M) < k(\omega)$ is a robust stability (performance) condition, then the condition $\mu(N_{11}) < k(\omega)$ is equivalent to requiring that the robust stability (performance) condition be satisfied for $T = 0$ at this frequency.

Condition (11.4–20) is necessary *and* sufficient for (11.4–19) if (11.4–19) is to be satisfied for *all* T's satisfying $\bar{\sigma}(T) \leq c_T$. (This follows directly from the proof of the theorem.) In most cases we are interested only in a specific M (and a specific T), and condition (11.4–20) is only sufficient for (11.4–19).

However, the value of c_T solving (11.4–21) provides the *least conservative* bound which may be derived on $\bar{\sigma}(T)$ such that (11.4–19) is satisfied.

Note that (11.4–21) is computed based on the structure of Δ *and of* T. The least restrictive bound on $\bar{\sigma}(T)$ (c_T large) is found when $T = tI$ is assumed, where t is a scalar, and the most restrictive bound (c_T small) when T is a full matrix. The reason is that by requiring $T = tI$, the class of perturbations is restricted, and the magnitude of the perturbations is allowed to be larger.

Theorem 11.4-1 may be used to derive a bound on any transfer matrix T which is related to M through a linear fractional transformation (LFT). Note that these bounds (e.g., on $\bar{\sigma}(\tilde{H})$ and $\bar{\sigma}(\tilde{E})$) may be *combined* over different frequency ranges since Thm. 11.4-1 applies on a frequency-by-frequency basis. This provides a powerful method for deriving simple robustness bounds.

Table 11.4-1.

Case		$\mu_{\tilde{\Delta}}^2 \begin{pmatrix} 0 & A \\ B & 0 \end{pmatrix}$
1	Δ and T full	$\bar{\sigma}(A)\bar{\sigma}B$
2	$T = tI$	$\mu_\Delta(AB)$
3	$\Delta = \delta I$	$\mu_T(BA)$
4	$\Delta = \delta I, T = tI$	$\rho(AB) = \rho(BA)$
5	$B = I$	$\mu_{T\Delta}(A)$

The theorem below which follows directly from Thms. 11.4-1 and 11.4-2 is useful in specific applications.

Theorem 11.4-3: *Let $\tilde{\Delta} = \text{diag}\{\Delta, T\}$. Then*

$$\mu_\Delta(ATB) \le \bar{\sigma}(T)\mu_{\tilde{\Delta}}^2 \begin{pmatrix} 0 & A \\ B & 0 \end{pmatrix} \qquad (11.4-23)$$

(Note that ATB and T are square matrices, while A and B may be non-square.) In special cases $\mu_{\tilde{\Delta}}^2 \begin{pmatrix} 0 & A \\ B & 0 \end{pmatrix}$ may be evaluated in terms of other quantities as shown in Table 11.4-1.

Proof. From Thms. 11.4-1 and 11.4-2 for the case $N_{11} = N_{22} = 0$:

$$\mu_\Delta(N_{12}TN_{21}) < k \quad \text{if} \quad \bar{\sigma}(T)\mu_{\tilde{\Delta}}^2 \begin{pmatrix} 0 & N_{12} \\ N_{21} & 0 \end{pmatrix} < k \qquad (11.4-24)$$

Since (11.4–24) holds for *any* choice of k it is equivalent to

$$\mu_\Delta(N_{12}TN_{21}) \le \bar{\sigma}(T)\mu_{\tilde{\Delta}}^2 \begin{pmatrix} 0 & N_{12} \\ N_{21} & 0 \end{pmatrix}$$

Inequality (11.4–23) follows by choosing $N_{12} = A, N_{21} = B$.

The relations in Table 11.4-1 are proved next. Let Δ_1 and Δ_2 have the same *structure* as Δ and T in the theorem. Define $\tilde{\Delta} = \text{diag}\{\Delta_1, \Delta_2\}$, $\bar{\sigma}(\Delta_i) \le 1$. Then

$$\mu_{\tilde{\Delta}}^2 \begin{pmatrix} 0 & A \\ B & 0 \end{pmatrix} < 1/k$$

$$\Leftrightarrow \mu_{\tilde{\Delta}} \begin{pmatrix} 0 & kA \\ B & 0 \end{pmatrix} < 1$$

$$\Leftrightarrow \det \left[I + \begin{pmatrix} 0 & kA \\ B & 0 \end{pmatrix} \begin{pmatrix} \Delta_1 & 0 \\ 0 & \Delta_2 \end{pmatrix} \right] \neq 0 \qquad \forall \Delta_1, \Delta_2$$

$$\Leftrightarrow \det \begin{pmatrix} I & kA\Delta_2 \\ B\Delta_1 & I \end{pmatrix} \neq 0 \qquad \forall \Delta_1, \Delta_2$$

$$\Leftrightarrow \det(I - kA\Delta_2 B\Delta_1) = \det(I - kB\Delta_1 A\Delta_2) \neq 0 \qquad \forall \Delta_1, \Delta_2$$

$$\Leftrightarrow \mu_{\Delta_1}(A\Delta_2 B) < \frac{1}{k}, \qquad \forall \Delta_2 \tag{11.4 - 25}$$

$$\Leftrightarrow \mu_{\Delta_2}(B\Delta_1 A) < \frac{1}{k}, \qquad \forall \Delta_1 \tag{11.4 - 26}$$

$$\Leftrightarrow \rho(A\Delta_2 B\Delta_1) = \rho(B\Delta_1 A\Delta_2) < \frac{1}{k} \qquad \forall \Delta_1, \Delta_2 \tag{11.4 - 27}$$

1: From the basic properties of norms $\rho(A\Delta_2 B\Delta_1) \leq \bar{\sigma}(A)\bar{\sigma}(B)$. We now have to show that this holds as an equality for some choice of Δ_1, Δ_2. Let $A = U_A \Sigma_A V_A^H$ and $B = U_B \Sigma_B V_B^H$. Since Δ_1 and Δ_2 are full we may choose them such that $\Delta_2 U_B = V_A$ and $V_B^H \Delta_1 = U_A^H$. Then $\rho(A\Delta_2 B\Delta_1) = \rho(U_A \Sigma_A \Sigma_B U_A^H) = \rho(\Sigma_A \Sigma_B) = \bar{\sigma}(A)\bar{\sigma}(B)$.

2: From (11.4-25); 3: From (11.4-26); 4,5: From (11.4-27). □

11.4.3 Examples

Example 11.4-2 (Input Uncertainty.) If there is only multiplicative input uncertainty of magnitude $\bar{\sigma}(\Delta_I) < w(\omega)$ we find [see (11.4–11)] the necessary and sufficient condition for robust stability

$$\mu(\tilde{P}^{-1}\tilde{H}\tilde{P}) \leq w^{-1}(\omega) \tag{11.4 - 28}$$

Here μ is computed with respect to the structure of Δ_I which may be a diagonal matrix. The least conservative bound on $\bar{\sigma}(\tilde{H})$ which may be derived from (11.4–28) is found using Thm. 11.4-3:

$$\bar{\sigma}(\tilde{H}) \leq 1/\mu^2 \begin{pmatrix} 0 & \tilde{P}^{-1} \\ \tilde{P} & 0 \end{pmatrix} w(\omega) \Rightarrow \text{Robust Stability} \tag{11.4 - 29}$$

μ in (11.4–29) is computed with respect to the structure diag$\{\Delta_I, H\}$. Note the following special cases:

(i) Δ_I and H are both full matrices: $\mu^2 \begin{pmatrix} 0 & \tilde{P}^{-1} \\ \tilde{P} & 0 \end{pmatrix} = \kappa(\tilde{P})$, where $\kappa(\tilde{P})$ is the condition number of the plant (compare Sec. 11.1.4).

(ii) $H = hI : \mu^2 \begin{pmatrix} 0 & \tilde{P}^{-1} \\ \tilde{P} & \tilde{P} \end{pmatrix} = 1$ $\square$

Example 11.4-3 (Robust Performance for SISO Plant.) For multiplicative uncertainty ($|\ell(i\omega)| \leq w_O$) we find the necessary and sufficient condition for robust performance ($\bar{\sigma}(Ew_p) < 1$)

$$\mu(M) < 1 \tag{11.4 – 30}$$

where (see Sec. 11.3.2)

$$M = \begin{pmatrix} w_O \tilde{H} & w_O \tilde{H} \\ w_P \tilde{E} & w_P \tilde{E} \end{pmatrix} \tag{11.4 – 31}$$

The SSV in (11.4-30) is computed with respect to the diagonal 2×2-matrix diag$\{\Delta_O, \Delta_P\}$. Bounds on $\bar{\sigma}(\tilde{H}) = |\tilde{H}|$ and $\bar{\sigma}(\tilde{E}) = |\tilde{E}|$ are easily derived using Thm. 11.4-1 with $k = 1$. Write M as a LFT of H:

$$N_{11}^H = \begin{pmatrix} 0 & 0 \\ w_P & w_P \end{pmatrix} \quad N_{12}^H = \begin{pmatrix} w_O \\ -w_P \end{pmatrix} \quad N_{21}^H = (1 \quad 1) \quad N_{22}^H = 0$$

Theorem 12.4-1 provides the sufficient condition for robust performance: $|\tilde{H}| \leq c_H$, $\forall\omega$ where c_H solves at each frequency

$$\mu \begin{pmatrix} 0 & 0 & w_O \\ w_P & w_P & -w_P \\ c_H & c_H & 0 \end{pmatrix} = 1 \tag{11.4 – 32}$$

The SSV μ in (11.4–32) is computed with respect to the structure diag$\{\Delta_O, \Delta_P, \tilde{H}\}$ — i.e., a diagonal 3×3 matrix. Note that (11.4–32) is independent of the plant model $\tilde{P}$. However, M (and therefore $\tilde{H}$) must be *stable*, and this implicitly makes the allowable $\tilde{H}$'s dependent on $\tilde{P}$. An analytic expression for c_H may be derived for this simple case.

$$|\tilde{H}| \leq c_H = \frac{1 - |w_P|}{|w_O| + |w_P|} \Rightarrow \text{Robust Performance} \tag{11.4 – 33}$$

Similarly, a condition in terms of $\tilde{E}$ is derived

$$|\tilde{E}| \leq c_E = \frac{1 - |w_O|}{|w_O| + |w_P|} \Rightarrow \text{Robust Performance} \qquad (11.4-34)$$

The expressions for c_H and c_E in (11.4–33) and (11.4–34) are most easily derived from the identity (see Sec. 11.3.2)

$$\mu \begin{pmatrix} w_O\tilde{H} & w_O\tilde{H} \\ w_P\tilde{E} & w_P\tilde{E} \end{pmatrix} = |w_O\tilde{H}| + |w_P\tilde{E}| \qquad (11.4-35)$$

combined with the triangle inequality (e.g., use $|\tilde{E}| = |1 - \tilde{H}| \leq 1 + |\tilde{H}|$ to derive (11.4–33)). Note that (11.4–33) is impossible to satisfy at low frequencies where tight performance is desired and $|w_P|$ is larger than one (corresponds to $\mu(N_{11}) > k$ in Thm. 11.4-1). Similarly, (11.4–34) is impossible to satisfy at high frequencies where the uncertainty exceeds 100% and $|w_O|$ is larger than one. However, we may combine the bounds: (11.4–30) is satisfied if (11.4–34) is satisfied at low frequencies and (11.4-33) at high frequencies. The bounds (11.4–33) and (11.4–34) (even when combined) tend to be conservative around cross-over where $|\tilde{H}|$ and $|\tilde{E}|$ have similar magnitude. This means that there will be systems which satisfy (11.4–30), but do not satisfy (11.4–33) and (11.4–34).

Conditions (11.4–33) and (11.4–34) are shown graphically in Fig. 11.4-3A for the choice $w_O(s) = 0.2(0.5s + 1)$ and $w_P(s) = 0.5(1 + s^{-1})$. Assume that the plant is minimum phase such that $\tilde{H} = (s + 1)^{-1}$ is an allowable (stable) closed-loop transfer function. This corresponds to a nominal first-order response with time constant one. This choice is seen to satisfy (11.4–33) for $\omega > 1.2$ and (11.4–34) for $\omega < 2$ (Fig. 11.4-3B). Consequently, (11.4–30) is satisfied at all frequencies and robust performance is guaranteed. □

11.5 Summary

The simplest uncertainty description for MIMO systems is in terms of a single norm-bounded perturbation matrix with the same dimensions as the plant. Typical examples are multiplicative output (L_O) and multiplicative input (L_I) uncertainty:

$$P = (I + L_O)\tilde{P}; \quad L_O = (P - \tilde{P})\tilde{P}^{-1}; \quad \bar{\sigma}(L_O(i\omega)) \leq \bar{\ell}_O(\omega), \forall \omega \qquad (11.1-2,5)$$

$$P = \tilde{P}(I + L_I); \quad L_I = \tilde{P}^{-1}(P - \tilde{P}); \quad \bar{\sigma}(L_I(i\omega)) \leq \bar{\ell}_I(\omega), \forall \omega \qquad (11.1-3,5)$$

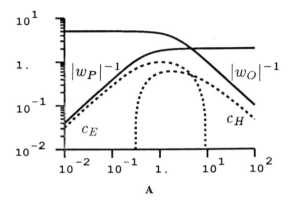

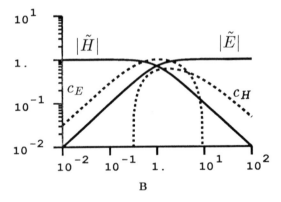

Figure 11.4-3. Graphical representation of conditions (11.4–33) and (11.4–34). Robust performance is guaranteed since $|\tilde{E}| < c_E$ for $\omega < 2$ and $|\tilde{H}| < c_H$ for $\omega > 1.4$.

The advantage of these uncertainty descriptions is that they lead to very simple necessary and sufficient robust stability conditions:

$$\bar{\sigma}(\tilde{P}C(I + \tilde{P}C)^{-1})\bar{\ell}_O = \bar{\sigma}(\tilde{H})\bar{\ell}_O < 1 \quad \forall\omega \tag{11.1 – 11}$$

$$\bar{\sigma}(C(I + \tilde{P}C)^{-1}\tilde{P})\bar{\ell}_I = \bar{\sigma}(\tilde{H}_I)\bar{\ell}_I < 1 \quad \forall\omega \tag{11.1 – 13}$$

It follows directly that robust integral control is possible for an open loop stable system if and only if $\bar{\ell}_O(0) < 1$ or $\bar{\ell}_I(0) < 1$. This implies that the gain matrix must remain nonsingular for all perturbations (Thm. 11.1-3).

It is often difficult to model the physical uncertainty accurately and non-conservatively with single perturbations. Therefore an uncertainty description involving multiple norm-bounded perturbations Δ_i was introduced. Employing weighting matrices W_1 and W_2 the actual perturbation is

$$L_i = W_2\Delta_i W_1 \tag{11.2 – 2}$$

where Δ_i may be any rational transfer matrix satisfying $\bar{\sigma}(\Delta_i) \leq 1$, $\forall\omega$. The individual perturbations are combined into one large block diagonal perturbation matrix

$$\Delta = \text{diag}\{\Delta_1, \ldots, \Delta_m\}; \quad \bar{\sigma}(\Delta) \leq 1 \tag{11.2 – 3, 4}$$

Many practical uncertainty problems can be cast into the $M - \Delta$ structure shown in Fig. 11.1-2 where Δ is of the form (11.2–3,4). Assuming that M is stable a necessary and sufficient condition for robust stability can be established via the Structured Singular Value (SSV) μ

$$\mu_\Delta(M(i\omega)) < 1 \qquad \forall\omega \tag{11.2 – 10}$$

where the subscript Δ indicates that μ is computed with respect to the *structure* of Δ. Among the problems which can be treated in this framework are simultaneous input and output uncertainty (Sec. 11.2.3), and independent uncertainty in the transfer matrix elements (Sec. 11.2.5). The latter type imposes severe constraint on performance if the plant is ill-conditioned: if the nominal response is decoupled — i.e., $\tilde{H} = \text{diag}\{\tilde{\eta}_i\}$ — then robust stability is guaranteed if

$$|\tilde{\eta}_i| < \frac{1}{r_{max}\sqrt{n}\kappa^*(P)} \qquad \forall\omega, \forall i \tag{11.2 – 36}$$

where r_{max} is the maximum relative element uncertainty, n the dimension of the system and κ^* the minimized condition number. Tighter conditions can be derived for 2×2 systems. The minimized condition number κ^* is closely related to the Relative Gain Array Λ:

$$\kappa^*(P) \cong \|\Lambda\|_m \qquad \text{for } \kappa^*(P) \text{ large}$$

where

$$\|\Lambda\|_m = 2 \cdot \max\left\{\|\Lambda\|_1, \|\Lambda\|_\infty\right\} \qquad (11.2-46)$$

Thus, systems with large RGA are very sensitive to *independent* element uncertainty. In practice, however, the variations of the transfer matrix elements are usually highly *correlated*.

The major advantage of the H_∞ performance objective is that it allows us to express the robust *performance* test as a robust *stability* test in the presence of a structured perturbation (Thm. 11.3-1): the nominally stable system G (Fig. 11.3-1) subjected to the block diagonal uncertainty Δ_u $(\bar{\sigma}(\Delta_u) \leq 1)$ satisfies the H_∞ robust performance condition if and only if

$$\mu_\Delta(G) < 1 \qquad \forall \omega \qquad (11.3-8)$$

For multiplicative output uncertainty the SSV in (11.3–8) can be approximated by

$$\mu_\Delta(G) \leq \bar{\sigma}(w\tilde{E}) + \bar{\sigma}(\bar{\ell}_O\tilde{H}) \qquad (11.3-15)$$

and for multiplicative input uncertainty by

$$\mu_\Delta(G) \leq \bar{\sigma}(w\tilde{E}) + \kappa(\tilde{P})\bar{\sigma}(\bar{\ell}_O\tilde{H}) \qquad (11.3-20)$$

or

$$\mu_\Delta(G) \leq \bar{\sigma}(w\tilde{E}) + \kappa(C)\bar{\sigma}(\bar{\ell}_O\tilde{H}) \qquad (11.3-22)$$

Thus, multiplicative output uncertainty does not cause any robust performance difficulties: if both the nominal performance $(\bar{\sigma}(w\tilde{E}) < 1)$ and the robust stability $(\bar{\sigma}(\bar{\ell}_O\tilde{H}) < 1)$ conditions are satisfied with some margin, then robust performance is guaranteed automatically. On the other hand input uncertainty causes robust performance problems when either the plant or the controller is ill-conditioned.

Sometimes, it is attractive to express robust stability and performance conditions in terms of bounds on transfer functions of direct engineering significance (e.g., $\tilde{H}$ or $\tilde{E}$) rather than in an implicit manner (11.3–8). This is possible if the transfer matrix G is related to the transfer matrix T of interest through a linear fractional transformation (Sec. 11.4). However, contrary to condition (11.3–8) these bounds are only sufficient for robust stability and performance.

11.6 References

Sections 11.1 through 11.3 closely follow the paper by Skogestad & Morari (1987b). All the examples in these sections are also taken from that paper.

11.1.2. The general robust stability theorem is covered in the paper by Doyle & Stein (1981). They emphasize, in particular, the *necessity* of the theorem. The proof presented here is patterned after that by Lehtomaki (1981). A detailed explanation of what is meant by a "connected" set of plants is provided by Vidyasagar, et al. (1982) and Postlethwaite & Foo (1985).

11.1.5. Postlethwaite & Foo (1985) show that Cor. 11.1-1 and 11.1-3 can be combined over different frequency ranges.

11.1.7. A condition for robust integral control similar to (11.1–33) was proved by Garcia & Morari (1985a).

11.2.2. The Structured Singular Value was introduced by Doyle (1982) who also discussed its properties and a generalized gradient search procedure to minimize its upper bound. Osborne (1960) developed the iterative scheme to minimize $\|DMD^{-1}\|_F$. An efficient procedure for computing the SSV based on its lower bound was proposed by Fan & Tits (1986).

11.2.5. Alternative *sufficient* stability conditions in the presence of element by element uncertainty were derived by Kantor & Andres (1983) and Kouvaritakis & Latchman (1985). The claim about necessity in the latter paper is incorrect.

11.2.6. Morari (1983a), Morari et al. (1985) and Grosdidier, Morari & Holt (1985) argued in a somewhat qualitative manner that for robust stability the minimized condition number is a measure of sensitivity with respect to uncertainty. Bristol (1966), in his original paper on the RGA, pointed out its similarity with the condition number. A quantitative relationship between the two was first established by Grosdidier et al. (1985) and then extended by Nett & Manousiouthakis (1987). Tighter but more complicated conditions for robust stability involving the condition number were derived by Skogestad & Morari (1987b).

11.3. Doyle & Wall (1982) and Doyle (1984) pointed out the equivalence between robust stability and robust performance and proposed the SSV as a tool to assess robust performance.

11.3.2, 11.3.3. The sufficient robust performance conditions in the presence of multiplicative input and output uncertainty were derived by Stein (1985).

11.3.4. The method in this section was proposed by Zafiriou and Morari (1986b).

11.3.5. The example is taken from the paper by Skogestad, Morari & Doyle (1988). The procedure described by Doyle (1985) was used to find the "μ-optimal" controller.

11.4. This section follows closely the paper by Skogestad & Morari (1988c). Postlethwaite & Foo (1985) also derive robustness conditions in terms of bounds on transfer matrices of interest. However, in particular for structured uncertainty, their bounds are not as tight as the bounds derived here.

11.4.1. Doyle (1984) showed that $N_{22} = 0$ when T is a *closed loop* transfer function.

Chapter 12

MIMO IMC DESIGN

We assume that the reader has mastered Chaps. 3 through 6 and is familiar with the IMC concept and its implications. Thus, the MIMO treatment of this topic will be much briefer. Some concepts which extend to the MIMO case in a straightforward manner (e.g., two-degree-of-freedom design) will not be covered at all. Issues for which analogies in the SISO case are lacking (e.g., structured uncertainty, ill-conditioned systems) will be treated in depth.

For SISO systems it was possible to design the IMC controllers analytically and to obtain insight into the effect of RHP zeros, RHP poles and robustness issues in the process. For tutorial reasons a separate derivation for stable (Chap. 4) and unstable (Chap. 5) systems was justified. Except in trivial cases (MP systems) the derivation of the IMC controllers for MIMO systems and the effect of RHP zeros and poles is quite complex. Therefore, only one derivation for both stable and unstable systems will be presented.

12.1 IMC Structure

The block diagram of the IMC structure is shown in Fig. 12.1-1A. Here P denotes the plant and P_m the measurement device transfer function. In general, neither P nor P_m are known exactly, only their approximate models $\tilde{P}$ and $\tilde{P}_m$ are available. The process transfer function P_d describes the effect of the disturbance d on the process output y. The measurement of y is corrupted by measurement noise n. The controller Q determines the value of the input (manipulated variable) u. The control objective is to keep y close to the reference (setpoint) r.

Commonly we will use the simplified block diagram in Fig. 12.1-1B. Here d denotes the effect of the disturbance on the output. Exact knowledge of the output y is assumed ($P_m = I, n = 0$). Note that the complete control system to be implemented through computer software or analog hardware is contained in the shaded box in Fig. 12.1-1B.

If the IMC controller Q and the classic controller C (Fig. 10.2-1B) are related

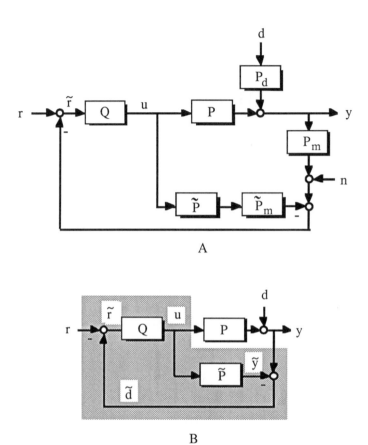

A

B

Figure 12.1-1. General (A) and simplified (B) IMC block diagram. Shaded portion indicates control system.

through

$$C = Q(I - \tilde{P}Q)^{-1} \qquad (12.1-1)$$

$$Q = C(I + \tilde{P}C)^{-1} \qquad (12.1-2)$$

then the IMC structure and the classic feedback structure are *equivalent* in the sense that to each input pair $\{r, d\}$ there corresponds the *same* output pair $\{y, u\}$.

For the IMC structure the sensitivity E and complementary sensitivity H are defined through

$$e = y - r = (I - \tilde{P}Q)(I + (P - \tilde{P})Q)^{-1}(d - r) \overset{\triangle}{=} E(d - r) \qquad (12.1-3)$$

$$y = PQ(I + (P - \tilde{P})Q)^{-1}r \overset{\triangle}{=} Hr \qquad \text{(for } d = 0\text{)} \qquad (12.1-4)$$

For the case that the model is perfect $(P = \tilde{P})$ these expressions yield

$$E = I - PQ \qquad (12.1-5)$$

$$H = PQ \qquad (12.1-6)$$

12.2 Conditions for Internal Stability

In order to test for internal stability we examine the transfer matrices between all possible system inputs and outputs. From the block diagram in Fig. 12.2-1 we note that there are three independent system inputs r, u_1, and u_2 and three independent outputs y, u, and $\tilde{y}$. If there is no model error $(P = \tilde{P})$, they are related through the following transfer matrix:

$$\begin{pmatrix} y \\ u \\ \tilde{y} \end{pmatrix} = \begin{pmatrix} PQ & (I - PQ)P & P \\ Q & -QP & 0 \\ PQ & -PQP & P \end{pmatrix} \begin{pmatrix} r \\ u_1 \\ u_2 \end{pmatrix} \qquad (12.2-1)$$

Theorem 12.2-1 follows trivially by inspection.

Theorem 12.2-1. *Assume that the model is perfect $(P = \tilde{P})$. Then the IMC system in Fig. 12.1-1B is internally stable if and only if both the plant P and the controller Q are stable.*

Thus the structure in Fig. 12.1-1B cannot be used to control plants which are open-loop unstable. Nevertheless, even for unstable plants we will exploit the features of the IMC structure for control system *design* and then *implement* the controller C (obtained from (12.1-1)) in the classic manner. For internal stability of the classic feedback structure the expression (10.2-14) has to be stable. We substitute for C from (12.1-1) to obtain for $P = \tilde{P}$

$$\begin{pmatrix} y \\ u \end{pmatrix} = \begin{pmatrix} PQ & (I - PQ)P \\ Q & -QP \end{pmatrix} \begin{pmatrix} r \\ u' \end{pmatrix} \overset{\triangle}{=} S \begin{pmatrix} r \\ u' \end{pmatrix} \qquad (12.2-2)$$

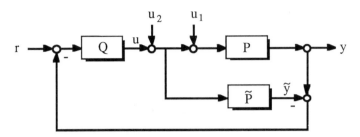

Figure 12.2-1. Block diagram for derivation of conditions for internal stability.

This implies that Q has to be stable and that in the elements of S the factors Q and $(I - PQ)$ have to cancel any unstable poles of P. Thus both Q and $(I - PQ)$ must have RHP zeros at the plant RHP poles. Special care has to be taken to cancel these common RHP zeros when the controller $C = Q(I - PQ)^{-1}$ is constructed. Minimal or balanced realization software can be used to accomplish that.

12.3 Parametrization of All Stabilizing Controllers

For SISO systems we derived a parametrization of all stabilizing controllers in terms of a stable transfer function q_1 (Thm. 5.1-2). A similar development involving coprime factorization is possible for MIMO systems. However, the resulting MIMO parameter Q cannot be interpreted as nicely as in the SISO case. Therefore, the IMC design procedure where Q is augmented by a low-pass filter for robustness and on-line adjustment would not be meaningful. We will derive an alternate parametrization which preserves the physical meaning of Q and allows us to compute the H_2-optimal controller for both stable and unstable MIMO systems. This controller will then again be a possible starting point (Step 1) for the IMC design procedure. We will make the following two assumptions.

Assumption A1. *If π is a pole of $\tilde{P}$ in the open RHP, then (a) the order of π is equal to 1 and (b) $\tilde{P}$ has no zeros at $s = \pi$.*

Assumption A1a is made solely to simplify the notation. Assumption A1b is not very restrictive because the presence of a zero at $s = \pi$ implies an exact cancellation in $\det(\tilde{P}(s))$, which is a nongeneric property — i.e., it does not happen after a slight perturbation in the coefficients of $\tilde{P}$ is introduced.

Assumption A1a is not made for poles at $s = 0$ because more than one such pole may appear in an element of $\tilde{P}$, introduced by capacitances present in the process. However, the following assumption is made which is true for all practical process control problems.

Assumption A2. *Any poles of $\tilde{P}$ or P on the imaginary axis are at $s = 0$. Also $\tilde{P}$ has no finite zeros on the imaginary axis.*

These two assumptions allow us to derive the following theorem.

Theorem 12.3-1. *Assume that Assns. A1 and A2 hold and that $Q_0(s)$ stabilizes P — i.e., that S in (12.2–2) is stable for $Q = Q_0$. Then all Q's which make S stable are given by*

$$Q(s) = Q_0(s) + Q_1(s) \tag{12.3 - 1}$$

where Q_1 is any stable transfer matrix such that PQ_1P is stable.

Proof. $\Rightarrow$ Assume that Q_0 makes S stable and that Q_1 and PQ_1P are stable. From substitution of (12.3-1) into (12.2–2) it follows that $S(Q)$ is stable if and only if $S' = (PQ_1 \quad Q_1P \quad PQ_1P)$ is stable. By assumption the third element of S' is stable. Stability of the other two elements follows by pre- and post-multiplication of PQ_1P by P^{-1}, since according to Assns. A1 and A2, P has no zeros at the location of its unstable poles and these are the only possible unstable poles of S'.

$\Leftarrow$ Assume that Q and Q_0 make S stable. Then the difference matrix

$$\Delta S = S(Q) - S(Q_0) = \begin{pmatrix} P(Q - Q_0) & -P(Q - Q_0)P \\ (Q - Q_0) & -(Q - Q_0)P \end{pmatrix}$$

is stable. This implies that $(Q - Q_0) = Q_1$ and PQ_1P are stable. $\qquad\square$

Note that for SISO systems we can choose

$$Q_1 = b_p^2 s^{2\ell} Q_1', \quad Q_1' \text{ stable} \tag{12.3 - 2}$$

as we did for Thm. 5.1-2. Then clearly Q_1 and PQ_1P are stable. Poles of MIMO systems have a "structure" (matrix of residues) which is not reflected in (12.3–2). While even for MIMO systems any controller of the form (12.3–1 and 12.3–2) would indeed be a stabilizing controller there are many other controllers which are not of this form but are also stabilizing. Thus (12.3–1 and 12.3–2) would not constitute a *complete* parametrization of all MIMO stabilizing controllers.

12.4 Asymptotic Properties of Closed-Loop Response

"System types" were defined in Sec. 10.4.2 to classify the asymptotic closed-loop behavior.

Type m:

$$\lim_{s \to 0} s^{-k} E(s) = 0; \qquad 0 \le k < m \qquad\qquad (12.4-1)$$

Using (12.1–3) this definition becomes

Type m:

$$\lim_{s \to 0} s^{-k} (I - \tilde{P}Q)(I + (P - \tilde{P})Q)^{-1} = 0; \quad 0 \le k < m \qquad (12.4-2)$$

This is equivalent to requiring

Type m:

$$\lim_{s \to 0} \frac{d^k}{ds^k}(I - \tilde{P}Q) = 0; \qquad 0 \le k < m \qquad\qquad (12.4-3)$$

Specifically we obtain

Type 1:

$$\lim_{s \to 0} \tilde{P}Q = I \qquad\qquad (12.4-4)$$

Type 2:

$$\lim_{s \to 0} \tilde{P}Q = I \quad \text{and} \quad \lim_{s \to 0} \frac{d}{ds}(\tilde{P}Q) = 0 \qquad (12.4-5)$$

Thus, in order to track error free asymptotically constant inputs the controller gain has to be the inverse of the model steady state gain. Note that (12.4–4) and (12.4–5) are necessary and sufficient for off-set free tracking of steps and ramps, respectively, even when model error is present.

12.5   Outline of the IMC Design Procedure

In Chaps. 10 and 11 we discussed the objectives of control system design.

Nominal Performance:

Specification:

$$\|W_2 \tilde{E} W_1\|_\alpha < 1; \qquad \alpha = 2, \infty \qquad\qquad (12.5-1)$$

Optimal Design Problem:

$$\min_Q \|W_2 \tilde{E} W_1\|_\alpha = \min_Q \|W_2(I - \tilde{P}Q)W_1\|_\alpha; \qquad \alpha = 2, \infty \qquad (12.5-2)$$

In Sec. 10.4.3 we concluded that from a deterministic point of view the H_2-objective (12.5–1) is somewhat artificial. As an alternative we suggested considering the ISE for a single input or for a finite set of inputs.

Robust Stability:

Multiplicative Output Uncertainty:

$$\bar{\sigma}(\tilde{H}\bar{\ell}_O) < 1; \quad \forall \omega \quad \Leftrightarrow \quad \|\tilde{H}\bar{\ell}_O\|_\infty = \|\tilde{P}Q\bar{\ell}_O\|_\infty < 1 \qquad (12.5-3)$$

Multiplicative Input Uncertainty:

$$\bar{\sigma}(\tilde{P}^{-1}\tilde{H}\tilde{P}\bar{\ell}_I) < 1; \quad \forall \omega \quad \Leftrightarrow \quad \|\tilde{P}^{-1}\tilde{H}\tilde{P}\bar{\ell}_I\|_\infty = \|Q\tilde{P}\bar{\ell}_I\|_\infty < 1 \qquad (12.5-4)$$

General (block diagonal) Uncertainty (Δ_u):

$$\mu_{\Delta_u}(G_{11}^F(\tilde{P}, Q)) < 1 \qquad \forall \omega \qquad (12.5-5)$$

Robust Performance:

Specification:

$$\|W_2 E W_1\|_\infty < 1 \qquad \forall P \in \Pi \qquad (12.5-6)$$

or

$$\mu_\Delta(G^F(\tilde{P}, Q)) < 1 \qquad \forall \omega, \quad \Delta = \text{diag}\{\Delta_u, \Delta_p\} \qquad (12.5-7)$$

Optimal Design Problem:

$$\min_Q \sup_\omega \mu_\Delta(G^F(\tilde{P}, Q)) \qquad (12.5-8)$$

where the superscript ^{F} in G^F is used to indicate that at the robustness stage, the design variables are the IMC filter parameters [see (12.5-11)].

For good performance a weighted norm of the sensitivity function $\tilde{E}$ should be made small (12.5-1). Usually the 2-norm, ∞-norm or the ISE for a finite set of inputs is minimized (12.5-2). Robust stability imposes constraints on the controller design: multiplicative output uncertainty constrains the maximum singular value of the complementary sensitivity function (12.5-3) but in general the constraint takes on a complex form (12.5-5). The robust performance constraint is expressed through the SSV. As an alternative to (12.5-8) one could minimize for a specific input the worst ISE that can result from any plant in the set Π.

Often what we need in practice is robust performance. However, at present, efficient and reliable techniques for the solution of (12.5-8) are not available and the specification of the performance weights which yield a practically useful controller is an art. Furthermore, sometimes it is more meaningful to consider an H_2-type performance objective [and not the H_∞-type implicit in 12.5-8] and to optimize nominal performance subject to the constraint of "robust stability" or "robust performance." Also, it is desirable to provide for convenient on-line robustness adjustment as the plant changes and the model quality degrades with time. For these reasons, we adopt the two-step IMC design approach which is capable of generating good engineering solutions to the robust control problem.

Step 1: Nominal Performance

$\tilde{Q}$ is selected to yield a "good" system response for the input(s) of interest, without regard for constraints and model uncertainty. Several H_2-type objective functions will be considered. Define

$$\Phi(v) = \|We\|_2^2 = \|W\tilde{E}v\|_2^2 = \|W(I - \tilde{P}\tilde{Q})v\|_2^2 \qquad (12.5-9)$$

Then one objective could be

Objective O1:

$$\min_{\tilde{Q}} \Phi(v)$$

for a particular input $v = (v_1, v_2, \ldots, v_n)^T$, where $\tilde{Q}$ satisfies the internal stability requirements. However, minimizing the weighted ISE for just one vector v is not very meaningful, because of the different directions in which the disturbances enter the process or the setpoints are changed. It is desirable to find a $\tilde{Q}$, that minimizes $\Phi(v)$ for every single v in a set of external inputs v of interest for the particular process. For an $n \times n$ P this set can be defined as

$$\mathcal{V} = \{v^i(s) : i = 1, \ldots, n\} \qquad (12.5-10)$$

where $v^1(s), \ldots, v^n(s)$ are vectors that describe the expected directions and frequency content of the external system inputs — e.g., steps, ramps, or other types of inputs.

The objective can then be written as

Objective O2:

$$\min_{\tilde{Q}} \Phi(v) \qquad \forall v \in \mathcal{V}$$

under the constraint that $\tilde{Q}$ satisfies the internal stability requirements. A linear time invariant $\tilde{Q}$ that meets Obj. O2 does not necessarily exist. In Sec. 12.6.3 it will be shown that it exists for a large class of $\mathcal{V}$'s. An alternative objective would be:

Objective O3:

$$\min_{\tilde{Q}}[\Phi(v^1) + \Phi(v^2) + \ldots + \Phi(v^n)]$$

In this case the objective is to minimize the sum of the ISE's that each of the inputs v^i would cause when applied to the system separately. This was the objective discussed in Sec. 10.4.3.

Step 2: Robust Stability and Performance

The aggressive controller $\tilde{Q}$ obtained in Step 1 is detuned to satisfy the robustness requirements. For that purpose $\tilde{Q}$ is augmented by a filter F of fixed structure

$$Q = Q(\tilde{Q}, F) \qquad (12.5-11)$$

and the filter parameters are adjusted to solve (12.5-8). Generally, the weights W_i chosen for nominal performance (12.5-1, 12.5-2) and robust performance (12.5-6 to 12.5-8) are not the same. As in the SISO case the solution of the nominal performance problem suggests an ideal sensitivity operator $\bar{E}$ which cannot be realized in practice because of robustness problems. The robust performance design objective is to make the sensitivity operator as similar to $\bar{E}$ as possible while satisfying all robustness constraints. Thus, generally the performance weight for (12.5-6 to 12.5-8) is derived from the optimal $\bar{E}$ obtained in Step 1. For example, one could choose $W_2 = w_2 I$ where $w_2^{-1} > \bar{\sigma}(\bar{E})$ which implies that the magnitude (maximum singular value) of the sensitivity function is allowed to increase up to w_2^{-1}.

Sometimes F enters Q (12.5-11) in a simple manner

$$Q = \tilde{Q}F \qquad (12.5-12)$$

Sometimes a more complicated form is required as we will discuss later. In general, it might not be possible to meet the robust performance requirement (12.5-7). The reason could be that our two-step design procedure fails to produce an acceptable Q. On the other hand, there might not exist any Q such that (12.5-7) is satisfied. Then the performance requirements have to be relaxed and/or the model uncertainty has to be reduced.

12.6 Nominal Performance

We will first make some assumptions on the input v and the set of inputs $\mathcal{V}$. Then we will find the controllers $\tilde{Q}$ which meet Objs. O1 through O3.

12.6.1 Assumptions

In Sec. 12.3 we made some assumptions on the RHP poles and zeros of $\tilde{P}$ to enable us to derive a simple controller parametrization. The following assumptions on v and $\mathcal{V}$ are of a similar nature.

Assumption A3. *Every nonzero element of v includes all open RHP poles of $\tilde{P}$, each of them with degree 1, and those are the only open RHP poles of v.*

Assumption A4. *Let ℓ_i be the maximum number of poles at $s = 0$ of any element in the ith row of $\tilde{P}$. Then the ith element of v, $v_i(s)$, has at least ℓ_i poles at $s = 0$. Also v has no other poles on the imaginary axis and its elements have no finite zeros on the imaginary axis.*

Assumptions A3 and A4 are not restrictive when v is an output disturbance d, generated by an input disturbance that has passed through the process. In that case d usually includes all the unstable process poles as postulated by Assns. A3 and A4. Note that the control system will also reject other disturbances with fewer unstable poles, without producing steady-state offset. Assumption A4 is different for poles at $s = 0$ because their number in each row of $\tilde{P}$ can be different (capacitancies may be associated only with certain process outputs). Also, the output disturbance may have more poles at $s = 0$ than the process.

Assumptions A3 and A4 may be restrictive for setpoints. Then, however, we can employ the two-degree-of-freedom structure (Sec. 5.2.4) which allows us to disregard the existence of any unstable poles of $\tilde{P}$ and Assns. A3 and A4 are not needed.

For Objs. O2 and O3 we are considering the set $\mathcal{V}$ (12.5–10) of n inputs v^i, $i = 1, n$. Define the square matrix

$$V(s) \triangleq (v^1(s) \quad v^2(s) \quad \dots \quad v^n(s)) \qquad (12.6 - 1)$$

Assumption A5. *The matrix V has no zeros at the location of its unstable poles and no finite zeros on the imaginary axis, and V^{-1} cancels the closed RHP poles of $\tilde{P}$ in $V^{-1}\tilde{P}$.*

Assumptions A3 and A4 for each column of V do not necessarily imply that Assn. A5 is satisfied. A matrix V which satisfies Assn. A5 can be easily constructed. One way is to obtain V as $\tilde{P}$ times a matrix with no open RHP poles and no finite zeros on the imaginary axis. This case corresponds to the physically meaningful situation, where the output disturbances v^i are generated by disturbances at the plant input. Another way is to use a diagonal V, in which case satisfaction of Assns. A3 and A4 by every column of V implies satisfaction of Assn. A5.

Note that Assns. A1-A5 are only relevant for unstable plants and impose no restrictions on stable plants.

12.6.2 H_2-Optimal Control for a Specific Input

The solution of Obj. O1 is also the first step toward the solution of Obj. O2 if such a solution exists. Let $v_0(s)$ be the scalar allpass that includes the *common*

RHP zeros of the elements of v. Factor v as follows:

$$v(s) = v_0(s)(\hat{v}_1(s) \ldots \hat{v}_n(s))^T \triangleq v_0(s)\hat{v}(s)^T \qquad (12.6-2)$$

The plant P can be factored into a stable allpass portion P_A and an MP portion P_M such that

$$P = P_A P_M \qquad (12.6-3)$$

Hence P_A and P_M^{-1} are stable and $P_A^H(i\omega)P_A(i\omega) = I$. The procedure for carrying out this "inner-outer" factorization is discussed in Sec. 12.6.4.

Theorem 12.6-1. *Assume that Assns. A1-A4 hold. Then the set of controllers $\tilde{Q}$ that solves Obj. O1 satisfies*

$$\tilde{Q}\hat{v} = P_M^{-1}W^{-1}\{WP_A^{-1}\hat{v}\}_* \qquad (12.6-4)$$

where the operator $\{\cdot\}_*$ denotes that after a partial fraction expansion of the operand, all terms involving the poles of P_A^{-1} are omitted. Furthermore, for $n \geq 2$ the number of stabilizing controllers that satisfy (12.6-4) is infinite. Guidelines for the construction of a controller are given in the proof.

We stress that not every controller that satisfies (12.6-4) is stabilizing. Improper stabilizing controllers that satisfy (12.6-4) are accepted because in the second step of the design procedure a filter with the appropriate roll-off will produce a proper $Q(s)$. Note that any *constant* weight W cancels in (12.6-4).

The relationship (12.6-4) should be compared with the H_2-optimal controller for SISO systems stated in Thm. 5.2-1. If we assume that the plant and the disturbance have the *same* open-loop RHP poles, then the expressions (5.2-6) and (12.6-4) are equivalent. For SISO systems we did not have to make Assn. A3 because we can easily factor out the unstable portions of p and v.

Proof of Theorem 12.6-1. The proof of (12.6-4) is somewhat lengthy and can be skipped by the result-oriented reader without loss of continuity. We will assume $W = I$. The proof of the weighted case if left as an exercise. Let V_0 be a diagonal matrix where each column satisfies Assn. A3 and every element has ℓ_v poles at $s = 0$ where ℓ_v is the maximum number of such poles in any element of v. Assume that there exists Q_0 which stabilizes $\tilde{P}$ in the sense of Thm. 12.3-1 and also makes $(I - PQ_0)V_0$ stable. Its existence will be proven by construction. Substitution of (12.3-1) into (12.5-9) and use of the fact that pre- or post-multiplication of a function with an allpass does not change its L_2-norm, yields:

$$\Phi(v) = \|P_A^{-1}(I - PQ_0)\hat{v} - P_M Q_1 \hat{v}\|_2^2 \triangleq \|f - P_M Q_1 \hat{v}\|_2^2 \qquad (12.6-5)$$

L_2, the space of functions square integrable on the imaginary axis, can be decomposed into two subspaces: H_2 the subspace of functions analytic in the RHP (stable functions) and its orthogonal complement $H_2^\perp$ that includes all strictly unstable functions (functions with *all* their poles in the RHP). The term f has no poles on the imaginary axis because $(I - PQ_0)V_0$ has no such poles. However, f may not be an L_2 function because of the existence of a constant term in its PFE. This term is equal to $\lim_{s \to +\infty} f(s) \triangleq f(\infty)$. Then $\hat{f}(s) \triangleq f(s) - f(\infty)$ is in L_2 and it can be uniquely decomposed into two orthogonal functions $\{\hat{f}\}_+ \in H_2$ and $\{\hat{f}\}_- \in H_2^\perp$:

$$\hat{f}(s) \triangleq f(s) - f(\infty) = \{\hat{f}\}_+ + \{\hat{f}\}_-$$

In order for $\Phi(v)$ to be finite, the optimal Q_1 has to make $f - P_M Q_1 \hat{v}$ strictly proper. Hence Q_1 has to satisfy

$$\lim_{s \to \infty}(P_M Q_1 \hat{v}) \triangleq (P_M Q_1 \hat{v})(\infty) = f(\infty) \qquad (12.6-6)$$

Next we want to show that $P_M Q_1 \hat{v}$ has to be stable. The fact that $(I - PQ_0)V_0$ is stable and the way V_0 is constructed imply that $(I - PQ_0)v$ is stable. Then $(I - PQ_0)\hat{v}$ will also be stable because of the definition of v_0. We require that $(I - PQ)v$ have no open RHP poles and therefore that $(I - PQ)\hat{v} = (I - PQ_0)\hat{v} - PQ_1\hat{v}$ have no open RHP poles. But since $(I - PQ_0)\hat{v}$ is stable, this requirement reduces to $PQ_1\hat{v}$ having no poles in the open RHP. Also in order for $\Phi(v)$ to be finite, Q_1 must be such that $(I - PQ)v$ or $(I - PQ)\hat{v}$ has no poles on the imaginary axis. But since $(I - PQ_0)\hat{v}$ is stable this is equivalent to $PQ_1\hat{v}$ having no poles on the imaginary axis. Hence the optimal Q_1 must be such that $PQ_1\hat{v}$ is stable. Then the only possible unstable poles of $P_M Q_1 \hat{v} = P_A^{-1} PQ_1\hat{v}$ are the poles of P_A^{-1}. But Assns. A1, A2 imply that the poles of P_A^{-1} are not among those of $P_M Q_1 \hat{v}$ and therefore $P_M Q_1 \hat{v}$ has to be stable. To proceed we will assume that Q_1 has this property. We will verify later that the solution has indeed this property.

The fact that $P_M Q_1 \hat{v}$ is stable and (12.6-6) imply that $P_M Q_1 \hat{v} - f(\infty)$ is in H_2. Thus, we can then write (12.6-5) as

$$\Phi(v) = \|\{\hat{f}\}_-\|_2^2 + \|\{\hat{f}\}_+ - (P_M Q_1 \hat{v} - f(\infty))\|_2^2$$

Because $\hat{f}$ does not depend on Q_1, the obvious solution to the minimization of $\Phi(v)$ is

$$P_M Q_1 \hat{v} - f(\infty) = \{\hat{f}\}_+$$

or

$$P_M Q_1 \hat{v} = \{\hat{f}\}_+ + f(\infty) \triangleq \{f\}_{\infty+}$$

or

$$Q_1 \hat{v} = P_M^{-1} \{f\}_{\infty+} \qquad (12.6-7)$$

where the notation $\{\cdot\}_{\infty+}$ is used to indicate that after a PFE all strictly proper stable terms as well as the constant term are retained.

Clearly such a Q_1 produces a stable $P_M Q_1 \hat{v}$ as was assumed and also satisfies (12.6–6). It should now be proved that Q_1's that satisfy the internal stability requirements exist among those described by (12.6–7) so that the obvious solution is a true solution. For $n = 1$, (12.6–7) yields a unique Q_1, which we have shown to satisfy the requirements in the proof of Thm. 5.2-1. For $n \geq 2$ write

$$Q_1 \triangleq (q_1 \quad q_2) \qquad (12.6-8)$$

$$\hat{V}_2 \triangleq (\hat{v}_2 \quad \ldots \quad \hat{v}_n)^T \qquad (12.6-9)$$

where without loss of generality the first element of v, and thus $\hat{v}_1$, is assumed to be nonzero. Also q_1 is $n \times 1$ and q_2 is $n \times (n-1)$. Then it follows from (12.6–7) that

$$Q_1 = \left(\hat{v}_1^{-1}(P_M^{-1}\{f\}_{\infty+} - q_2\hat{V}_2) \quad q_2 \right) \qquad (12.6-10)$$

We now need to show that a stable q_2 exists such that Q_1 is stable and produces a stable PQ_1P. Select a q_2 of the form:

$$q_2(s) = \hat{q}_2(s)s^{3\ell_v} \prod_{i=1}^{k}(s - \pi_i)^3 \qquad (12.6-11)$$

where $\hat{q}_2$ is stable and $\{\pi_1, \ldots, \pi_k\}$ are the poles of P in the open RHP. Then if follows from (12.6–10) that in order for PQ_1P to be stable it is sufficient that $P\hat{v}_1^{-1}P_M^{-1}\{f\}_{\infty+}\{P\}_{1^{st}row}$ have no closed RHP poles. But $PP_M^{-1} = P_A$ is stable and the only possible unstable poles of $\hat{v}_1^{-1}\{P\}_{1^{st}row}$ are open RHP poles of $\hat{v}_1^{-1}$ because of Assns. A3 and A4. These are also the only possible unstable poles of Q_1. Let α be such a pole (zero of $\hat{v}_1$). Then for stability we need to find $\hat{q}_2$ such that

$$\hat{q}_2(\alpha)\hat{V}_2(\alpha) = \alpha^{-3\ell_v} \prod_{i=1}^{k}(\alpha - \pi_i)^{-3} P_M^{-1}(\alpha)\{f\}_{\infty+}\Big|_{s=\alpha} \qquad (12.6-12)$$

The above equation always has a solution because the vector $\hat{V}_2(\alpha)$ is not identically zero since any common RHP zeros in v were factored out in v_0.

We shall now proceed to obtain an expression for $Q\hat{v}$. Equations (12.3–1) and (12.6–7) yield

$$Q\hat{v} = P_M^{-1}(P_A^{-1}PQ_0\hat{v} - \{P_A^{-1}PQ_0\hat{v}\}_{\infty+} + \{P_A^{-1}\hat{v}\}_{\infty+})$$

$$= P_M^{-1}(\{P_A^{-1}PQ_0\hat{v}\}_{0-} + \{P_A^{-1}\hat{v}\}_{\infty+}) \qquad (12.6-13)$$

where $\{\cdot\}_{0-}$ indicates that in the partial fraction expansion all poles in the closed RHP are retained. For (12.6–13), these poles are the poles of $\hat{v}$ in the closed RHP; $P_A^{-1}PQ_0 = P_M Q_0$ is strictly stable because of Assn. A1 and the fact that Q_0 is a stabilizing controller. The fact that $(I - PQ_0)V_0$ has no poles at $s = 0$ and the form of V_0 imply that $(I - PQ_0)$ and its derivatives up to and including the $(\ell_v - 1)^{th}$ are equal to zero at $s = 0$. Also, the fact that $(I - PQ_0)V_0$ is stable and that the columns of this diagonal V_0 satisfy Assn. A3, imply that $(I - PQ_0) = 0$ at $\pi_1, \ldots, \pi_k$. Hence (12.6–13) simplifies to (12.6–4).

We now need to show that a stabilizing controller Q_0 exists with the property that $(I - PQ_0)V_0$ is stable. The selection of a V_0 with the properties mentioned at the beginning of this section and its use instead of V in (12.6–15) yields such a controller.   □

12.6.3 H_2-Optimal Control for a Set of Inputs

We determined in Sec. 12.6.2 that the set of controllers which meets Obj. O1 for a MIMO system is infinite. Thus Obj. O1 is not a very interesting design objective. In this section we will address Objs. O2 and O3. Clearly any solution for Obj. O2 (if it exists) is also a solution for Obj. O3. Therefore, we will start by determining the controller which meets Obj. O3. If a solution for Obj. O2 exists it will be given by the solution for Obj. O3.

Factor V, defined in (12.6–1), similarly to P (see Sec. 12.6.4)

$$V = V_M V_A \qquad (12.6-14)$$

Then the following theorem holds:

Theorem 12.6-2. *Assume that Assns. A1-A5 hold. Then the controller*

$$\tilde{Q} = P_M^{-1}W^{-1}\{WP_A^{-1}V_M\}_*V_M^{-1} \qquad (12.6-15)$$

is the unique solution for Obj. O3. Here the operator $\{\cdot\}_*$ denotes that after a partial fraction expansion of the operand, all terms involving the poles of P_A^{-1} are omitted.

The formula is identical to that obtained in the SISO case when p and v have the same RHP poles.

Proof. Again we assume $W = I$ and leave the weighted case as an exercise. The L_2-norm of a matrix $G(s)$ analytic on the imaginary axis is given by

$$\|G\|_2 = \left(\frac{1}{2\pi} \int_{-\infty}^{\infty} \text{trace}[G^H(i\omega)G(i\omega)] \quad d\omega\right)^{1/2} \qquad (12.6-16)$$

Then from (12.5–9), (12.6–1) and (12.6–16) it follows that

$$\Phi(v^1) + \Phi(v^2) + \ldots + \Phi(v^n) = \|(I - P\tilde{Q})V\|_2^2 \overset{\Delta}{=} \Phi(V) \qquad (12.6-17)$$

The minimization of $\Phi(V)$ follows the steps in the proof of Thm. 12.6-1 up to (12.6–7), with V_M used instead of $\hat{v}$. In this case ℓ_v is the maximum number of poles at $s = 0$ in any element of V. From the equivalent to (12.6–7) we obtain

$$Q_1 = P_M^{-1}\{f\}_{\infty+}V_M^{-1} \qquad (12.6-18)$$

where V_M is used instead of $\hat{v}$ in f. This Q_1 makes $f - P_M Q_1 V_M$ strictly proper. We now have to establish that Q_1 is stable and produces a stable PQ_1P. In PQ_1P the unstable poles of the P on the left cancel with those of P_M^{-1}. As for the P on the right, the same follows from Assn. A5. Then in the same way that (12.6–4) follows from (12.6–13), (12.6–15) follows from (12.6–18). $\quad\square$

Let us now consider the more meaningful Obj. O2. Factor each of the v^i in the same way as in (12.6–2):

$$v^i(s) = v_0^i(s)\hat{v}^i(s) \qquad (12.6-19)$$

Define

$$\hat{V} \overset{\Delta}{=} (\hat{v}^1 \quad \hat{v}^2 \quad \ldots \quad \hat{v}^n) \qquad (12.6-20)$$

Theorem 12.6-3.

i) *If $\hat{V}(s)$ is non-minimum phase, then there exists no solution to O2.*

ii) *If $\hat{V}(s)$ is minimum phase, then use of $\hat{V}$ instead of V_M in (12.6–15) yields exactly the same $\tilde{Q}$, which also solves Obj. O2. In addition $\tilde{Q}$ minimizes $\Phi(v)$ for any v that is a linear combination of v^i's that have the same v_0^i's.*

Proof. $(W = I)$. A stabilizing controller that solves Obj. O2 has to solve Obj. O1 for all v^i, $\quad i = 1, \ldots n$. Satisfying (12.6–4) for every v^i is equivalent to

$$\tilde{Q} = P_M^{-1}\{P_A^{-1}\hat{V}\}_*\hat{V}^{-1} \qquad (12.6-21)$$

Hence the above $\tilde{Q}$ is the only potential solution for Obj. O2. However, it is not necessarily a stabilizing controller since not only stabilizing $\tilde{Q}$'s satisfy (12.6–4) for some v. Indeed, if $\hat{V}$ is non-minimum phase, $\hat{V}^{-1}$ is unstable and this results in an unstable $\tilde{Q}$, which is therefore unacceptable. Hence in such a case, there exists no solution for Obj. O2, which completes the proof of part (i) of the theorem.

In the case where $\hat{V}^{-1}$ is stable ($\hat{V}$ minimum phase), the controller given by (12.6–21) is stable and therefore it is the same as the one given by (12.6–15). This fact can be explained as follows. We have

$$V = \hat{V}V_0 \qquad\qquad (12.6-22)$$

where

$$V_0 = \text{diag}\left\{v_0^1, v_0^2, \ldots, v_0^n\right\} \qquad\qquad (12.6-23)$$

Since $\hat{V}^{-1}$ is stable, (12.6–22) represents a factorization of V similar to that in (12.6–14). From spectral factorization theory it follows that

$$\hat{V}(s) = V_M(s)A \qquad\qquad (12.6-24)$$

where A is a constant matrix, such that $AA^H = I$. Then (12.6–15) is not altered when $\hat{V}$ is used instead of V_M because A cancels.

Let us now assume without loss of generality that the first j v^i's have the same v_0^i's. Consider a v that is a linear combination of $v^1, \ldots, v^j$:

$$v(s) = \alpha_1 v^1(s) + \ldots + \alpha_j v^j(s) \qquad\qquad (12.6-25)$$

Then it follows that

$$v_0(s) = v_0^1(s) = \ldots = v_0^j(s) \qquad\qquad (12.6-26)$$

$$\hat{v}(s) = \alpha_1 \hat{v}^1(s) + \ldots + \alpha_j \hat{v}^j(s) \qquad\qquad (12.6-27)$$

One can easily check that a $\tilde{Q}$ that satisfies (12.6–4) for $\hat{v}^1, \ldots, \hat{v}^j$, will also satisfy (12.6–4) for the $\hat{v}$ given by (12.6–27) because of the property

$$\{\alpha_1 f_1(s) + \ldots + \alpha_j f_j(s)\}_* = \alpha_1 \{f_1(s)\}_* + \ldots + \alpha_j \{f_j(s)\}_* \qquad (12.6-28)$$

But then from Thm. 12.6-1 it follows that if a stabilizing controller $\tilde{Q}$ satisfies (12.6–4) for $\hat{v}$, then it minimizes the ISE $\Phi(v)$. $\square$

The following corollary to Thm. 12.6-3 holds for a specific choice of V.

Corollary 12.6-1. *Let*

$$V = \operatorname{diag} \{v_1, v_2, \ldots, v_n\} \qquad (12.6-29)$$

where $v_1(s), \ldots, v_n(s)$ are scalars. Then use of $\hat{V}$ instead of V_M in (12.6–15) yields exactly the same $\tilde{Q}$, which minimizes $\Phi(v)$ for the following n vectors:

$$v = \begin{pmatrix} v_1 \\ 0 \\ \vdots \\ 0 \end{pmatrix}, \begin{pmatrix} 0 \\ v_2 \\ \vdots \\ 0 \end{pmatrix}, \ldots, \begin{pmatrix} 0 \\ 0 \\ \vdots \\ v_n \end{pmatrix} \qquad (12.6-30)$$

and their multiples, as well as for the linear combinations of those directions that correspond to v_i's with the same open RHP zeros with the same degree.

In summary it is most popular to obtain $\tilde{Q}$ by solving Obj. O3. As we pointed out in Sec. 10.4.3 the $\tilde{Q}$ which solves Obj. O3 minimizes the 2-norm of the weighted sensitivity operator. Theorem 12.6-3 can be helpful for choosing a meaningful V.

12.6.4 Algorithm for "Inner-Outer" Factorization

The following theorem is the tool for obtaining the factorizations (12.6–3) and (12.6–14).

Theorem 12.6-4 (Chu, 1985). *Let* $G(s) = C(sI - A)^{-1}B + D$ *be a minimal realization of the square transfer matrix* $G(s)$, *and let* $G(s)$ *have no zeros on the* $i\omega$*-axis including infinity. Then we have*

$$G(s) = N(s)M(s)^{-1} \qquad (12.6-31)$$

where N, M *are stable and* $N(i\omega)^H N(i\omega) = I$. $N(s)$ *and* $M(s)^{-1}$ *are given by*

$$N(s) = (C - QF)(sI - (A - BR^{-1}F))^{-1}BR^{-1} + Q \qquad (12.6-32)$$

$$M(s)^{-1} = F(sI - A)^{-1}B + R \qquad (12.6-33)$$

where

$$D = QR \qquad (12.6-34)$$

is the QR factorization of D into an orthogonal matrix Q $(Q^T Q = I)$ *and an upper triangular matrix R, and*

$$F = Q^T C + (BR^{-1})^T X \qquad (12.6-35)$$

with X the stabilizing [i.e., it makes $(A - BR^{-1}F)$ *stable] real symmetric solution of the following algebraic Riccati equation (ARE):*

$$(A - BR^{-1}Q^T C)^T X + X(A - BR^{-1}Q^T C) - X(BR^{-1})(BR^{-1})^T X = 0 \quad (12.6-36)$$

Such a solution X exists for (12.6–36).

Methods for solving ARE's can be found in the cited literature. Many control software packages include programs for that purpose.

By comparing (12.6–31) to (12.6–3) we see that $P_A = N$ and $P_M = M^{-1}$. Note however that a $P(s)$ that represents a physical system, is usually strictly proper ($D = 0$). In order to apply Thm. 12.6-4 in such a case we need to add to $P(s)$ a small $D = \epsilon I$. Repeated applications have shown this to be a satisfactory approach. Also note that $N(s)$ in (12.6–32) is unique only up to premultiplication with a constant unitary matrix U ($U^H U = 1$). Although not necessary, one may wish for consistency with the SISO case, to choose $P_A(s)$ such that $P_A(0) = I$. This can be accomplished by premultiplying $N(s)$ (12.6–32) by $N(0)^{-1}$.

A comparison of (12.6–31) and (12.6–14) indicates that Thm. 12.6-4 cannot be directly applied to $G(s) = V(s)$. We can, however, apply Thm. 12.6-4 to $\tilde{V}(s) \triangleq V^T(s)$ to obtain

$$\tilde{V}(s) = V^T(s) = \tilde{V}_A(s)\tilde{V}_M(s) \qquad (12.6 - 37)$$

with $\tilde{V}_A, \tilde{V}_M^{-1}$ stable and $\tilde{V}_A(i\omega)^H \tilde{V}_A(i\omega) = I$. From (12.6–37) we find

$$V(s) = \tilde{V}_M^T(s)\tilde{V}_A^T(s) \qquad (12.6 - 38)$$

Since $\tilde{V}_A^T(s), (\tilde{V}_M^T(s))^{-1}$ are stable and

$$\tilde{V}_A^T(i\omega)(\tilde{V}_A^T(i\omega))^H = (\tilde{V}_A^H(i\omega)\tilde{V}_A(i\omega))^T = I$$

we can select V_A, V_M as

$$V_A(s) = \tilde{V}_A^T(s) \qquad (12.6 - 39)$$

$$V_M(s) = \tilde{V}_M^T(s) \qquad (12.6 - 40)$$

Similarly to the case for $P(s)$, we may have to introduce a small D matrix in $V(s)$. Also the factorization is only unique up to post-multiplication with a constant unitary matrix.

12.7 Robust Stability and Performance

The controller $\tilde{Q}$ is to be detuned through a lowpass filter F such that for the detuned controller $Q = Q(\tilde{Q}, F)$ both the robust stability (12.5–5) and the robust performance (12.5–7) conditions are satisfied. Because (12.5–7) implies (12.5–5) a procedure is proposed to minimize $\mu_\Delta(G^F(\tilde{P}, Q))$ as a function of the filter parameters for a filter with a fixed structure. First we will postulate reasonable filter structures. Then we will discuss an algorithm for carrying out the minimization.

12.7.1 Filter Structure

In principle the structure of F can be as complex as the designer wishes. However, in order to keep the number of variables in the optimization problem small, a simple structure like a diagonal F with first- or second-order terms is recommended. In most cases this is not restrictive because the controller $\tilde{Q}$ designed in the first step of the IMC procedure is a full matrix with high order elements. Some restrictions must be imposed on the filter in the case of an open-loop unstable plant. Also a more complex filter structure may be necessary in cases of highly ill-conditioned systems.

Open-Loop Unstable Plants.
   The filter $F(s)$ is chosen to be a diagonal rational function that satisfies the following requirements:

(a) Pole-zero excess. The controller $Q = \tilde{Q}F$ must be proper. Assume that the designer has specified a pole-zero excess of ν for the filter $F(s)$.

(b) Internal stability. The transfer matrix S in (12.2–2) must be stable.

(c) Asymptotic setpoint tracking and/or disturbance rejection. $(I-\tilde{P}\tilde{Q}F)v$ must be stable.

Write

$$F(s) = \text{diag}\,\{f_1(s),\ldots,f_n(s)\} \qquad (12.7-1)$$

Under Assns. A1-A5, (b) and (c) are equivalent to the following conditions. Let π_i $(i = 1,\ldots,k)$ be the open RHP poles of $\tilde{P}$. Let $\pi_0 = 0$ and m_{0l} be the largest multiplicity of such a pole in any element of the lth row of V. From Assns. A1-A5 and the fact that $\tilde{Q}$ makes S and $(I-P\tilde{Q})V$ stable, it follows that the lth element, f_l, of the filter F must satisfy:

$$f_l(\pi_i) = 1, \qquad i = 0, 1, \ldots, k \qquad (12.7-2)$$

$$\frac{d^j}{ds^j} f_l(s)\Big|_{s=\pi_0} = 0, \qquad j = 1, \ldots, m_{0l} - 1 \qquad (12.7-3)$$

Requirements (12.7–2) clearly show the limitation RHP poles place on the robustness properties of a control system designed for an open-loop unstable plant. Since because of (12.7–2) one cannot reduce the closed-loop bandwidth of the nominal system at frequencies corresponding to the RHP poles of the plant, only a relatively small model error can be tolerated at those frequencies.

Experience has shown that the following structure for a filter element $f_l(s)$ is reasonable:

$$f_l(s) = \frac{a_{\nu_l-1,l}s^{\nu_l-1} + \ldots + a_{1,l}s + a_{0,l}}{(\lambda s + 1)^{\nu+\nu_l-1}} \qquad (12.7-4)$$

where

$$\nu_l = m_{0l} + k \qquad (12.7-5)$$

For a specific tuning parameter λ the numerator coefficients can be computed to satisfy (12.7–2) and (12.7–3). This involves solving a system of ν_l linear equations with ν_l unknowns.

Example 12.7-1. Assume that we desire a pole-zero excess of ν and that there is only one pole π. Then

$$f(s) = \frac{(\lambda\pi + 1)^\nu}{(\lambda s + 1)^\nu} \qquad (12.7-6)$$

If $\pi = 0$, (12.7–6) reduces to the standard filter for stable systems $f(s) = (\lambda s + 1)^{-\nu}$. □

Example 12.7-2. Assume that $\nu = 2$ and the only pole is a double pole at $s = 0$. Then

$$f(s) = \frac{3\lambda s + 1}{(\lambda s + 1)^3} \qquad (12.7-7)$$

□

Ill-Conditioned Plants.
Problems arise because the optimal controller $\tilde{Q}$ designed for $\tilde{P}$ tends to be an approximate inverse of $\tilde{P}$ and as a result $\tilde{Q}$ is ill-conditioned when $\tilde{P}$ is. Although robust stability can generally be achieved by significant detuning of the diagonal filter, the robust performance condition is usually not satisfied. We have shown in Sec. 11.3.3 that an ill-conditioned P and C (and therefore an inverting ill-conditioned $\tilde{Q}$ as well) can cause problems when input uncertainty is present. This problem can be addressed through a filter that acts directly on the singular values of $\tilde{Q}$, at the frequency where the condition number of $\tilde{Q}$ is highest, say ω^*. Let

$$\tilde{Q}(i\omega^*) = U_Q\Sigma_Q V_Q^H \qquad (12.7-8)$$

be the SVD of $\tilde{Q}$ at ω^* and let R_u, R_v be real matrices that solve the pseudo-diagonalization problems:

$$U_Q^H R_u \cong I \qquad\qquad (12.7-9)$$

$$V_Q^H R_v \cong I \qquad\qquad (12.7-10)$$

Then the controller Q including the filter, is chosen to be of the form

$$Q(s) = R_u F_1(s) R_u^{-1} \tilde{Q}(s) F_2(s) \qquad\qquad (12.7-11)$$

or

$$Q(s) = \tilde{Q}(s) R_v F_1(s) R_v^{-1} F_2(s) \qquad\qquad (12.7-12)$$

where $F_1(s)$, $F_2(s)$ are diagonal filters, such that $F_1(0) = F_2(0) = I$ if integral action is desired. Note that for F_1, m_0 should be used in (12.7-3), (12.7-5), for all l, instead of m_{0l}, where $m_0 = \max_l m_{0l}$. This is required for internal stability and no steady-state offset.

It should be pointed out that the success of this approach depends on the quality of either one of the pseudo-diagonalizations (12.7-9) or (12.7-10). The diagonalization will be perfect if U_Q or V_Q is real. This will happen for example if $\omega^* = 0$, that is if the problems arise because the plant is ill-conditioned at steady-state, as for example high purity distillation columns are.

12.7.2 General Interconnection Structure with Filter

Consider the block diagram in Fig. 12.1-1B. Assume that $Q = \tilde{Q}F$. In order to use the SSV effectively for designing F, some rearrangement of the structure is necessary. Figure 12.1-1B can be transformed to Fig. 12.7-1A, where $v = d - r$, $e = y - r$ and

$$G = \begin{pmatrix} 0 & 0 & \tilde{Q} \\ I & I & \tilde{P}\tilde{Q} \\ -I & -I & 0 \end{pmatrix} \qquad\qquad (12.7-13)$$

where the blocks 0 and I have the appropriate dimensions.

We will assume throughout the following that the set of plants Π can be modeled in a way that allows Fig. 12.7-1A to be transformed into Fig. 12.7-1B where Δ is a block diagonal matrix with the additional property that $\bar{\sigma}(\Delta) \leq 1$, $\forall \omega$ (see Sec. 11.2.1). The superscript u in G^u denotes the dependence of G^u not only on G but also on the specific uncertainty description available for the model $\tilde{P}$. Next G^u is derived for some typical examples.
1. *Additive Uncertainty* (11.1–1), (11.1–5)

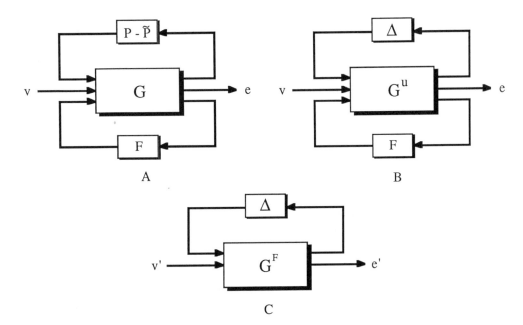

Figure 12.7-1. Model uncertainty block diagrams.

$$G^u = G^A = \begin{pmatrix} \bar{\ell}_A I & 0 & 0 \\ 0 & I & 0 \\ 0 & 0 & I \end{pmatrix} G \qquad (12.7-14)$$

where G is defined in (12.7–13)

2. *Multiplicative Output Uncertainty* (11.1–2), (11.1–5)

$$G^u = G^O = \begin{pmatrix} \bar{\ell}_O \tilde{P} & 0 & 0 \\ 0 & I & 0 \\ 0 & 0 & I \end{pmatrix} G \qquad (12.7-15)$$

3. *Multiplicative Input Uncertainty* (11.1–3), (11.1–5)

$$G^u = G^I = G \begin{pmatrix} \bar{\ell}_I \tilde{P} & 0 & 0 \\ 0 & I & 0 \\ 0 & 0 & I \end{pmatrix} \qquad (12.7-16)$$

4. *Independent Uncertainty in the Transfer Matrix Elements* (11.2–32), (11.2–33)

$$G^u = G^e = \begin{pmatrix} W_{e1} & 0 & 0 \\ 0 & I & 0 \\ 0 & 0 & I \end{pmatrix} G \begin{pmatrix} W_{e2} & 0 & 0 \\ 0 & I & 0 \\ 0 & 0 & I \end{pmatrix} \qquad (12.7-17)$$

Note that all the above relations yield a G^u already partitioned as

$$G^u = \begin{pmatrix} G^u_{11} & G^u_{12} & G^u_{13} \\ G^u_{21} & G^u_{22} & G^u_{23} \\ G^u_{31} & G^u_{32} & G^u_{33} \end{pmatrix} \qquad (12.7-18)$$

Then Fig. 12.7-1B can be written as Fig. 12.7-1C with $e' = W_2 e$ and $v = W_1 v'$

$$G^F = \begin{pmatrix} G^u_{11} & G^u_{12} W_1 \\ W_2 G^u_{21} & W_2 G^u_{22} W_1 \end{pmatrix} + \begin{pmatrix} G^u_{13} \\ W_2 G^u_{23} \end{pmatrix} (I - FG^u_{33})^{-1} F (G^u_{31} \quad G^u_{32} W_1)$$

$$\triangleq \begin{pmatrix} G^F_{11} & G^F_{12} \\ G^F_{21} & G^F_{22} \end{pmatrix} \qquad (12.7-19)$$

If the special filter structure for ill-conditioned systems is used, the procedure for deriving G^u is slightly modified. Define

$$F(s) = \text{diag}\{F_1(s), F_2(s)\} \qquad (12.7-20)$$

and

$$\tilde{Q}_A(s) = R_u; \qquad A(s) = R_u^{-1} \tilde{Q}(s) \qquad (12.7-21)$$

or

$$\tilde{Q}_A(s) = \tilde{Q}(s) R_v; \qquad A(s) = R_v^{-1} \qquad (12.7-22)$$

depending on whether (12.7–11) or (12.7–12) is chosen. Then in Fig. 12.7-1B use $G^{u,ill}$ instead of G^u, where

$$G^{u,ill} = \begin{pmatrix} G^u_{11} & G^u_{12} & G^u_{13} & 0 \\ G^u_{21} & G^u_{22} & G^u_{23} & 0 \\ 0 & 0 & 0 & A \\ G^u_{31} & G^u_{32} & G^u_{33} & 0 \end{pmatrix} \qquad (12.7-23)$$

and G^u is obtained from G (12.7–13) with $\tilde{Q}$ replaced by $\tilde{Q}_A$.

12.7.3 Robust Control: H_∞ Performance Objective

We can write

$$F \triangleq F(s; \Lambda) \qquad (12.7-24)$$

where Λ is an array of filter parameters. Then the design problem (12.5-8) has been converted into a nonlinear program with the constraint that the filter F must be stable. This is easily accomplished by expressing the filter denominator as a product of polynomials of degree 2 whose coefficients must be positive. If an odd degree is desired, an additional first-order term with positive coefficients can be included. The positivity constraint can be handled by writing the coefficients as squares of the independent variables Λ. The design objective (12.5–8) becomes

Objective O4:

$$\min_{\Lambda} \sup_{\omega} \mu_\Delta(G^F(P, \tilde{Q})) \qquad (12.7-25)$$

One should note that the objective function is not convex. Good initial guesses for the filter parameters (elements of Λ) can usually be obtained by matching them with the frequencies where the peaks of $\mu_\Delta(G^F)$ appear for $F = I$.

The SSV μ in Obj. O4 is computed by minimizing its upper bound (11.2–15). In the computation of the supremum in Obj. O4 only a finite number of frequencies is considered. Hence Obj. O4 is transformed into

Objective O4':

$$\min_{\Lambda} \max_{\omega \in \Omega} \inf_{D \in \mathcal{D}} \bar{\sigma}(DG^F D^{-1})$$

where Ω is a finite set of frequencies. Define

$$\Phi_\infty(\Lambda) \triangleq \max_{\omega \in \Omega} \inf_{D \in \mathcal{D}} \bar{\sigma}(DG^F D^{-1}) \qquad (12.7-26)$$

The gradient of Φ_∞ with respect to Λ can be computed analytically except when two or more of the largest singular values of $DG^F D^{-1}$ coincide. This is quite uncommon, however, and although the computation of a generalized gradient is possible, experience has shown the use of a mean direction to be satisfactory. A similar problem appears when the $\max_{\omega \in \Omega}$ is attained at more than one frequency, but again the use of a mean direction seems to be sufficient. We shall now proceed to obtain the expression for the gradient of $\Phi_\infty(\Lambda)$ in the general case.

Assume that for the value of Λ where the gradient of $\Phi_\infty(\Lambda)$ is computed, the $\max_{\omega \in \Omega}$ is attained at $\omega = \omega_0$ and that the $\inf_{D \in \mathcal{D}} \bar{\sigma}(DG^F(i\omega_0)D^{-1})$ is obtained at $D = D_0$, where only one singular value σ_1 is equal to $\bar{\sigma}$. Let the singular value decomposition be

$$D_0 G^F(i\omega_0)D_0^{-1} = (\mathrm{u}_1 \quad \mathrm{U}) \begin{pmatrix} \sigma_1 & 0 \\ 0 & \Sigma \end{pmatrix} \begin{pmatrix} \mathrm{v}_1^H \\ V^H \end{pmatrix} \qquad (12.7-27)$$

Then for the element of the gradient vector corresponding to the filter parameter λ_k we have under the above assumptions:

$$\frac{\partial}{\partial \lambda_k}\Phi_\infty = \frac{\partial}{\partial \lambda_k}\sigma_1(D_0 G^F(i\omega_0)D_0^{-1}) \qquad (12.7-28)$$

because $\nabla_{D_0}(\sigma_1) = 0$ since we are at an optimum with respect to the D's. To simplify the notation use

$$A \triangleq D_0 G^F(i\omega_0)D_0^{-1} = U_A \Sigma_A V_A^H \qquad (12.7-29)$$

Then the gradient can be computed as follows:

$$AA^H = U_A \Sigma_A^2 U_A^H$$

$$\Rightarrow \frac{\partial}{\partial \lambda_k}(AA^H) = \frac{\partial}{\partial \lambda_k}(U_A)\Sigma_A^2 U_A^H + U_A \frac{\partial}{\partial \lambda_k}(\Sigma_A^2)U_A^H + U_A \Sigma_A^2 \frac{\partial}{\partial \lambda_k}(U_A^H)$$

Multiply this expression by u_1^H on the left and u_1 on the right. Consider the first term on the RHS

$$u_1^H \frac{\partial}{\partial \lambda_k}(U_A)\Sigma_A^2 U_A^H u_1 = u_1^H \frac{\partial}{\partial \lambda_k}(U_A)(\sigma_1^2 \quad 0 \quad \cdots \quad 0)^T = \sigma_1^2 u_1^H \frac{\partial}{\partial \lambda_k} u_1 = 0$$

because u_1 is a vector on the unit sphere whose gradient is orthogonal to u_1. Thus both the first and the last term on the RHS vanish and we find

$$u_1^H \frac{\partial}{\partial \lambda_k}(AA^H)u_1 = u_1^H(\frac{\partial}{\partial \lambda_k}(A)A^H + A\frac{\partial}{\partial \lambda_k}(A^H))u_1 = u_1^H U_A(2\Sigma_A \frac{\partial}{\partial \lambda_k}(\Sigma_A))U_A^H u_1$$

$$\Rightarrow u_1^H \frac{\partial}{\partial \lambda_k}(A)v_1\sigma_1 + \sigma_1 v_1^H \frac{\partial}{\partial \lambda_k}(A^H)u_1 = 2\sigma_1 \frac{\partial}{\partial \lambda_k}(\sigma_1)$$

$$\Rightarrow \frac{\partial}{\partial \lambda_k}(\sigma_1) = \text{Re}\left[u_1^H\left(\frac{\partial}{\partial \lambda_k}(D_0 G^F(i\omega_0)D_0^{-1})\right)v_1\right] \qquad (12.7-30)$$

From (12.7–19) we get

$$\frac{\partial}{\partial \lambda_k}(D_0 G^F(i\omega_0)D_0^{-1}) = D_0 \begin{pmatrix} G_{13}^u \\ W_2 G_{23}^u \end{pmatrix} \frac{\partial}{\partial \lambda_k}\left[(I - FG_{33}^u)^{-1}F\right]\begin{pmatrix} G_{31}^u & G_{32}^u W_1 \end{pmatrix}D_0^{-1}$$

$$(12.7-31)$$

We use the identity

$$\frac{d}{dz}(M(z)^{-1}) = -M(z)^{-1}\frac{d}{dz}(M(z))M(z)^{-1} \qquad (12.7-32)$$

in

$$\frac{\partial}{\partial \lambda_k}((I - FG_{33}^u)^{-1}F) = \left(\frac{\partial}{\partial \lambda_k}(I - FG_{33}^u)^{-1}\right)F + (I - FG_{33}^u)^{-1}\frac{\partial F}{\partial \lambda_k} \quad (12.7-33)$$

to obtain

$$\frac{\partial}{\partial \lambda_k}((I-FG_{33}^u)^{-1}F) = -(I-FG_{33}^u)^{-1}\left(\frac{\partial}{\partial \lambda_k}(I-FG_{33}^u)\right)(I-FG_{33}^u)^{-1}F+(I-FG_{33}^u)^{-1}\frac{\partial F}{\partial \lambda_k}$$

$$= (I - FG_{33}^u)^{-1} \cdot \frac{\partial F}{\partial \lambda_k} \cdot (G_{33}^u(I - FG_{33}^u)^{-1}F + I)$$

$$= (I - FG_{33}^u)^{-1} \cdot \frac{\partial F}{\partial \lambda_k} \cdot (I - G_{33}^u F)^{-1} \quad (12.7-34)$$

Equation (12.7–28) can now be expressed in terms of (12.7–30), (12.7–31) and (12.7–34):

$$\frac{\partial}{\partial \lambda_k}\Phi_\infty = \mathrm{Re}\left[\mathbf{u}_1^H D_0 \begin{pmatrix} G_{13}^u \\ W_2 G_{23}^u \end{pmatrix} (I - FG_{33}^u)^{-1}\frac{\partial}{\partial \lambda_k}(F(i\omega_0))\right.$$

$$\left. (I - G_{33}^u F)^{-1}(G_{31}^u \quad G_{32}^u W_1)D_0^{-1}\mathbf{v}_1\right] \quad (12.7-35)$$

where F, G_{ij}^u, W_1, and W_2 are computed at $\omega = \omega_0$. The derivatives of F with respect to its parameters (elements of Λ) depend on the particular filter form selected by the designer and can be computed easily.

12.7.4   Robust Control: H_2-Type Performance Objective

In the previous section we outlined how to design the filter such that in the presence of model uncertainty the sensitivity operator remains "close" to its nominal value. If the performance for a *particular* external input v is of primary interest then it will be more appropriate to minimize the ISE for this particular input in the presence of model uncertainty — i.e., to minimize

$$\max_{P\in\Pi}\|W_2 Ev\|_2^2 = \frac{1}{2\pi}\int_{-\infty}^{\infty}\beta_0^2 \ d\omega \quad (11.3-32)$$

Hence the filter parameters are obtained by solving

   Objective O5:

$$\min_\Lambda \|\beta_0\|_2$$

Note that in Sec. 12.5 we defined H_2-type objective functions not only for a single input v but also for finite sets of inputs. For the robust performance case here we have to restrict our problem definition to a single input.

Objective O5 defines a nonconvex nonlinear program. The solution is simplified by the fact that the gradient $\partial\beta_0/\partial\lambda_k$ can be computed explicitly as we will show next.

Define G^β from G^F (12.7-19) as

$$G^\beta = \begin{pmatrix} 1 & 0 \\ 0 & \beta^{-1} \end{pmatrix} G^F \tag{12.7 - 36}$$

The function β_0 is defined through [see (11.3–28), (11.3–29)]

$$\mu(G^\beta(i\omega)) = 1 \quad \Leftrightarrow \quad \beta(\omega) = \beta_0(\omega) \tag{12.7 - 37}$$

First, β_0 has to be computed at a finite set Ω of frequencies. Theorem 11.3-2 implies that any basic descent method should be sufficient. To obtain the gradient of $\|\beta_0\|_2$ with respect to the filter parameters, we need to compute the gradient of $\beta_0(\omega)$ with respect to these parameters for every frequency $\omega \in \Omega$. From the definition of β_0 in (12.7-37) we see that as some filter parameter λ_k changes, $\beta_0(\omega)$ must also change so that $\mu(G^{\beta_0}(i\omega))$ remains constantly equal to 1. Hence we find

$$\frac{\partial\mu}{\partial\beta_0}\frac{\partial\beta_0}{\partial\lambda_k} + \frac{\partial\mu}{\partial\lambda_k} = 0 \quad \Rightarrow \quad \frac{\partial\beta_0}{\partial\lambda_k} = -\frac{\partial\mu}{\partial\lambda_k} / \frac{\partial\mu}{\partial\beta_0} \tag{12.7 - 38}$$

where μ is computed by minimizing its upper bound $\inf_{D\in\mathcal{D}} \bar{\sigma}(DG^\beta D^{-1})$ [see (11.2–15)]. We assume again that the two largest singular values of $DG^\beta D^{-1}$ for the optimal D's at the value of β where the gradient is computed, are not equal to each other. If this is not the case a mean direction can be used as mentioned above.

The numerator $\partial\mu/\partial\lambda_k$ of (12.7–38) can be evaluated in the same way as (12.7–28) and (12.7–35) but with G^β instead of G^F:

$$\frac{\partial}{\partial\lambda_k}(\mu(G^\beta(i\omega))) = \text{Re}\Big[u_1^H D_0 \begin{pmatrix} G_{13}^u \\ \beta^{-1}W_2 G_{23}^u \end{pmatrix} (I - FG_{33}^u)^{-1}$$

$$\frac{\partial F}{\partial\lambda_k}(I - G_{33}^u F)^{-1}(G_{31}^u \quad G_{32}^u W_1)D_0^{-1}v_1\Big] \tag{12.7 - 39}$$

where

$$W_1 = (v \quad 0) \tag{12.7 - 40}$$

For the computation of the denominator $\partial \mu / \partial \beta_0$ let the $\inf_{D \in \mathcal{D}} \bar{\sigma}(DG^\beta(i\omega)D^{-1})$ be attained for $D_0 = D_0(\omega; \beta)$ and let σ_1 be the maximum singular value and u_1, v_1 the corresponding singular vectors of $D_0 G^\beta D_0^{-1}$. Then the same steps as for obtaining (12.7–30) are valid and we find after some algebra

$$\frac{\partial}{\partial \beta}(\mu(G^\beta(i\omega))) = -\frac{1}{\beta^2} \operatorname{Re}\left[u_1^H D_0 \begin{pmatrix} 0 & 0 \\ G_{21}^\beta & G_{22}^\beta \end{pmatrix} D_0^{-1} v_1\right] \qquad (12.7-41)$$

Substitution of (12.7–39) and (12.7–41) into (12.7–38) yields the desired gradient.

12.8 Application: High-Purity Distillation

Consider again the distillation column described in the Appendix where the overhead composition is to be controlled at $y_D = 0.99$ and the bottom composition at $x_B = 0.01$ using the reflux L and boilup V as manipulated inputs. Approximating the dynamics by a first-order system we find the linear model

$$\tilde{P}(s) = \frac{1}{75s+1}\begin{pmatrix} 0.878 & -0.864 \\ 1.082 & -1.096 \end{pmatrix} \qquad (12.8-1)$$

Problems arise from the fact that high purity distillation columns tend to be ill-conditioned. For (12.8–1) the condition number is 142. Hence any controller Q based on the inverse of $\tilde{P}$ will also be ill conditioned and this might result in a control system which is not robust. For simplicity we will use the (conservative) uncertainty description (11.1–22)

$$\bar{\ell}_I(s) = 0.2 \frac{5s+1}{0.5s+1} \qquad (12.8-2)$$

The performance weights

$$W_2^{-1}(s) = \frac{20s}{10s+1} I \qquad (12.8-3)$$

$$W_1 = I \qquad (12.8-4)$$

require a closed loop time constant of about 20 and restrict the maximum peak of the sensitivity operator to less than two.

The plant $\tilde{P}$ is MP and therefore

$$\tilde{Q}(s) = \tilde{P}(s)^{-1} \qquad (12.8-5)$$

First a diagonal filter structure is chosen. A one-parameter search based on the analytic gradient expression derived in Sec. 12.7.3 yields

$$F(s) = \frac{1}{7.28s + 1} I \qquad (12.8-6)$$

Plots of μ for robust stability and robust performance are shown in Fig. 12.8-1A. Clearly, although the system is guaranteed to remain stable in the presence of modeling error within the bound defined by (12.8–2), the performance is expected to deteriorate. This is confirmed by the simulations shown in Fig. 12.8-2. For the nominal case ($P = \tilde{P}$), the outputs are decoupled and the performance is acceptable (Fig. 12.8-2A). However in the case where

$$P(s) = \tilde{P}(s) \begin{pmatrix} 1.2 & 0 \\ 0 & 0.8 \end{pmatrix} \qquad (12.8-7)$$

the performance deteriorates to the point where it is totally unacceptable (Fig. 12.8-2). Note that the plant in (12.8-7) includes a 20% error in each plant input and is within the bound (12.8–2). The same plant is used in all other simulations in this section when model-plant mismatch is assumed. Higher order filters with different elements were also tried but were found not to improve the performance substantially. The reason is that, in general, a diagonal filter cannot affect the condition number of $\tilde{Q}$ significantly.

We shall proceed with the filter structure suggested in Sec. 12.7.1 for ill-conditioned systems. For our example $\omega^* = 0$ and therefore the diagonalizations (12.7–9) and (12.7–10) are exact. Hence (12.7–11) and (12.7–12) yield the same Q. Objective O4' was solved with a gradient search method using the analytic gradient expression of Sec. 12.7.3. Different filter orders were tested and a few different initial guesses were tried to avoid local minima. The final result for filters with two parameters in each element was:

$$F_1(s) = \begin{pmatrix} \frac{0.244s+1}{(0.184s+1)^2} & 0 \\ 0 & \frac{0.00284s+1}{(8.72s+1)^2} \end{pmatrix} \qquad (12.8-8)$$

$$F_2(s) = \begin{pmatrix} \frac{0.164s+1}{(0.446s+1)^2} & 0 \\ 0 & \frac{0.213s+1}{(0.476s+1)^2} \end{pmatrix} \qquad (12.8-9)$$

The values of μ for robust stability and performance are shown in Fig. 12.8-1B. The clear improvement over the diagonal filter is verified by the simulations in Fig. 12.8-3. It is interesting to note that settling times for the nominal case are similar as with the diagonal filter (Fig. 12.8-2 and 3). Hence, as expected, the robustness improvement was not the result of additional detuning, but of the two-filter structure which allowed us to affect the singular values of $\tilde{Q}$ directly and reduce its condition number in the critical frequency range.

Finally, a comparison will be made between the performance obtained by the two-filter IMC controller and the "true" μ-optimum controller, defined as the

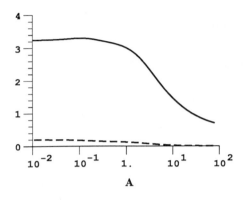

A

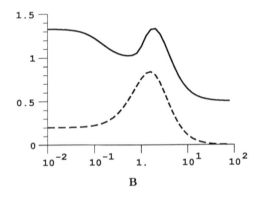

B

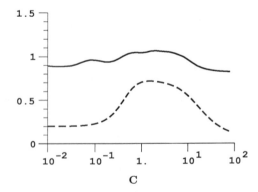

C

Figure 12.8-1. Solid lines: μ for robust performance. Dashed lines. μ for robust stability. (A) One-filter IMC controller, (B) Two-filter IMC controller, (C) μ-optimal controller.

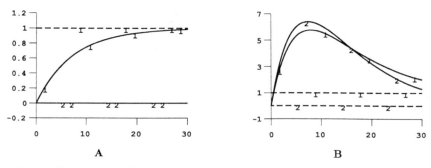

Figure 12.8-2. Time response for the one-filter IMC controller for unit step setpoint change for distillate (output 1). Dashed line: Setpoint; Solid lines: Outputs. (A) $P = \tilde{P}$, (B) $P \neq \tilde{P}$.

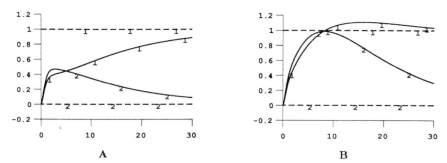

Figure 12.8-3. Time response for the two-filter IMC controller for unit step setpoint change for distillate (output 1). Dashed line: Setpoint; Solid lines: Outputs. (A) $P = \tilde{P}$, (B) $P \neq \tilde{P}$.

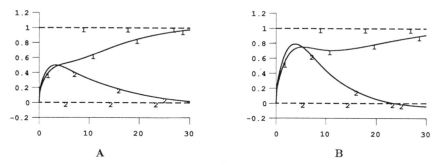

Figure 12.8-4. Time response for the μ-optimal controller for unit step setpoint change for distillate (output 1). Dashed line: Setpoint; Solid lines: Outputs. (A) $P = \tilde{P}$, (B) $P \neq \tilde{P}$.

result of minimizing $\sup_\omega \mu_\Delta(G^F)$ over *all* stabilizing controllers Q (or C). However, the iterative approach presently available for solving this problem is not guaranteed to converge and indeed, in our experience it has often failed to converge. For this particular example though, we were able to obtain a μ-optimal controller. The values of μ for robust performance and stability are shown in Fig. 12.8-1C. Clearly the difference is not significant and this is verified by the simulations shown in Fig. 12.8-4.

12.9 Summary

For internal stability of the IMC structure (12.1–1) both the plant P and the IMC controller Q have to be stable. For open-loop unstable plants it is convenient to *design* the IMC controller Q but for *implementation* the classic feedback structure must be chosen. Under some mild assumptions about pole-zero cancellation (Sec. 12.3) *all* stabilizing controllers Q for the plant P are parametrized by

$$Q_1(s) = Q_0(s) + Q_1(s) \qquad (12.3-1)$$

where $Q_0(s)$ is an arbitrary stabilizing controller for P and Q_1 is any stable transfer matrix such that PQ_1P is stable.

The IMC design procedure consists of two steps. In the *first step*, $\tilde{Q}$ is selected to yield a good system response for the inputs of interest, without regard for constraints and model uncertainty. As one option we propose to define a set of n inputs

$$\mathcal{V} = \left\{ v^i(s): \quad i = 1, \ldots n \right\} \qquad (12.5-10)$$

and to minimize the *sum* of the ISE's that each of the inputs v^i would cause when applied to the system separately:

Objective O3:

$$\min_{\tilde{Q}} \left[\Phi(v^1) + \Phi(v^2) + \ldots + \Phi(v^n) \right]$$

where

$$\Phi(v^i) = \|We^i\|_2^2 = \|W\tilde{E}v^i\|_2^2 = \|W(I - \tilde{P}\tilde{Q})v^i\|_2^2 \qquad (12.5-9)$$

The unique controller $\tilde{Q}$ which meets Obj. O3 is given by

$$\tilde{Q} = P_M^{-1}W^{-1}\left\{WP_A^{-1}V_M\right\}_* V_M^{-1} \qquad (12.6-15)$$

where the operator $\{\cdot\}_*$ denotes that after a partial fraction expansion of the operand, all terms involving the poles of P_A^{-1} are omitted. The factorization of the plant

$$P = P_A P_M \qquad (12.6-3)$$

into an allpass portion P_A and an MP portion P_M can be accomplished via inner-outer factorization (Sec. 12.6.4). The input matrix $V = (v^1 \quad v^2 \quad \ldots \quad v^n)$ can be factored similarly

$$V = V_M V_A \qquad (12.6-14)$$

In some special cases, $\tilde{Q}$ defined by (12.6-15) is also H_2-optimal for *each* one of the inputs v^i separately and their linear combinations.

In the *second step* of the design procedure, the controller $\tilde{Q}$ is detuned through a lowpass filter F such that for the detuned controller $Q = Q(\tilde{Q}, F)$ the robust performance condition

$$\mu_\Delta(G^F(\tilde{P}, Q)) < 1 \qquad \forall \omega, \qquad \Delta = \text{diag}\{\Delta_u, \Delta_p\} \qquad (12.5-7)$$

is satisfied. A nonlinear program was formulated (Sec. 12.7.3) to minimize $\mu_\Delta(G^F(\tilde{P}, Q))$ as a function of the filter parameters for a filter with a fixed diagonal structure. For unstable plants the filter has to be identity at all the unstable plant poles. For ill conditioned plants two filters can be necessary to meet the requirement (12.5-7).

12.10 Discussion and References

12.2. The idea of a "balanced realization" as a framework for model reduction is described by Moore (1981).

12.3. A general stable parametrization of all stabilizing controllers for a particular plant was developed by Youla, Jabr & Bongiorno (1976). It involves matrix coprime factors.

12.6, 12.7. The MIMO procedure for obtaining the controller for optimal nominal performance is patterned after the SISO case. For SISO systems the tradeoff between robustness and performance is straightforward because there are very few degrees of freedom. It makes sense to start from (almost) "perfect control" and to detune for robustness. Usually, the robust performance obtained in this manner is only marginally inferior to what can be accomplished via an integrated procedure which optimizes robust performance directly.

For MIMO systems this is not true anymore. There are tradeoffs between the different outputs and between the different inputs. One or even two diagonal filters might not offer enough degrees of freedom to reach the desirable performance-robustness compromise when the starting point is the (improper) H_2-optimal controller as proposed in Sec. 2.6. Short of going to the integrated procedure of optimizing robust performance directly, there are two options: a more complex filter or a different design for nominal performance. Because of the nonconvex nonlinear programming problem (Sec. 12.7.3 and 4) necessary to determine the filter parameters, the first option is likely to cause many difficulties. It appears more promising to investigate alternate nominal-performance designs. One possibility is to include a penalty term for the control action in the H_2-objective. The linear quadratic optimal controller minimizing the modified objective can be found by solving two Riccati equations (Kwakernaak & Sivan, 1972). The controller is always proper and should be better conditioned because of the control action penalty. The other possibility is to postulate $\tilde{H}(= \tilde{P}\tilde{Q})$ and to determine $\tilde{Q}$ from $\tilde{H}$. The advantage is that the *structure* of $\tilde{H}$ (decoupled, one-way decoupled, etc.) can be postulated directly by the designer. The limitations on $\tilde{H}$ imposed by NMP elements in $\tilde{P}$ have been explored by Holt & Morari (1985a,b), Zafiriou & Morari (1987) and Morari, Zafiriou & Holt (1987).

12.6-1-12.6-3. These sections follow the development in Chap. V of Zafiriou (1987).

12.6.4. Theorem 12.6-4 is Thm. 4 of Chu (1985, p. 40), applied to square systems and with some notational changes. Methods for solving ARE's can be found in Laub (1979) and Molinari (1973). Details on the QR factorization can be found in many books — e.g., Sec. 5.7.2 of Dahlquist and Björck (1974).

12.7.1. An algorithm for pseudo-diagonalization is described by Rosenbrock (1974).

12.7.3, 12.7.4. These sections follow the discussion in Zafiriou and Morari (1986b).

12.8. For computing the μ-optimal controller the procedure proposed by Doyle (1985) was used.

Chapter 13

PERFORMANCE LIMITATIONS FOR MIMO SYSTEMS

We know from practical experience that for a SISO system with a small gain, a long deadtime or severe "inverse response" it is impossible to achieve good closed-loop performance. Our thereotical analysis in Sec. 3.3.4 corroborated these observations. NMP characteristics are detrimental to closed-loop performance regardless of the employed controllers, in particular long delays and RHP zeros close to the origin should be avoided. Also, when the plant gain is small, only small disturbances can be controlled effectively without saturation of the manipulated variable.

Unlike SISO systems, MIMO systems display "directionality" which makes the assessment of their inherent performance limitations much more difficult. There is no single "gain" but the MIMO "gain" depends on the *direction* of the input vector. This is most apparent from the singular value decomposition which was discussed in Sec. 10.1.5. Whether a MIMO RHP zero leads to poor closed-loop performance or not, depends not only on its location in the RHP but also on its "structure." If it affects primarily a process output which is of minor importance, its presence might be irrelevant, even when it is located close to the origin. Finally, in Chap. 11 we pointed out that model uncertainty is a very critical component in the design of MIMO controllers. SISO systems do not display a similar sensitivity to model uncertainty.

The objective of this chapter is to develop a quantitative understanding of the factors which limit the achievable closed loop performance and to demonstrate how they can be used to screen alternate choices of "control structures" — i.e., manipulated variables.

13.1   Effect of Plant Gain

We will use the process description

$$y = Pu + P_d d' \qquad\qquad (13.1-1)$$

as suggested by Fig. 10.2-1. P_d is the disturbance model expressing the relationship between the physical disturbances d_i' and their effects d_i on the output. For a distillation column the components of $d' = (d_1', \ldots, d_i', \ldots, d_n')^T$ may correspond to disturbances in feed rate, feed composition, boilup rate, etc. The column vector p_{di} of P_d represents the disturbance model for the disturbance d_i'. The effect of a particular disturbance d_i' on the process output is d_i,

$$d_i = p_{di} d_i' \qquad\qquad (13.1-2)$$

The direction of the vector d_i will be referred to as the *direction of the disturbance i*. The overall effect of all disturbances d_i' on the output is d,

$$d = \sum_i d_i = \sum_i p_{di} d_i' = P_d d' \qquad\qquad (13.1-3)$$

In most cases we will consider the effect of one particular disturbance d_i'. To simplify notation we will usually drop the subscript i, and $d = p_d d'$ will denote the effect of this *single* disturbance $d_i' = d'$ on the outputs. We will be referring to d as a "disturbance," although, in general, it represents the *effect* of the physical disturbance. The "disturbance" d can also represent a setpoint change $(-r)$ as indicated in Fig. 10.2-1.

We assume that the disturbance model and the process model have been scaled such that at steady state $-1 \le d_i' \le 1$ corresponds to the expected range of each disturbance and $-1 \le u_j \le 1$ corresponds to the acceptable range for each manipulated variable. For process control $u_j = -1$ may correspond to a closed valve and $u_j = 1$ to a fully open valve.

13.1.1   Constraints on Manipulated Variable

In this section we investigate the magnitude of the manipulated variables necessary to cancel the influence of a disturbance on the process output at steady state. It should be obvious that this magnitude is independent of the controller. This analysis will allow us to identify problems with *actuator* (e.g., valve) *constraints at steady state*. However, we should point out that the issue of constraints at steady state is not really a *control* problem, but rather a plant *design* problem. Any well designed plant should be able to reject disturbances at steady state.

For complete disturbance rejection ($y = 0$) at steady state

$$u = -P(0)^{-1}p_d(0)d' \qquad (13.1-4)$$

To avoid actuator saturation we must require

$$|u|_\infty \leq 1 \qquad \forall d' \ni |d'|_\infty \leq 1$$

or equivalently

$$|P(0)^{-1}p_d(0)|_\infty \leq 1 \qquad (13.1-5)$$

where $|\cdot|_\infty$ denotes the ∞-vector norm which was defined in Sec. 10.1. Whether (13.1–5) is violated and saturation causes problems depends both on the process P and the disturbance p_d. Even if $\|P(0)^{-1}\|_\infty$ is large (implying a small plant "gain"), $|P(0)^{-1}p_d(0)|_\infty$ can still be small if p_d is "aligned" with P^{-1} in a certain manner. This is the topic of the next section.

13.1.2 Disturbance Condition Number

When saturation is not an issue it is more reasonable to use the Euclidean (2-) norm as a measure of magnitude because it "sums up" the deviations of all manipulated variables rather than accounting for the maximum deviation only (like the ∞-norm). Consider a particular disturbance $d = p_d d'$. For complete rejection of this disturbance

$$u = -P^{-1}d \qquad (13.1-6)$$

The quantity

$$\frac{|u|_2}{|d|_2} = \frac{|P^{-1}d|_2}{|d|_2} \qquad (13.1-7)$$

depends only on the direction of the disturbance d but not on its magnitude. It measures the magnitude of u needed to reject a disturbance d of unit magnitude which enters in a particular direction expressed by $d/|d|_2$.

From the discussion of the SVD in Sec. 10.1.5 we note that the RHS of (13.1–7) is minimized for

$$d = \underline{v}(P^{-1}) = \bar{u}(P) \qquad (13.1-8)$$

For d defined by (13.1-8), (13.1-7) becomes

$$\frac{|u|_2}{|d|_2} = |P^{-1}\underline{v}(P^{-1})|_2 = \underline{\sigma}(P^{-1}) = \frac{1}{\bar{\sigma}(P)} \qquad (13.1-9)$$

Thus, the best disturbance direction requiring the *least* action by the manipulated variables, is that of the singular vector $\bar{u}(P)$ associated with the largest singular value of P.

By normalizing (13.1–7) with $|u|_2/|d|_2$ for the "best" disturbance defined by (13.1–8) we obtain the *disturbance condition number of the plant P*

$$\kappa_d(P) = \frac{|P^{-1}d|_2}{|d|_2}\bar{\sigma}(P)$$

$$= \frac{|P^{-1}p_d|_2}{|p_d|_2}\bar{\sigma}(P) \tag{13.1 – 10}$$

It expresses the magnitude of the *manipulated variable* needed to reject a disturbance in the direction d relative to rejecting a disturbance with the same magnitude, but in the "best" direction $[\bar{u}(P)]$.

The "worst" disturbance direction is

$$d = \bar{v}(P^{-1}) = \underline{u}(P)$$

In this case we get

$$\kappa_d(P)_{max} = \bar{\sigma}(P^{-1})\bar{\sigma}(P) = \kappa(P)$$

and therefore for all disturbance directions

$$1 \leq \kappa_d(P) \leq \kappa(P) \tag{13.1 – 11}$$

Thus, $\kappa_d(P)$ may be viewed as a generalization of the condition number $\kappa(P)$ of the plant, which also takes into account the direction of the disturbances. It measures how well the disturbance direction d is aligned with the direction of maximum effectiveness of the manipulated variables. A large value of $\kappa(P)$ indicates a large degree of directionality in the plant P. If this directionality is not compensated by the controller the closed-loop performance for different disturbance directions is vastly different as we will show next.

13.1.3   Implications of κ_d for Closed-Loop Performance

The objective of the control system is to minimize the effect of the disturbances on the outputs y. Consider a particular disturbance $d(s) = p_d(s)d'(s)$. The closed-loop relationship between this disturbance and the outputs is

$$y(s) = (I + P(s)C(s))^{-1}d(s) = E(s)d(s) \tag{13.1 – 12}$$

Let $|y(i\omega)|_2$ denote the Euclidean norm of y evaluated at each frequency. The quantity

$$\alpha(\omega) = \frac{|Ed(i\omega)|_2}{|d(i\omega)|_2} \tag{13.1 – 13}$$

depends only on the disturbance direction but not on its magnitude. $\alpha(\omega)$ measures the magnitude of the output vector $y(i\omega)$ resulting from a sinusoidal disturbance $d(i\omega)$ of unit magnitude and frequency ω.

The "best" disturbance direction causing the smallest output deviation is that of the right singular vector $\underline{v}(E)$ associated with the smallest singular value $\underline{\sigma}(E)$. By normalizing $\alpha(\omega)$ with this best disturbance we obtain the *disturbance condition number of E^{-1}*

$$\kappa_d(E^{-1}) = \frac{|Ed|_2}{|d|_2} \frac{1}{\underline{\sigma}(E)} = \frac{|Ed|_2}{|d|_2} \bar{\sigma}(E^{-1}) \qquad (13.1-14)$$

Again

$$1 \leq \kappa_d(E^{-1}) \leq \kappa(E^{-1}) = \kappa(E) \qquad (13.1-15)$$

At low frequencies, where the controller gain is high we have $E(i\omega) \cong (PC(i\omega))^{-1}$. In particular, this expression is exact at steady state ($\omega = 0$) when the controller includes integral action. Based on this approximation we derive the *disturbance condition number of PC.*

$$\kappa_d(PC) = \frac{|(PC)^{-1}d|_2}{|d|_2} \bar{\sigma}(PC) \qquad (13.1-16)$$

As stated above this quantity has physical significance only when $\underline{\sigma}(PC) \gg 1$. To avoid problems when this measure is evaluated at $\omega = 0$ write

$$C(s) = c(s)D(s) \qquad (13.1-17)$$

where $c(s)$ is a scalar transfer function which includes any integral action present in C. $D(s)$ may be viewed as a "decoupler." Then we have

$$\kappa_d(PC) = \frac{|(PD)^{-1}d|_2}{|d|_2} \bar{\sigma}(PD) \qquad (13.1-18)$$

For $D = I$ (i.e., the controller $c(s)I$) we find from (13.1–18) the disturbance condition number of P (13.1–10) derived previously. Thus $\kappa_d(P)$ can be interpreted in terms of closed loop performance as follows: If a scalar controller $C = c(s)I$ is chosen (which keeps the directionality of the plant unchanged), then $\kappa_d(P)$ measures the magnitude of the *output y* for a particular disturbance d, compared to the magnitude of the output if the disturbance were in the "best" direction (corresponding to the large plant gain). If $\kappa_d(P) = \kappa(P)$, the disturbance has all its components in the "bad" direction corresponding to low plant gain and low bandwidth. If $\kappa_d(P) = 1$, the disturbance has all its components in the "good" direction corresponding to high plant gain and high bandwidth.

Though a large value of $\kappa_d(P)$ does not necessarily imply bad performance, it usually does. In principle we could choose a compensator C which makes $\kappa_d(PC) = 1$ for all disturbances. However, such a controller can lead to serious robustness problems as we will show in Sec. 13.3.

13.1.4 Decomposition of d along Singular Vectors

The objective here is to gain insight into the type of dynamic response which is to be expected when disturbances along a particular direction affect a system with a high degree of directionality ($\kappa(E^{-1})$ is "large"). The singular vectors $v_j(E)$ form an orthonormal basis. The disturbance vector d can be represented in terms of this basis

$$d = \sum_{j=1}^{n} (v_j(E)^T \cdot d) v_j(E) \qquad (13.1-19)$$

Then the output y is described by

$$y(i\omega) = Ed(i\omega) = \sum_{j=1}^{n} Ev_j(E)(v_j(E)^T \cdot d)(i\omega)$$

$$= \sum_{j=1}^{n} \sigma_j(E)u_j(E)(v_j(E)^T \cdot d) = \sum_{j=1}^{n} \sigma_j(E)d^j(i\omega) \qquad (13.1-20)$$

where we have defined the new "disturbance components"

$$d^j = (v_j(E)^T \cdot d)u_j(E) \qquad (13.1-21)$$

(13.1-20) shows that the response to a particular disturbance can be viewed as the sum of responses to the disturbances d^j passing through the scalar transfer function $\sigma_j(E)$. The magnitude of d^j depends on the alignment of the disturbance d with the singular vector $v_j(E)$. The characteristics of the (speed of) response to d^j depend on $\sigma_j(E)$.

For the controller $C = c(s)D(s)$ with integral action in $c(s)$ the approximation $E(i\omega) \cong c^{-1}(PD)^{-1}(i\omega)$ is valid for small ω. Then with $\ell = n - j + 1$ (13.1-20) becomes

$$y(i\omega) = \sum_{\ell=1}^{n} \frac{1}{c\sigma_\ell(PD)}\tilde{d}^\ell \qquad (13.1-22)$$

where

$$\tilde{d}^\ell = d^{n-\ell+1} = (u_\ell^T(PD) \cdot d)v_\ell(PD) \qquad (13.1-23)$$

The magnitude of $\tilde{d}^\ell$ is given by the component of d in the direction of the singular vector $u_\ell(PD)$ and $\tilde{d}^\ell$ affects the output along the direction of the singular vector $v_\ell(PD)$. If the loop transfer matrix PD has a high gain in this direction [i.e., $\sigma_\ell(PD)$ is large] then the control will be quick and good. If the gain is low the response will be slow and poor. If PD is ill-conditioned [$\kappa(PD)$ is large], the widely different response characteristics for different disturbance components will result in unusual overall system responses. This will be observed in the following example.

Example 13.1-1. LV-Distillation Column. Consider the distillation column described in the Appendix with L and (-V) as manipulated variables and the product compositions y_d and x_B as controlled outputs. The model is given by

$$P(s) = \frac{1}{75s + 1} \begin{pmatrix} 0.878 & 0.864 \\ 1.082 & 1.096 \end{pmatrix} \qquad (13.1-24)$$

We assume there are no problems with constraints. We want to study how well the system rejects various disturbances using a diagonal controller $C(s) = c(s) \cdot I$. Since we are only concerned about the outputs (y_D and x_B), the scaling does not matter provided the outputs are scaled such that an output of magnitude one is equally "bad" for both y_D and x_B. We have

$$\bar{\sigma}(P) = 1.972, \quad \underline{\sigma}(P) = 0.0139, \quad \kappa(P) = 141.7$$

Consider disturbances d' of unit magnitude in feed composition, x_F, feed flowrate, F, feed liquid fraction, q_F, and boilup rate, $-V_d$. The linearized steady state disturbance models are

$$d = P_d d' = \begin{pmatrix} 0.881 \\ 1.119 \end{pmatrix} x_F + \begin{pmatrix} 0.394 \\ 0.586 \end{pmatrix} F + \begin{pmatrix} 0.868 \\ 1.092 \end{pmatrix} q_F + \begin{pmatrix} 0.864 \\ 1.096 \end{pmatrix} (-V_d) \quad (13.1-25a)$$

Also consider setpoint changes in y_D and x_B of magnitude one. These are mathematically equivalent to disturbances with

$$d = \begin{pmatrix} 1 \\ 0 \end{pmatrix} \text{ and } d = \begin{pmatrix} 0 \\ 1 \end{pmatrix} \qquad (13.1-25b)$$

The steady state values of the disturbance condition number, $\kappa_d(P)$, are given for these disturbances in Table 13.1-1. The disturbance condition number of E^{-1}, with the controller described below, is shown as a function of frequency in Fig. 13.1-1. From these data we see that disturbances in x_F, q_F and V are very well aligned with the plant, and there is little need for using a decoupler to change the direction of P. The feed flow disturbance is the "worst" disturbance, but even it has its dominant effect in the "good" direction.

A "decoupler" is desirable if we want to follow setpoint changes which have a large component in the "bad" direction corresponding to low plant gains. However, a decoupler is *not* recommended for this distillation column because of severe

Table 13.1.1. Disturbance condition numbers for distillation example ($\kappa(P) = 141.7$).

	Disturbance/Setpoint Change d'					
	x_F	F	q_F	$-V_d$	y_{Ds}	x_{Bs}
d	$\begin{pmatrix} 0.881 \\ 1.119 \end{pmatrix}$	$\begin{pmatrix} 0.394 \\ 0.586 \end{pmatrix}$	$\begin{pmatrix} 0.868 \\ 1.092 \end{pmatrix}$	$\begin{pmatrix} 0.864 \\ 1.096 \end{pmatrix}$	$\begin{pmatrix} 1 \\ 0 \end{pmatrix}$	$\begin{pmatrix} 0 \\ 1 \end{pmatrix}$
$\kappa_d(P)$	1.48	11.75	1.09	1.41	110.7	88.5

robustness problems caused by uncertainty (see Sec. 13.3). Therefore we cannot expect good setpoint tracking for this LV-configuration. If setpoint changes are of little or no interest, the LV-configuration with a diagonal controller may be a good choice. The response to a feed-rate disturbance is then expected to be somewhat sluggish because of the high value of $\kappa_d(P)$.

We will now confirm the predictions based on the data in Table 13.1-1 by studying some time responses. We use a diagonal controller of the form $C(s) = c(s)I$ where $c(s)$ is the PI controller $c(s) = 0.1(75s + 1)s^{-1}$.

Time domain simulations are shown for "disturbances" in x_F and F and for setpoint changes in y_D in Figs. 13.1-2 to 13.1-4. We have simulated all responses as step setpoint changes of size d (13.1–25) to make comparisons easier. Dynamics have not been included in the "disturbances" for x_F and F (which is clearly unrealistic) to make the example simpler. The time responses confirm the predictions based on Table 13.1-1. The rather odd-looking response can be explained easily by decomposing the disturbances along the singular vector directions of the closed-loop system, as shown before. For each disturbance, the closed-loop frequency response at low frequencies can be approximated by $y(i\omega) \cong c^{-1}P^{-1}d(i\omega)$. By decomposing d along the "directions" of P as in (13.1-22) and (13.1-23) we may write this response as the sum of two SISO responses

$$y(i\omega) \cong \left[\frac{1}{c\bar{\sigma}(P)}\tilde{d}^1 + \frac{1}{c\underline{\sigma}(P)}\tilde{d}^2 \right] \qquad (13.1-26)$$

where

$$\tilde{d}^1 = (\bar{u}^T \cdot d)\bar{v}(P) \qquad (13.1-27a)$$

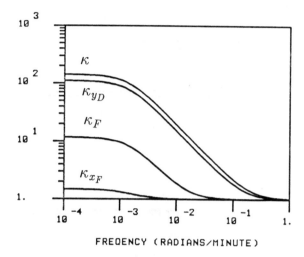

Figure 13.1-1. Disturbance condition number of E^{-1} for disturbances in feed rate F, feed composition x_F, and setpoint change in y_D. $C(s) = 0.1(75s + 1)s^{-1} \cdot I$. (Reprinted with permission from *Ind. Eng. Chem. Res.*, **26**, 2033 (1987), American Chemical Society.)

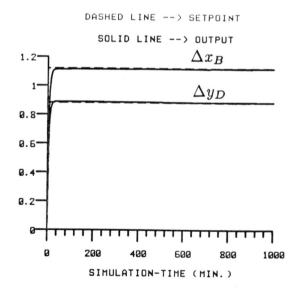

Figure 13.1-2. Step change in setpoint $(0.881, 1.119)^T$. (Closed-loop response to "disturbance" in x_F.) (Reprinted with permission from *Ind. Eng. Chem. Res.*, **26**, 2034 (1987), American Chemical Society.)

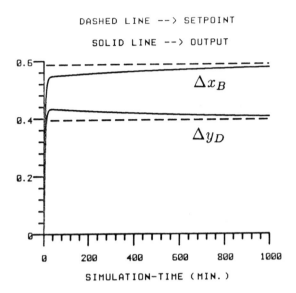

Figure 13.1-3. Step change in setpoint $(0.394, 0.586)^T$. (Closed-loop response to "disturbance" in F.) (Reprinted with permission from *Ind. Eng. Chem. Res.*, **26**, 2034 (1987), American Chemical Society.)

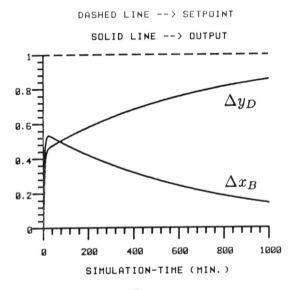

Figure 13.1-4. Step change in setpoint $(1,0)^T$. (Closed-loop response to setpoint change for y_D.) (Reprinted with permission from *Ind. Eng. Chem. Res.*, **26**, 2034 (1987), American Chemical Society.)

Table 13.1-2. $\tilde{d}^1$ and $\tilde{d}^2$ (13.1-27) for distillation example.

| | Disturbance/Setpoint Change | | |
	x_F	F	y_{Ds}
d	$\begin{pmatrix} 0.881 \\ 1.119 \end{pmatrix}$	$\begin{pmatrix} 0.394 \\ 0.586 \end{pmatrix}$	$\begin{pmatrix} 1 \\ 0 \end{pmatrix}$
$\tilde{d}^1$	$\begin{pmatrix} 1.00 \\ 1.00 \end{pmatrix}$	$\begin{pmatrix} 0.50 \\ 0.50 \end{pmatrix}$	$\begin{pmatrix} 0.44 \\ 0.44 \end{pmatrix}$
$\tilde{d}^2$	$\begin{pmatrix} -0.008 \\ 0.008 \end{pmatrix}$	$\begin{pmatrix} -0.04 \\ 0.04 \end{pmatrix}$	$\begin{pmatrix} 0.55 \\ -0.55 \end{pmatrix}$

$$\tilde{d}^2 = (\underline{u}^T \cdot d)\underline{v}(P) \qquad\qquad (13.1-27b)$$

Thus, each disturbance response is the sum of two responses: one fast in the direction $\tilde{d}^1$ and one slow in the direction $\tilde{d}^2$ (see Table 13.1-2). The singular value decomposition $P = U\Sigma V^H$ gives

$$\Sigma = \begin{pmatrix} \bar{\sigma} & 0 \\ 0 & \underline{\sigma} \end{pmatrix} = \begin{pmatrix} 1.972 & 0 \\ 0 & 0.01391 \end{pmatrix}$$

$$U = (\bar{u} \quad \underline{u}) = \begin{pmatrix} 0.625 & -0.781 \\ 0.781 & 0.625 \end{pmatrix}$$

$$V = (\bar{v} \quad \underline{v}) = \begin{pmatrix} 0.707 & -0.708 \\ 0.708 & 0.707 \end{pmatrix}$$

The decomposition expressed through (13.1–26) and (13.1–27) which holds for low frequencies, explains the actual responses very well: Initially there is a very fast response in the direction of $\bar{v}^T = (0.707, \quad -0.708)$. This response arises from the overall open-loop transfer function $c\bar{\sigma}(P) = 0.197/s$ corresponding to a first-order response with time constant $(0.1\bar{\sigma}(P))^{-1} = 5.1$ min. Added to this is a slow first-order response with time constant $(0.1\underline{\sigma}(P))^{-1} = 720$ min in the direction of $\underline{v}^T = (-0.708, 0.707)$.

Note that the slow disturbance component $\tilde{d}^2$ is the "error" at $t \approx 40$ min, because the fast response has almost settled at this time. As an example consider the disturbance in feed rate F (Fig. 13.1-3). At $t \approx 40$ min the deviation from the desired setpoint, $(0.394, 0.586)^T$, is approximately equal to $\tilde{d}^2 = (-0.04, 0.04)^T$. Similarly, for the setpoint change in y_D (Fig. 13.1-4) the deviation from the desired setpoint, $(1, 0)^T$, at $t \approx 40$ min is approximately equal to $\tilde{d}^2 = (0.55, -0.55)^T$. □

13.1.5 Summary

The disturbance condition number of a matrix A with respect to a disturbance with direction d is defined as follows:

$$\kappa_d(A) = \frac{|A^{-1}d|_2}{|d|_2}\bar{\sigma}(A)$$

For non-square A (e.g., plants with more inputs than outputs) one should replace $|A^{-1}d|_2$ by $\{\min|m|_2 \text{ s.t. } Am = d\}$, that is — one should replace A^{-1} by $A^{\#} \triangleq A^T(AA^T)^{-1}$ (the pseudo-inverse of A).

We have introduced the disturbance condition number of the plant P (13.1–10), of E^{-1} (13.1–14) and of PC (13.1–16). The disturbance condition number $\kappa_d(E^{-1})$ measures the performance (error) for a disturbance in direction $d/|d|$ relative to the performance for a disturbance which enters in the "optimal" direction. For high gain feedback ($\underline{\sigma}(PC) >> 1$) $E^{-1} \cong PC$, which justifies the use of $\kappa_d(PC)$ and also $\kappa_d(P)$ (when the controller C is scalar) instead of $\kappa_d(E^{-1})$. The disturbance condition number is a measure of control *performance* and therefore *must* be scaling *dependent*: We define performance in terms of a weighted average of output deviations. Any measure of performance must depend on the relative importance of the outputs which is expressed through scaling factors. (On the other hand, the issue of *stability* is *independent* of scaling, and any measure used as a tool for evaluating a system's stability should be independent of scaling.)

If $\kappa_d(P)$ is large then for good performance in all directions a controller is needed which makes $\kappa_d(PC)$ small. As we will discuss in Sec. 13.3 robustness problems might prevent us from choosing such a controller. On the other hand, if $\kappa_d(P)$ is small then a *scalar* controller C can be expected to yield good performance. Thus we can use $\kappa_d(P)$ in two different ways:

1. Discriminating between process alternatives and selecting manipulated inputs. Plants with large values of $\kappa_d(P)$ are not necessarily bad, but if other factors are equal [e.g., RGA-values (Sec. 13.3), RHP-zeros (Sec. 13.2)], then we should prefer a design with a low value of $\kappa_d(P)$. This measure may therefore be used as *one* criterion for selecting manipulated variables and discriminating among alternatives.

2. Selecting controller structure (e.g., diagonal versus multivariable inverse-based controller). Guidance can be obtained from $\kappa_d(P)$ when it is used together with the RGA, as discussed in Sec. 13.3.

13.2 NMP Characteristics

In our discussions we stressed repeatedly that NMP characteristics limit the achievable closed loop performance. For example, the complementary sensitivity resulting from an H_2-optimal controller (Sec. 12.6.3) includes the factor P_A, the allpass incorporating all NMP characteristics of P. If $P_A = I$, perfect control is theoretically possible. Also, Thm. 4.1-4 shows that RHP zeros close to the origin are more detrimental to the performance than zeros far out in the RHP. In this section we want to capture quantitatively the "directional" effect of RHP zeros — i.e., which outputs are most and which are least affected by a particular zero. We will make the following three assumptions:

Assumption 13.2-1: *$P(s)$ is square $(n \times n)$ and nonsingular except for isolated values of s.*

Assumption 13.2-2. *All RHP zeros $\zeta_1, \ldots, \zeta_m$ are of degree one — i.e., the rank of $P(\zeta_i)$ is $n - 1$.*

Assumption 13.2-3. *No RHP poles are located at $\zeta_1, \ldots, \zeta_m$.*

Usually, these assumptions are not restrictive. They can be easily relaxed at the expense of a more involved notation.

13.2.1 Zero Direction

Definition 13.2-1: *Let ζ_i be a zero of $P(s)$. The vector z_i $(z_i \neq 0)$ satisfying $z_i^T P(\zeta_i) = 0$ is called the direction of the zero ζ_i.*

Note that $P(\zeta_i)$ is of rank $n - 1$ because the zero was assumed to be of degree one. The vector z_i is the eigenvector of $P(\zeta_i)^T$ associated with the eigenvalue zero. It is called zero direction because for any system input of the form $ke^{\zeta_i t}$, where k is an arbitrary complex vector, the output in the direction of z_i is identically equal to zero (given appropriate initial conditions).

Because the IMC controller Q has to be stable it cannot cancel the RHP zeros of P and they will appear unchanged in the complementary sensitivity $H = PQ$. Furthermore from Def. 13.2-1 we find the following relation for the sensitivity

$$z_i^T E(\zeta_i) = z_i^T (I - PQ(\zeta_i)) = z_i^T \qquad (13.2-1)$$

which implies that the magnitude of any disturbance component $z_i e^{\zeta_i t}$ entering along the zero direction z_i is passing through to the output y unaffected by feedback (given appropriate initial conditions). Thus RHP zeros affect both the achievable H and E.

RHP poles also impose limitations on the choice of Q as discussed in Sec. 12.2. With a two-degree-of-freedom structure, however, the achievable H is not constrained by the presence of RHP plant poles (Sec. 5.2.4).

We would like to characterize in a convenient manner all complementary sensitivity operators which can be achieved for a plant P. The following theorem is self-evident from the preceding discussion.

Theorem 13.2-1. *For a plant P the complementary sensitivity H is achievable by a two-degree-of-freedom controller (such that the closed-loop system is internally stable) if and only if there exists a stable Q such that $H = PQ$.*

A direct test for the existence of a stable Q is provided by the following theorem.

Theorem 13.2-2. *There exists a stable Q such that the complementary sensitivity is equal to a desired transfer matrix H if and only if H satisfies*

$$z_i^T H(\zeta_i) = 0 \qquad\qquad (13.2 - 2)$$

for all RHP zeros ζ_i of the plant $P(s)$ where z_i is the direction of the zero ζ_i $(z_i^T P(\zeta_i) = 0)$.

Proof.

$\Rightarrow$ Assume there exists a stable Q such that $H = PQ$. Then $z_i^T H(\zeta_i) = z_i^T PQ(\zeta_i) = 0$.

$\Leftarrow$ Find the partial fraction expansion of P^{-1}.

$$P(s)^{-1} = \frac{1}{s - \zeta_i} R_i + P_\zeta(s) \qquad\qquad (13.2 - 3)$$

where R_i is the matrix of residues and $P_\zeta(s)$ is a remainder term with no poles at $s = \zeta_i$. Postmultiply both sides of (13.2–3) by $P(s)$

$$I = \frac{1}{s - \zeta_i} R_i P(s) + P_\zeta(s) P(s) \qquad\qquad (13.2 - 4)$$

Because the LHS of (13.2–4) is the identity the RHS cannot have a pole at $s = \zeta_i$. $P(s)P_\zeta(s)$ does not have a pole at $s = \zeta_i$ and therefore it must be that

$$R_i P(\zeta_i) = 0 \qquad (13.2-5)$$

Because ζ_i is of degree one, $P(\zeta_i)$ is of rank $(n-1)$. Hence R_i is of rank 1 and as a result of Def. 13.2-1, the rows of R_i are multiples of z_i^T. Therefore $z_i^T H(\zeta_i) = 0$ implies

$$R_i H(\zeta_i) = 0 \qquad (13.2-6)$$

Now postmultiply both sides of (13.2–3) by H.

$$Q = P^{-1} H = \frac{1}{s - \zeta_i} R_i H(s) + P_\zeta(s) H(s) \qquad (13.2-7)$$

$P_\zeta(s) H(s)$ does not have a pole at $s = \zeta_i$. Hence (13.2–6) implies that Q does not have a pole at $s = \zeta_i$. $\qquad \square$

Example 13.2-1. Consider the plant P_1

$$P_1 = \frac{1}{s+1} \begin{pmatrix} 1 & 1 \\ 1+2s & 2 \end{pmatrix}$$

which has a zero at $s = \zeta = \frac{1}{2}$. The zero direction $z = (2, -1)$ satisfies

$$z^T P_1(\zeta) = z^T \begin{pmatrix} 1 & 1 \\ 2 & 2 \end{pmatrix} = 0$$

We will use condition (13.2–2) to construct decoupled and one-way decoupled H's for which stable Q's exist. Trivially for the decoupled plant

$$H = \begin{pmatrix} \frac{-2s+1}{2s+1} & 0 \\ 0 & \frac{-2s+1}{2s+1} \end{pmatrix}$$

$H(\zeta) = 0$ and therefore (13.2–2) is satisfied. Let us postulate

$$H^{LT} = \begin{pmatrix} 1 & 0 \\ x_1 & \frac{-2s+1}{2s+1} \end{pmatrix}$$

and

$$H^{UT} = \begin{pmatrix} \frac{-2s+1}{2s+1} & x_2 \\ 0 & 1 \end{pmatrix}$$

where x_1 and x_2 are to be determined. We find from (13.2–2)

$$2 - x_1(\zeta) = 0 \quad \text{or} \quad x_1(\zeta) = 2$$

$$2x_2(\zeta) - 1 = 0 \quad \text{or} \quad x_2(\zeta) = \frac{1}{2}$$

If we postulate x_1 and x_2 to be of the form $2\beta s/(2s+1)$ (with the constant β to be determined) so that there are no steady state interactions ($H(0) = I$) then we find

$$x_1(\zeta) = \frac{2\beta}{4} = 2$$

and

$$x_1(s) = \frac{8s}{2s+1}$$

Similarly

$$x_2(s) = \frac{2s}{2s+1}$$

□

13.2.2 Implications of Zero Direction for Achievable Performance

The vector z_i is constant and possibly complex. Condition (13.2–2) requires that each column of H, evaluated at the plant zero ζ_i, is orthogonal to z_i. For a particular element of the input vector v the elements of the output vector y cannot be selected independently but have to satisfy the linear "interpolation conditions" (13.2–2). The presence of RHP plant zeros requires some relationship between the elements of a column of H but the columns themselves can be selected independently of each other.

By assumption, ζ_i is a zero of $P(s)$ of degree one. Assume that H is selected to be diagonal – i.e., completely decoupled. Then according to the theorem the degree of ζ_i in H has to be at least equal to the number of nonzero entries in z_i. Usually this number is larger than one, generally equal to n. Thus, requiring a decoupled response generally leads to the introduction of RHP zeros not originally present in the plant $P(s)$. This is the price to be paid for decoupling.

The zero ζ_i is "pinned" to the outputs corresponding to nonzero entries in z_i: the zero has to affect at least one of these outputs and it cannot affect any of the outputs corresponding to zero entries in z_i. This is illustrated in the following example.

Example 13.2-2. The system

$$P_2(s) = \frac{1}{s+2} \begin{pmatrix} -s+1 & -s+1 \\ 1 & 2 \end{pmatrix}$$

has a zero at $\zeta = 1$ with the direction $z^T = (1,0)$. Because $z_2 = 0$, it can only affect the first output of H — i.e., it is "pinned" to the first output. Pinned zeros are nongeneric and therefore somewhat of a mathematical artifact. In the preceding example an infinitesimal change in the polynomial coefficients of p_{11} or p_{12} removes the RHP zero altogether.

Corollary 13.2-1. *Assume that the kth element z_{ik} of the zero direction z_i is nonzero. Then it is possible to obtain "perfect" control on all outputs $j \neq k$ with the remaining output exhibiting no steady state offset.*

This and the next result follow trivially from Thm. 13.2-2.

Corollary 13.2-2. *Assume that $P(s)$ has a single zero ζ and that the kth element z_k of the zero direction z is nonzero. Then H can be chosen of the form*

$$H = \begin{pmatrix} 1 & 0 & \cdot & 0 & 0 & 0 & \cdot & 0 \\ 0 & 1 & \cdot & 0 & 0 & 0 & \cdot & 0 \\ \cdot & \cdot & \cdot & & \cdot & \cdot & \cdot & \cdot \\ \cdot & \cdot & \cdot & & \cdot & \cdot & \cdot & \cdot \\ \frac{\beta_1 s}{s+\zeta} & \frac{\beta_2 s}{s+\zeta} & \cdot & \frac{\beta_{k-1} s}{s+\zeta} & \frac{-s+\zeta}{s+\zeta} & \frac{\beta_{k+1} s}{s+\zeta} & \cdot & \frac{\beta_n s}{s+\zeta} \\ \cdot & \cdot & \cdot & \cdot & \cdot & \cdot & \cdot & \cdot \\ \cdot & \cdot & \cdot & \cdot & \cdot & \cdot & \cdot & \cdot \\ 0 & 0 & \cdot & 0 & 0 & 0 & \cdot & 1 \end{pmatrix} \qquad (13.2-8)$$

where

$$\beta_j = -\frac{2z_j}{z_k} \text{ for } j \neq k \qquad (13.2-9)$$

The interaction terms will be insignificant if $z_k >> z_j$ ($\forall j \neq k$) – i.e., when the zero is "aligned" predominantly with output k. If for some j, $z_j >> z_k$ then the zero is aligned predominantly with output j. It can be pushed to output k only at the cost of generating significant interactions (large β's).

As a demonstration of the alignment effect recall Ex. 13.2-1 with the zero aligned with the first output ($z^T = (2, -1)$). Pushing the effect of the zero to the second output (H^{LT}) leads to strong interactions ($\beta = 4$), while aligning the zero with the first output (H^{UT}) is much more favorable ($\beta = 1$). Thus, if one-way decoupling is contemplated the zero direction should be used as a guideline.

13.2.3 Summary

The concept of zero direction is a convenient tool to judge the feasibility of alternate forms of decouplers. In the generic case when all elements of the zero direction vector are nonzero all kinds of decouplers are feasible, in principal. However, if the zero is very close to the origin complete decoupling is not advisable because it introduces more RHP zeros at this same location into the complementary sensitivity. Also, though generically the effect of a zero can be "pushed" to any arbitrary output, this can cause large interactions and violent moves in the manipulated variables unless the zero direction is aligned with this output — i.e., it has a large component in the direction of this output.

13.3   Sensitivity to Model Uncertainty

We have learned that model uncertainty limits the achievable closed-loop performance. The SSV introduced in Chap. 11 can be used to measure the performance deterioration caused by uncertainty. The extent of the deterioration depends on the (nominal) system, the controller and the type (structure) of the uncertainty. It is desirable that the system be of such a kind that even with a simplistic controller the sensitivity of closed loop performance to model uncertainty is small. Then the modelling and controller design effort necessary to achieve good performance will be small.

We know from Sec. 11.3.2 that in the presence of multiplicative output uncertainty alone simple SISO-type design techniques suffice to achieve good performance. On the other hand multiplicative input uncertainty (Sec. 11.3.3) can cause serious performance problems if the plant condition number is high: with a low condition number controller poor nominal performance is expected; when the controller condition number is high, robust performance is generally bad. Thus, high condition number plants should be avoided by appropriate *process design* and in particular by the appropriate choice of actuators.

In Secs. 11.2.5 and 11.2.6 we studied "element-by-element" uncertainty. A large *minimized* condition number $\kappa^*(P)$ is an indication that closed-loop performance is sensitive to this type of uncertainty. We were able to correlate $\kappa^*(P)$ with the RGA $\Lambda(P)$ and concluded that the stability and performance of plants with large RGA elements is strongly affected by independent element uncertainty. Example 11.2-2 illustrated, however, that this type of uncertainty description is usually inappropriate because the transfer matrix elements do not vary independently. In the following we will show that large RGA elements are also an indication of sensitivity to *diagonal* multiplicative input uncertainty, which is a good model of actuator uncertainty. Actuator uncertainty is clearly always present to some extent and therefore plants with large RGA elements should generally be avoided.

13.3.1   Sensitivity to Diagonal Input Uncertainty

Let u denote the actual plant input and u_c the input computed by the controller. Let Δ_i represent the relative uncertainty on the ith manipulated input. Then $u_i = u_{ci}(1+\Delta_i)$ or in vector form $u = u_c(I+\Delta_I)$ where $\Delta_I = \text{diag}\{\Delta_i\}$. Alternatively we may define the perturbed plant

$$P = \tilde{P}(I + \Delta_I), \qquad \Delta_I = \text{diag}\{\Delta_i\} \qquad (13.3-1)$$

The loop gain matrix PC which is closely related to performance, may be written in terms of the nominal $\tilde{P}C$

$$PC = \tilde{P}C(I + C^{-1}\Delta_I C) \qquad (13.3-2a)$$

$$= (I + \tilde{P}\Delta_I \tilde{P}^{-1})\tilde{P}C \qquad (13.3-2b)$$

For SISO plants, a relative input error of magnitude Δ results in the same relative change in $PC = \tilde{P}C(1 + \Delta)$, but for multivariable plants the effect of the input uncertainty on PC may be amplified significantly as we will show.

For 2×2 plants the error term in (13.3–2a) may be expressed in terms of the RGA of the *controller C* as follows

$$C^{-1}\Delta_I C = \begin{pmatrix} \lambda_{11}(C)\Delta_1 + \lambda_{21}(C)\Delta_2 & \lambda_{11}(C)\frac{c_{12}}{c_{11}}(\Delta_1 - \Delta_2) \\ -\lambda_{11}(C)\frac{c_{21}}{c_{22}}(\Delta_1 - \Delta_2) & \lambda_{12}(C)\Delta_1 + \lambda_{22}(C)\Delta_2 \end{pmatrix} \qquad (13.3-3)$$

where

$$\Lambda(C) = C \times (C^{-1})^T \qquad (13.3-4)$$

For $n \times n$ plants the *diagonal* elements of the error matrix $C^{-1}\Delta_I C$ may be written as a straightforward generalization of the 2×2 case

$$(C^{-1}\Delta_I C)_{ii} = \sum_{j=1}^{n} \lambda_{ji}(C)\Delta_j \qquad (13.3-5)$$

That is, the diagonal elements of $C^{-1}\Delta_I C$ depend on the elements of $\Lambda(C)$ in the same column and the magnitude of the uncertainty. Similarly, for 2×2 plants the error term in (13.3–2b) may be expressed in terms of the RGA of the *plant*

$$\tilde{P}\Delta_I \tilde{P}^{-1} = \begin{pmatrix} \lambda_{11}\Delta_1 + \lambda_{12}\Delta_2 & -\lambda_{11}\frac{\tilde{p}_{12}}{\tilde{p}_{22}}(\Delta_1 - \Delta_2) \\ \lambda_{11}\frac{\tilde{p}_{21}}{\tilde{p}_{11}}(\Delta_1 - \Delta_2) & \lambda_{21}\Delta_1 + \lambda_{22}\Delta_2 \end{pmatrix} \qquad (13.3-6)$$

For $n \times n$ plants the diagonal elements in $(P\Delta_I P^{-1})$ depend on the elements of $\Lambda(P)$ in the same row:

$$(\tilde{P}\Delta_I \tilde{P}^{-1})_{ii} = \sum_{j=1}^{n} \lambda_{ij}(\tilde{P})\Delta_j \qquad (13.3-7)$$

Controllers with large RGA-elements will lead to large elements in the matrix $C^{-1}\Delta_I C$, and plants with large RGA-elements will lead to large elements in the matrix $\tilde{P}\Delta_I \tilde{P}^{-1}$. Equations (13.3–2) seem to imply that either of these cases will lead to large elements in PC and therefore poor performance when there is input uncertainty ($\Delta_I \neq 0$). However, this interpretation is generally not correct since the "directionality" of PC may be such that the elements in PC remain small even though $C^{-1}\Delta_I C$ or $\tilde{P}\Delta_I \tilde{P}^{-1}$ have large elements. This should be clear from the following two extreme cases

1. Assume the controller has a small RGA. In this case the elements in the error term $C^{-1}\Delta_I C$ are similar to Δ_I in magnitude (13.3–2a and 13.3–5). Consequently, PC is not particularly influenced by input uncertainty, even though the plant itself may be strongly ill-conditioned with a large RGA.

2. Assume the plant has a small RGA. In this case PC is not particularly influenced by input uncertainty (13.3–2b and 13.3–6), even though the controller itself may have a large RGA. (From a practical point of view, one can argue that it is unlikely that anyone would design a controller with large RGA-elements for a plant with small RGA-elements.)

From items 1 and 2 we conclude that for a system to be sensitive to diagonal input uncertainty, *both* the controller and the plant must have large RGA-elements. This is consistent with Sec. 11.3.3 where we found that for block input uncertainty to cause robust performance problems the condition numbers of *both* the plant and the controller have to be large. Thus the RGA plays a similar role for *diagonal* input uncertainty as the condition number does for *block* input uncertainty.

13.3.2   Sensitivity with Different Controller Structures

Inverse-based controller. For "tight" control it is desirable to use an inverse-based controller $C(s) = P^{-1}(s)K(s)$ (where $K(s)$ is *diagonal*). A special case of such an inverse-based controller is a decoupler. With $C(s) = P^{-1}(s)K(s)$ we find $PC = K(s)$ and $\Lambda(C) = \Lambda(P^{-1}K) = \Lambda(P^{-1}) = \Lambda(P)^T$. Thus, if the elements of $\Lambda(P)$ are large, so will be the elements of $\Lambda(C)$ and from the discussion above we expect high sensitivity to input uncertainty. We also see directly from

$$PC = K(s)(I + \tilde{P}\Delta_I \tilde{P}^{-1}) = K(s)(I + C^{-1}\Delta_I C) \qquad (13.3-8)$$

that large elements in $\tilde{P}\Delta_I \tilde{P}^{-1}$ (or equivalently large elements in $C^{-1}\Delta_I C$) imply that the loop transfer matrix PC is very different from nominal ($\tilde{P}C$) and poor response or even instability is expected when $\Delta_I \neq 0$.

Decouplers have been discussed extensively in the chemical engineering literature, in particular in the context of distillation columns. The purpose of the decoupler (D) is to take care of the multivariable aspects and to reduce the tuning of the control system to a series of single-loop problems. Let $K(s)$ denote these "single-loop" controllers. Then the overall controller C including the decoupler is $C(s) = DK(s)$. Typical decoupler forms are:

Ideal Decoupling:

$$D_I = \tilde{P}^{-1}\tilde{P}_{diag}$$

Simplified Decoupling:

$$D_S = \tilde{P}^{-1}((\tilde{P}^{-1})_{diag})^{-1}$$

Steady State Decoupling:

$$D_0 = D_I(0) \text{ or } D_0 = D_S(0)$$

where we used the notation $A_{diag} = \text{diag}\{a_{11}, \ldots a_{nn}\}$. One-way decouplers are triangular and chosen such that $\tilde{P}C = \tilde{P}DK$ is triangular.

For systems with large RGA the closed loop performance has been reported to be sensitive to errors in the decoupler. This can be easily explained from the results in Sec. 11.2.5 and 11.2.6. However, the most important reason for the robustness problems encountered with decouplers is probably diagonal (actuator) and full-block *input* uncertainty. Because all decouplers (except one-way decouplers) lead to inverse-based controllers the closed-loop performance is highly sensitive to input uncertainty when the plant is ill-conditioned or has a large RGA.

Diagonal Controllers. Closed-loop systems with diagonal controllers are insensitive to diagonal input uncertainty because $\Lambda(C) = I$ when C is diagonal. However, when P is ill-conditioned or has a large RGA, then the performance with a diagonal controller is generally poor, except when for all inputs of interest the disturbance condition number $\kappa_d(PC)$ is close to unity.

In one special case a diagonal controller can lead to good performance for *arbitrary* input directions even for an ill-conditioned plant: when the plant is naturally decoupled at the input. Let the SVD of P be $P = U\Sigma V^H$ and assume $V = I$ (or more generally that V has only one nonzero element in each row and column, so that the inputs can be rearranged to give $V = I$). Then we can choose the diagonal controller $C(s) = c(s)\Sigma^{-1}$ to get $PC = c(s)U$ which has $\kappa_d(PC) = 1$ for *all* inputs d — i.e., good disturbance rejection independent of direction. Note, however that the response is not decoupled (unless U is diagonal).

Also note that in this case

$$\kappa^*(P) = \min_{D_1, D_2} \kappa(D_1 P D_2)$$

$$= \min_{D_1,D_2} \kappa(D_1 U \Sigma D_2)$$

$$= \min_{D_1,D_2} \kappa(D_1 U D_2) = 1$$

because Σ is diagonal and U is unitary. Furthermore, because of (11.2–43), (11.2–45) and (11.2–46) all elements of Λ have to be positive and smaller than one.

General Controller Structure. We have learned that an inverse-based controller is generally bad for plants with a large RGA. (In this case $\Lambda(C)$ is also large.) We also know that for diagonal controllers ($\Lambda(C) = I$) the closed-loop system is always insensitive to diagonal input uncertainty. This suggests that a plot of the magnitude of the elements of $\Lambda(C)$ as a function of frequency may be useful for evaluating a system's sensitivity to input uncertainty. In particular, a large $\Lambda(C)$ around the crossover frequency is undesirable.

We stress that a large $\Lambda(C)$ does not necessarily imply sensitivity to input uncertainty. For example, if $\Lambda(P)$ is small there are no sensitivity problems as we discussed. It is unlikely, however, that a controller with large $\Lambda(C)$ would lead to good *nominal* performance for a plant with small $\Lambda(P)$. Thus, if the controller is designed for good nominal performance, then $\Lambda(C)$ should be a good indicator of sensitivity to input uncertainty.

13.3.3 "Worst-Case" Uncertainty

It is of interest to know the "worst" possible input uncertainty — i.e., the "worst-case" combination of Δ_j's. Consider (13.3–7). If all Δ_j have the same magnitude ($|\Delta_j| < r_I$) then the largest possible magnitude (worst case) of any diagonal element in $P\Delta_I P^{-1}$ is given by $r_I \|\Lambda(P)\|_\infty$. To obtain this value the signs of the Δ_j's should be the same as those in the row of $\Lambda(P)$ with the largest elements.

Example 13.1-3. Consider a plant with steady-state gain matrix

$$P(0) = \begin{pmatrix} 1 & 0.1 & -2 \\ 1 & 2 & -3 \\ -0.1 & -1 & 1 \end{pmatrix}$$

The RGA is

$$\Lambda(P(0)) = \begin{pmatrix} -1.89 & -0.13 & 3.02 \\ 3.59 & 3.02 & -5.61 \\ -0.7 & -1.89 & 3.59 \end{pmatrix}$$

Assume the relative uncertainties Δ_1, Δ_2 and Δ_3 on each manipulated input have the same magnitude $|\Delta|$. Then the second row of $\Lambda(P)$ has the largest row sum ($\|\Lambda(P)\|_\infty = 12.21$) and the worst combination of input uncertainty for an inverse-based controller is

$$\Delta_1 = \Delta_2 = -\Delta_3 = \Delta$$

We find

$$\text{diag}(P\Delta_I P^{-1}) = \text{diag}\{-5.0\Delta, 12.2\Delta, -6.2\Delta\}$$

Note that in this specific example we would arrive at the same worst case by considering the first or third row. Therefore the worst case will always be obtained with Δ_1 and Δ_2 with the same sign and Δ_3 with a different sign even if their magnitudes are different. □

In some cases we may arrive at a different conclusion by considering other frequencies. Also note that, unless an inverse-based controller is used, it is not guaranteed that the worst case uncertainties are deduced using this approach.

13.3.4 Example

The high-purity distillation column described by the simple one-time constant model (see Appendix) will be used to demonstrate the effects of input (actuator) uncertainty. Two control configurations will be compared. In one case reflux (L) and boilup (V) are used for composition control, in the other case the distillate flow (D) and the boilup (V). We will study the responses for two setpoint changes: $r_1 = (1,0)^T$ and $r_2 = (0.4, 0.6)^T$, where r_2 is roughly equivalent to the effect of a change in feed flow to the column. The two configurations are compared in Table 13.3-1. Two controllers were designed for each one of these configurations, an inverse-based and a diagonal controller. The controller gains were adjusted to guarantee robust stability for diagonal input uncertainty with a magnitude bound $w_I(s) = 0.2(5s+1)/(0.5s+1)$. Robust stability is guaranteed for this kind of uncertainty if and only if (Ex. 11.2-1)

$$\mu(CP(I + CP)^{-1}) \leq 1/|w_I|, \qquad \forall \omega$$

where the structured singular value μ is computed with respect to a diagonal matrix. The condition is shown graphically in Fig 13.3-1 for the controllers used in this example.

Responses are obtained both for the nominal case ($\Delta_I = 0$) and with 20% actuator uncertainty: $\Delta_I = \text{diag}\{0.2, -0.2\}$ which yields the following error terms (13.3–6) for PC when an inverse-based controller is used:

$$(\tilde{P}\Delta_I\tilde{P}^{-1})_{LV} = \begin{pmatrix} 35.1\Delta_1 - 34.1\Delta_2 & -27.7(\Delta_1 - \Delta_2) \\ 43.2(\Delta_1 - \Delta_2) & -34.1\Delta_1 + 35.1\Delta_2 \end{pmatrix} = \begin{pmatrix} 13.8 & -11.1 \\ 17.2 & -13.8 \end{pmatrix}$$

$$(\tilde{P}\Delta_I\tilde{P}^{-1})_{DV} = \begin{pmatrix} 0.45\Delta_1 + 0.55\Delta_2 & 0.45(\Delta_1 - \Delta_2) \\ -0.55(\Delta_1 - \Delta_2) & 0.55\Delta_1 + 0.45\Delta_2 \end{pmatrix} = \begin{pmatrix} -0.02 & 0.18 \\ -0.22 & 0.02 \end{pmatrix}$$

The simulations shown in Figs. 13.3-2 to 13.3-5 illustrate the following points:

Table 13.3-1. Comparison of LV and DV configurations.

	LV	DV
$\left(\begin{array}{c} dy_d \\ dx_B \end{array}\right) = \frac{1}{75s+1} \times$	$\left(\begin{array}{cc} 0.878 & -0.864 \\ 1.082 & -1.096 \end{array}\right)\left(\begin{array}{c} dL \\ dV \end{array}\right)$	$\left(\begin{array}{cc} -0.878 & 0.014 \\ -1.082 & -0.014 \end{array}\right)\left(\begin{array}{c} dD \\ dV \end{array}\right)$
RGA : λ_{11}	35.1	0.45
RGA : $\Sigma_{i,j}\,|\lambda_{ij}|$	138.3	2
$\kappa(P)$	141.7	70.8
$\kappa_{r_1}, r_1 = (1,0)^T$	110.7	54.9
$\kappa_{r_2}, r_2 = (0.4, 0.6)^T$	11.8	4.3

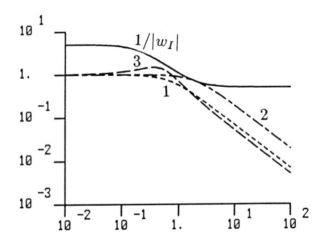

Figure 13.3-1. Robust stability test: $\mu(CP((I+CP)^{-1})$ is shown for 1: Inverse-controller for LV- and DV-configurations; 2: Diagonal LV-controller; 3: Diagonal DV-controller. (Reprinted with permission from *Ind. Eng. Chem. Res.*, **26**, 2329 (1987), American Chemical Society.)

- An inverse-based controller gives poor response when the plant has large RGA-elements (λ_{11} is large) and there is input uncertainty (Fig. 13.3-2).

- A diagonal controller cannot correct for the strong directionality of a plant with large RGA-elements. This results in responses which are strongly dependent on the disturbance (or setpoint) directions (Fig. 13.3-3). The response to r_2 (disturbance in F) is acceptable, but the response to the setpoint change r_1 is extremely sluggish. This system may be acceptable despite the large value of λ_{11}, provided setpoint changes are not important.

- An inverse-based controller may give very good response for an ill conditioned plant even with diagonal input uncertainty, provided λ_{11} is small (Fig. 13.3-4).

- A diagonal controller may remove most of the directionality in the plant if $V \approx I$. However, "interactions" are still present because U is not diagonal (Fig. 13.3-5).

13.3.5 Summary

To some extent actuator (diagonal input) uncertainty is present in every physical system. From the RGA information can be obtained on the sensitivity of the closed loop performance to input uncertainty for different controller structures.

1. Closed loop systems with diagonal controllers are insensitive to diagonal input uncertainty. However when P is ill-conditioned or has a large RGA, then the performance is poor, except when for all inputs of interest the disturbance condition number $\kappa_d(PC)$ is close to one.

2. An inverse-based (e.g., decoupling) controller should *never* be used for a plant with large RGA because it leads to extreme sensitivity to input uncertainty.

3. Inverse-based controllers may give poor response even if the RGA is small. This may happen in the case of a 2×2 system if p_{12}/p_{22} or p_{21}/p_{11} is large (see 13.3–6). One example is a triangular plant which always has $\lambda_{11} = 1$, but where the response obtained with an inverse-based controller may display large "interactions" in the presence of uncertainty.

4. One-way decouplers are generally much less sensitive to input uncertainty.

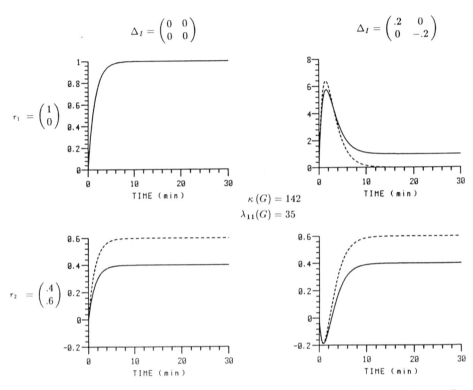

$$\Delta_I = \begin{pmatrix} 0 & 0 \\ 0 & 0 \end{pmatrix} \qquad \Delta_I = \begin{pmatrix} .2 & 0 \\ 0 & -.2 \end{pmatrix}$$

$$r_1 = \begin{pmatrix} 1 \\ 0 \end{pmatrix}$$

$$\kappa\,(G) = 142$$
$$\lambda_{11}(G) = 35$$

$$r_2 = \begin{pmatrix} .4 \\ .6 \end{pmatrix}$$

Figure 13.3-2. LV-configuration. Closed-loop responses y_1 and y_2 for inverse-based controller:

$$C(s) = \frac{0.7(75s + 1)}{s}G_{LV}^{-1} = \frac{0.7(75s + 1)}{s}\begin{pmatrix} 39.94 & -31.49 \\ 39.43 & -32.00 \end{pmatrix}$$

(Reprinted with permission from *Ind. Eng. Chem. Res.*, **26**, 2329 (1987), American Chemical Society.)

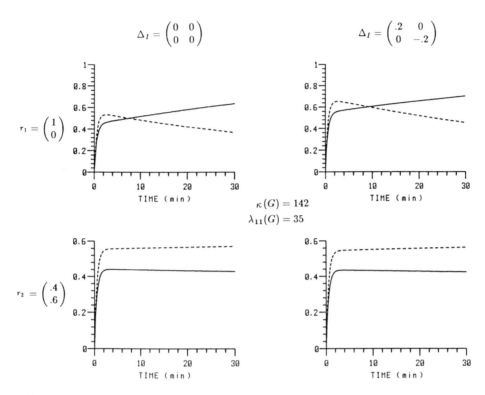

Figure 13.3-3. LV-configuration. Closed-loop responses y_1 and y_2 for diagonal controller:

$$C(s) = \frac{(75s + 1)}{s} \begin{pmatrix} 1 & 0 \\ 0 & -1 \end{pmatrix}$$

(Reprinted with permission from *Ind. Eng. Chem. Res.*, **26**, 2329 (1987), American Chemical Society.)

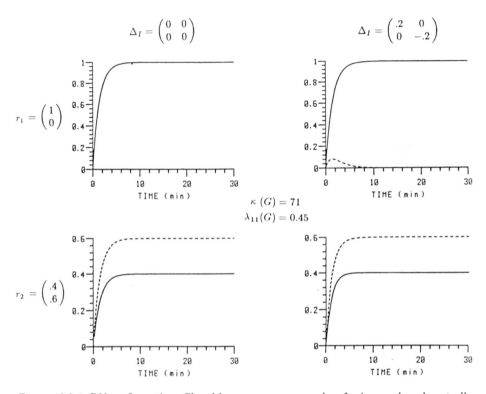

Figure 13.3-4. DV-configuration. Closed-loop responses y_1 and y_2 for inverse-based controller:

$$C(s) = \frac{0.7(75s + 1)}{s} G_{DV}^{-1} = \frac{0.7(75s + 1)}{s} \begin{pmatrix} -0.5102 & -0.5102 \\ 39.43 & -32.00 \end{pmatrix}$$

(Reprinted with permission from *Ind. Eng. Chem. Res.*, **26**, 2330 (1987), American Chemical Society.)

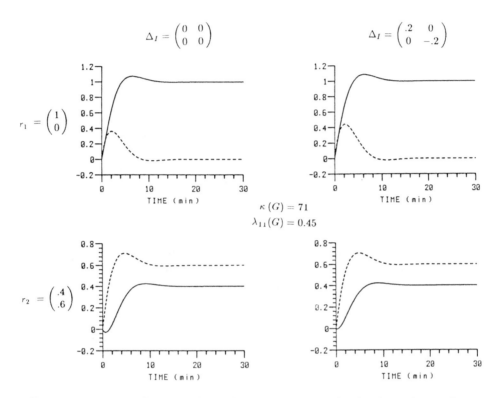

Figure 13.3-5. DV-configuration. Closed-loop responses y_1 and y_2 for diagonal controller:

$$C(s) = \frac{-0.2(75s+1)}{s}\Sigma^{-1} = \frac{0.2(75s+1)}{s}\begin{pmatrix} -0.718 & 0 \\ 0 & -50.8 \end{pmatrix}$$

(Reprinted with permission from *Ind. Eng. Chem. Res.*, **26**, 2330 (1987), American Chemical Society.)

Table 13.3-2. Guidelines for choice of preferred multivariable controller structure ("large" implies a comparison with one, typically > 10), $\|\Lambda\|_A = \sum_{i,j} |\lambda_{ij}|$.

		$max_d \kappa_d(P)$	
		Large	Small
$\|\Lambda(P)\|_A$	Large	–	Diagonal
	Small	Inverse-based ($V = I$: diagonal)	Inverse-based (diagonal)

5. For a general multivariable controller designed for good performance a plot of the magnitude of the elements of $\Lambda(C)$ vs. frequency is a good indicator of sensitivity to input uncertainty. In particular, a large $\Lambda(C)$ in the crossover frequency range is undesirable.

6. Table 13.3-2 is helpful for choosing between inverse-based (decoupling) and diagonal controllers in extreme situations.

13.4   References

13.1. The ideas in this section were first presented by Skogestad & Morari (1987d). Stanley, Marino-Galarraga & McAvoy (1985) also observed the dependence of the performance on the disturbance direction and developed the Relative Disturbance Gain as an indicator. Their observations regarding the behavior of high purity distillation columns parallel those made for Ex. 13.1-1.

13.2. This section follows the paper by Morari, Zafiriou & Holt (1987). Zafiriou & Morari (1987) presented a more general treatment removing some of the assumptions. A discrete framework was used in this paper, which allows one to handle multiple delays in exactly the same way as NMP zeros in one single step. The transmission blocking interpretation of zeros is due to Desoer & Schulman (1974).

13.2.2. Holt & Morari (1985b) give a general overview of the effect of RHP zeros on closed-loop performance. They review the concept of "pinned zeros" introduced by Bristol (1980). Desoer & Gündes (1986) point out in a very general context that additional RHP zeros can be introduced by decouplers.

13.3. Skogestad & Morari (1987c) discovered the RGA as an indicator of sensitivity to diagonal input (i.e., actuator) uncertainty.

13.3.2. The use of decouplers for distillation control is discussed by Luyben (1970) and Arkun, Manousiouthakis & Palozoglu (1984). Toijala & Fagervik (1982) report high sensitivity to decoupler errors for systems with large RGA.

13.3.3. The method developed by Fan & Tits (1986) for computing the SSV yields also the "worst-case" uncertainty.

Chapter 14

DECENTRALIZED CONTROL

14.1 Motivation

Let $P(s)$ be an $n \times n$ rational transfer function matrix relating the vector of system inputs y to the vector of system outputs u. Let r be the vector of reference signals or setpoints for the closed loop system. Assume that u, y, and r have been partitioned in the same manner: $u = (u_1, u_2, \ldots u_m)^T, y = (y_1, y_2, \ldots y_m)^T, r = (r_1, r_2, \ldots, r_m)^T$. In this book decentralized control means that the controller C is block diagonal (Fig. 14.1-1)

$$u_i = C_i(y_i - r_i) \qquad (14.1-1)$$

(It should be obvious that any control system where every input u_i and every output y_j are processed by a single controller block C_k, can be rearranged such that C is diagonal.) The constraints on the controller structure invariably lead to performance deterioration when compared to the system with a full controller matrix. This sacrifice has to be weighed against the following two factors:

1. *Hardware simplicity.* If u_i, y_i are physically close but u_i, y_j $(i \neq j)$ are far apart, a full controller could require expensive communication links. Also, the controller hardware costs could be high if an implementation through analog circuitry is required. These considerations are relevant, for example, for large networks of power stations where the distances between the stations can be significant. Hardware issues are generally irrelevant in the context of process control; in all modern plants all measurement signals are sent into a central control room from where all the actuator signals originate.

2. *Design simplicity.* If all the blocks $P_{ij} = 0$ $(i \neq j)$ then each controller C_i can be designed for the isolated subsystem P_{ii} without any loss of performance. If P_{ij} $(i \neq j)$ is "small" then it should still be possible to design the controller for the essentially independent subsystem P_{ii}. The advantage is that fewer controller parameters need to be chosen than for the full system. This is particularly relevant in process control where often thousands of variables have to be controlled, which

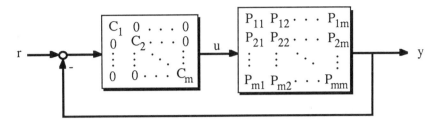

Figure 14.1-1. General decentralized control structure.

could lead to an enormously complex controller. It is also important that at least stability but hopefully also performance is preserved to some degree when individual sensors or actuators fail. This failure tolerance is generally easier to achieve with decentralized control systems, where parts can be turned off without significantly affecting the rest of the system, provided that the controller blocks are designed appropriately.

The designer of decentralized controllers is faced with two issues: (1) The control structure or "pairing" problem — i.e., which set of measurements should be used to affect which set of inputs. (2) The controller design problem — i.e., how to tune the individual controller blocks. The pairing problem can be formidable. Even for relatively small systems, there are many distinct decentralized control system structures to choose from: For a 4×4 system there are 130 possibilities, for a 5×5 system 1495, and so on. Thus efficient screening techniques are needed which are capable of eliminating quickly all the control structures which are definitely inappropriate according to certain criteria like failure tolerance.

All controller design techniques discussed in this book so far yield controllers whose structure and order are determined by the structure and order of the system to be controlled. In this section we will develop guidelines on how to design controllers with a fixed (decentralized) structure which is usually different from the structure of the system to be controlled. We will also develop a series of so-called "Interaction Measures" (IM's). Their purpose is, in general, to help in the screening of different control structures and to guide the design of the controller blocks. Some IM's indicate, for example, if it is possible to design a fault tolerant control system for a specific control structure. Others express the performance degradation caused by the decentralized control structure.

After defining the decentralized control problem more rigorously, we will derive a number of *necessary* conditions the plant and the associated control structure

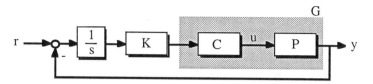

Figure 14.2-1. System with integrator and diagonal compensator K.

have to satisfy for the design of a fault tolerant decentralized control system to be possible. These conditions can be used to screen alternate control structures. Then we will present a decentralized design procedure which guarantees that the overall closed-loop system satisfies certain robust performance conditions. To simplify the notation we will often assume that the controller is fully decentralized — i.e., that the controller matrix C is diagonal rather than block diagonal. Most results can be extended in a obvious manner to the block diagonal case. Usually we will also assume that the square $n \times n$ plant P is stable. If we do so, then the results do *not* carry over easily to systems which are open-loop unstable.

14.2 Definitions

Consider the block diagram in Fig. 14.2-1 where the diagonal controller $C(s)$ is stable and where K is a diagonal gain matrix with positive entries:

$$K = \text{diag}\{k_i\} \qquad k_i > 0, \quad i = 1, n \qquad (14.2 - 1)$$

The factor s^{-1} implies that the control system features integral control in all channels. The following property is very desirable from a practical point of view.

Definition 14.2-1. *A plant P is Decentralized Integral Controllable (DIC) if there exists a diagonal controller CKs^{-1} with integral action such that (a) the closed loop system shown in Fig. 14.2-1 is stable and (b) the gains of any subset of loops can be reduced to $K_\epsilon = \text{diag}\{k_i\epsilon_i\}$, $0 \le \epsilon_i \le 1$ without affecting the closed loop stability.*

Condition (b) implies that any subset of loops can be detuned or taken out of service (put on "manual") while maintaining the stability of the rest of the system. If a system is DIC then it is possible to achieve stable closed-loop performance of the overall system by tuning every loop separately. DIC is a property of the system and in particular of the selected control structure. It is desirable to select a control structure such that the system is DIC. Unfortunately, no necessary *and* sufficient conditions for DIC are available which can be used to screen alternate control structures. There are some sufficient conditions but they should be used for screening only with great caution because they might eliminate attractive alternatives erroneously. Next we will define some properties which are weaker than DIC. They will lead in turn to *necessary* conditions for DIC. These necessary conditions are usually powerful enough to reduce the number of alternatives drastically.

Definition 14.2-2. *The system $G = PC$ is Integral Controllable (IC) if there exists a $k > 0$ such that (a) the closed-loop system shown in Fig. 14.2-1 is stable for $K = kI$ and (b) the gains of the loops can be reduced to $K_\epsilon = \epsilon k I$, $0 < \epsilon \leq 1$ without affecting the closed-loop stability.*

For a system to be IC, stability must be preserved when the gains k are reduced *simultaneously*, while for it to be DIC one must be able to change the gains *independently*. We will show later than when P is not IC with a certain class of compensators C then it cannot be DIC. Thus, control structures which fail the IC test can be eliminated when searching for a system which is DIC.

Definition 14.2-3. *The system $G = PC$ is Integral Stabilizable (IS) if there exists a $k > 0$ such that the closed-loop system shown in Fig. 14.2-1 is stable for $K = kI$.*

Clearly IS is necessary for IC. Easy tests for IS are available which can help to eliminate systems which cannot be DIC.

14.3   Necessary Conditions for Controllability

14.3.1   Results

We will start with some necessary conditions for IS and IC which will then lead to necessary conditions for DIC. In order not to interrupt the flow of the presentation all proofs are collected in Sec. 14.3.2.

Theorem 14.3-1. *Assume G is a proper rational transfer matrix. G is IS only if $\det G(0) > 0$.*

Theorem 14.3-2. *Assume G is a rational transfer matrix. G is IC if all the eigenvalues of $G(0)$ lie in the open RHP. G is not IC if any of the eigenvalues of $G(0)$ lie in the open LHP (the test is inconclusive if any of the eigenvalues are purely imaginary).*

Corollary 14.3-1. *Assume $g(s)$ is a proper rational transfer function. The SISO system described by $g(s)$ is IS and IC if and only if $g(0) > 0$.*

Corollary 14.3-1 simply states the well known fact that positive feedback leads to instability. An SISO system with positive steady state gain can only be controlled with a negative feedback loop. Theorems 14.3-1 and 14.3-2 are generalizations of the negative feedback condition to multivariable systems. These theorems involve steady state information only, the dynamic components of $P(s)$ and $C(s)$ are not important. Because $P(s)$ and $C(s)$ are assumed to be stable it is always possible to choose the dynamic elements of $C(s)$ conservatively enough for the overall system to be stable and to remain stable with controller gain changes — as long as the steady state requirements — expressed, for example, through Thm. 14.3-2 – are satisfied.

For a system to be DIC all the individual loops have to be stable. Therefore, according to Cor. 14.3-1 the controller gains $c_i(0)$ have to be chosen such that $p_{ii}(0)c_i(0) > 0$. Alternatively we can define the matrix $P^+(0)$ which is derived from $P(0)$ by multiplying each column with $+1$ or -1 such that all diagonal elements are positive. Then for the individual loops to be stable we require $c_i(0) > 0$ so that $p_{ii}^+(0)c_i(0) > 0$. Now we can state necessary conditions for DIC in terms of $P^+(0)$ as corollaries of Th. 14.3-1 and 14.3-2.

Corollary 14.3-2. *Let $P(s)C(s)$ be proper and rational. The system $P(s)$ is DIC only if $\det(P^+(0)) > 0$.*

Corollary 14.3-3. *The rational system $P(s)$ is DIC only if all eigenvalues of the matrix product $P^+(0)C(0)$ are in the closed RHP for all nonnegative diagonal gain matrices $C(0)$.*

These corollaries follow from the fact that for DIC it must be possible to adjust the controller gains $k_i > 0$ or equivalently $c_i(0) > 0$ arbitrarily without affecting closed loop stability: if for some gain matrix $C(0)$ the system is not IS or not IC then it cannot be DIC.

Finally it can be shown that the system is not DIC if any of the diagonal elements of the RGA are negative.

Corollary 14.3-4. *If the RGA element $\lambda_{jj}(P) < 0$ then for any diagonal compensator $C(s)$ chosen so that $P(s)C(s)$ is proper and any K the closed loop system shown in Fig. 14.2-1 has at least one of the following properties. (a) The closed-loop system is unstable. (b) Loop j is unstable by itself — i.e., with all the other loops opened. (c) The closed loop system is unstable as loop j is removed.*

Conditions for DIC. In summary, a system $P(s)$ is DIC with a controller $C(s)$ selected such that PC is proper *only if all* the following conditions are met:

(a) $\det(P^+(0)) > 0$

(b) $\text{Re}\{\lambda_i(P^+(0)C(0))\} \geq 0$ $\forall i$, for all diagonal $C(0) \geq 0$

(c) $\text{Re}\{\lambda_i(P^+(0))\} \geq 0$ $\forall i$

(d) $\text{Re}\{\lambda_i(L(0))\} \geq -1$ $\forall i$, where $L = (P - \bar{P})\bar{P}^{-1}$, $\bar{P} = \text{diag}\{p_{11}, \dots, p_{nn}\}$

(e) RGA: $\lambda_{ii}(P(0)) > 0$ $\forall i$

Conditions (a), (b) and (e) follow from Cor. 14.3-2, 14.3-3 and 14.3-4 respectively. Condition (c) and (d) are special cases of (b) for $C(0) = I$ and $P^+(0)C(0) = P(0)\bar{P}(0)^{-1}$. Condition (a) is implied by (c) and is therefore redundant. Condition (b) is difficult to test. Condition (c), (d), and (e) are popular tools for screening control structures in terms of DIC. They are independent as the following examples illustrate.

Example 14.3-1.

$$P(0) = \begin{pmatrix} 10 & 0 & 20 \\ 0.2 & 1 & -1 \\ 11 & 12 & 10 \end{pmatrix}$$

$$\lambda_i(P^+(0)) = \{24.7, -3.0, -0.65\}$$

$$\lambda_i(L(0)) = \{1.19, -0.59 \pm 0.232i\}$$

$$\text{RGA}: \quad \lambda_{ii} = \{4.6, -2.5, 2.1\}$$

Here $\lambda_i(L(0))$ is inconclusive, the other two tests indicate that the system is not DIC.   □

Example 14.3-2.

$$P(0) = \begin{pmatrix} 8.72 & 2.81 & 2.98 & -15.80 \\ 6.54 & -2.92 & 2.50 & -20.79 \\ -5.82 & 0.99 & -1.48 & -7.51 \\ -7.23 & 2.92 & 3.11 & 7.86 \end{pmatrix}$$

$$\lambda_i(P^+(0)) = \{-9.7, 4.7, 6.1, 19.9\}$$

$$\lambda_i(L(0)) = \{-3.3, 1.9, 0.7 \pm i\}$$

$$\text{RGA} \quad \lambda_{ii} = \{0.41, 0.45, 0.17, 0.04\}$$

Here the RGA is inconclusive, the other two tests indicate that the system is not DIC.   □

Example 14.3-3.

$$P(0) = \begin{pmatrix} 0.5 & 0.5 & -0.005 \\ 1 & 2 & -0.01 \\ -30 & -250 & 1 \end{pmatrix}$$

$$\lambda_i(P^+(0)) = \{3.43, 0.036 \pm 0.32i\}$$

$$\lambda_i(L(0)) = \{1.7, -0.85 \pm 0.38i\}$$

$$\text{RGA} \quad \lambda_{ii} = \{-0.71, 2, 1.43\}$$

Here only the RGA allows one to conclude that the system is not DIC. $\qquad\square$

Note that sometimes it is not possible to find any control structure for which a system is DIC. This is illustrated in the next example.

Example 14.3-4.

$$P(0) = \begin{pmatrix} 1 & 1 & -0.1 \\ 0.1 & 2 & -1 \\ -2 & -3 & 1 \end{pmatrix}$$

$$\text{RGA} = \begin{pmatrix} -1.89 & 3.59 & -0.7 \\ -0.13 & 3.02 & -1.89 \\ 3.02 & -5.61 & 3.59 \end{pmatrix}$$

Clearly it is not possible to permute the rows and columns of $P(0)$ and equivalently the RGA such that all elements on the diagonal are positive. Thus the system is not DIC for any pairing of variables. $\qquad\square$

For 2×2 systems much stronger results are available.

Corollary 14.3-4. *For 2×2 systems all tests (a)-(e) are equivalent and necessary and sufficient for DIC. Moreover, for every 2×2 system, there is always a pairing such that the system is DIC.*

14.3.2 Proofs

Proof of Theorem 14.3-1. The proof is based on the Routh test. The characteristic equation (CE) for the closed-loop system in Fig. 14.2-1 is given by

$$\phi(s) \cdot \det(I + G(s)\frac{k}{s}) = 0 \qquad (14.3-1)$$

where $\phi(s)$ is the open-loop characteristic polynomial of $G(s) = P(s)C(s)$. Express $G(s)$ as $G(s) = N(s)d^{-1}(s)$ where $d(s)$ is the common denominator of the elements of $G(s)$ and $N(s)$ is a polynomial matrix. Equation (14.3–1) can then be expressed as

$$\frac{\phi(s)}{sd(s)} \cdot \det(sd(s)I + kN(s)) = 0 \qquad (14.3-2)$$

Upon expansion of the determinant, this expression becomes

$$\frac{\phi(s)}{sd(s)} \cdot (s^n d^n(s) + \ldots + k^n \det N(0)) = 0 \qquad (14.3-3)$$

If $G(s)$ is proper, the coefficient of the highest power of s in (14.3–3) will be the coefficient of the highest power of s in $d(s)$. This coefficient will be positive because of the stability assumption. The closed-loop system will be stable only if all the coefficients in $\det(sd(s)I + kN(s))$ are positive. The constant coefficient is $\det(kN(0))$ and therefore for closed-loop stability it is required that $\det(N(0)) > 0$ and $\det(G(0)) > 0$.   □

Proof of Theorem 14.3-2. Let the Nyquist D-contour be indented at the origin to the right to exclude the pole of $s^{-1}G(s)$ at the origin. The system will be closed-loop stable if none of the characteristic loci (CL) encircles the point $(-1/k, 0)$. For IC it is necessary and sufficient that the CL intersect the negative real axis only at *finite* values. An intersection at $(-\infty, 0)$ could only occur because of the pole of $1/s\ G(s)$ at the origin. Along the indentation, the small semicircle with radius ϵ around the origin, the CL can be described by

$$\lambda_j(G(0))\frac{1}{\epsilon}e^{i\phi} \quad -\frac{\pi}{2} \le \phi \le \frac{\pi}{2}; \quad j = 1, n$$

for small ϵ. Let $\lambda_j(G(0)) = r_j e^{i\theta_j}$; then the expression for the CL can be rewritten as

$$\frac{r_j}{\epsilon}e^{i(\theta_j+\phi)} \quad -\frac{\pi}{2} \le \phi \le \frac{\pi}{2}; \quad j = 1, n$$

The CL do not cross the negative real axis if $-\pi < \theta_j + \phi < \pi$ or $-\pi/2 < \theta_j < \pi/2$, which means $\lambda_j(G(0))$, $j = 1, n$ has to be in the open RHP. The characteristic locus j crosses the negative real axis if $\pi/2 < \theta_j < 3\pi/2$, which means $\lambda_j(G(0))$ is in the open LHP. Nothing can be said from this proof about the systems for which the spectrum of $G(0)$ is constrained to the closed right-half plane and includes eigenvalues on the imaginary axis.   □

Proof of Corollary 14.3-4. Because the RGA element λ_{ij} is invariant under input and output scaling we have for any diagonal $C(0)$

$$\lambda_{ij} = (-1)^{i+j}p_{ij}\frac{\det(P^{ji}(0))}{\det(P(0))} \qquad (14.3-4)$$

$$\lambda_{ij} = (-1)^{i+j}g_{ij}\frac{\det(G^{ji}(0))}{\det(G(0))} \qquad (14.3-5)$$

where the notation A^{ij} indicates that row i and column j of matrix A have been deleted.

If $\lambda_{jj} < 0$ then one or three of the terms in (14.3-5) is negative. For property (a) $\det(G(0)) < 0$; for property (b) $g_{jj} < 0$; for property (c) $\det(G^{jj}(0)) < 0$. □

The proofs of the other corollaries are straightforward and are left as an exercise.

14.4 Stability Conditions – Interaction Measures

Let us define the term Interaction Measure (IM) more precisely with reference to Fig. 14.4-1.

A controller

$$C = \mathrm{diag}\{C_1, C_2, \ldots, C_m\} \tag{14.4 – 1}$$

is to be designed for the system

$$\bar{P} \triangleq \mathrm{diag}\{P_{11}, P_{22}, \ldots, P_{mm}\} \tag{14.4 – 2}$$

such that the block diagonal closed-loop system with the transfer matrices

$$\bar{H} \triangleq \bar{P}C(I + \bar{P}C)^{-1} \tag{14.4 – 3}$$

$$\bar{E} \triangleq (I + \bar{P}C)^{-1} \tag{14.4 – 4}$$

is stable ($\delta = 0$ in Fig. 14.4-1). An IM expresses the constraints imposed on the choice of the closed-loop transfer matrix $\bar{H}$ or $\bar{E}$ for the block diagonal system, which guarantee that the full closed-loop system

$$H = PC(I + PC)^{-1} \tag{14.4 – 5}$$

$$E = (I + PC)^{-1}$$

is stable (i.e., $\delta = 1$) in Fig. 14.4-1.

This definition of an IM has its limitations and therefore the results should be interpreted with caution. The reason is that the IM is based on the block diagonal $\bar{H}$ or $\bar{E}$ which might or might not be indicative of the actual full closed-loop transfer matrix H or E. Though this definition of the IM guarantees the system to be stable it can be very badly behaved. Even if the IM indicates "small" interactions, the performance can be arbitrarily poor. The following matrices play central roles in interaction analysis:

$$L_H = (P - \bar{P})\bar{P}^{-1} \tag{14.4 – 6}$$

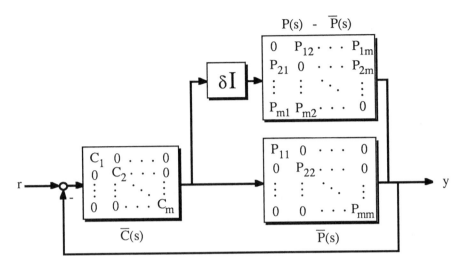

Figure 14.4-1. Block diagram representation of interactions as additive uncertainty.

$$L'_H = \bar{P}^{-1}(P - \bar{P}) \tag{14.4 - 7}$$

$$L_E = (P - \bar{P})P^{-1} \tag{14.4 - 8}$$

They can be viewed as "relative errors" arising from the "approximation" of the full system P by $\bar{P}$.

14.4.1 Necessary and Sufficient Stability Conditions

The following two stability conditions follow from the multivariable Nyquist criterion. Let us denote by $N(k, g(s))$ the net number of clockwise encirclements of the point $(k, 0)$ by the image of the Nyquist D contour under $g(s)$. Then we can state

Theorem 14.4-1. *Assume that P and $\bar{P}$ have the same number of RHP poles and that $\bar{H}$ is stable. Then the closed-loop system H is stable if and only if*

$$N(0, \ \det(I + L_H \bar{H})) = 0 \tag{14.4 - 9}$$

Proof. The return difference operator for the full system H can be factored as

$$(I + PC) = (I + L_H \bar{H})(I + \bar{P}C) \tag{14.4 - 10}$$

Let the number of open loop unstable poles of P and $\bar{P}$ be p_0. H is stable if and only if

$$N(0, \det(I+PC)) = N(0, \det(I+\bar{P}C)) + N(0, \det(I+L_H\bar{H})) = -p_0 \quad (14.4-11)$$

Because $\bar{H}$ is stable by assumption

$$N(0, \det(I + \bar{P}C)) = -p_0 \qquad (14.4-12)$$

Substituting (14.4–12) into (14.4–11), (14.4–9) follows immediately. □

Theorem 14.4-2. *Assume that P and $\bar{P}$ have the same number of RHP zeros and that $\bar{H}$ is stable. Then the closed-loop system H is stable if and only if*

$$N(0, \det(I - \bar{E}L_E)) = 0 \qquad (14.4-13)$$

Proof. Consider the following identity

$$P^{-1}(I + PC) = \bar{P}^{-1}(I + \bar{P}C)(I - \bar{E}L_E) \qquad (14.4-14)$$

Let the number of RHP zeros of P and $\bar{P}$ be z_0. H is stable if and only if

$$N(0, \det(P^{-1}(I + PC))) = N(0, \det(\bar{P}^{-1}(I + \bar{P}C))) + N(0, \det(I - \bar{E}L_E)) = -z_0$$
$$(14.4-15)$$

Because $\bar{H}$ is stable by assumption and the number of RHP zeros of $\bar{P}$ is z_0

$$N(0, \det(\bar{P}^{-1}(I + \bar{P}C))) = -z_0 \qquad (14.4-16)$$

Substituting (14.4–16) into (14.4–15), (14.4–13) follows immediately. □

Theorems 14.4-1 and 14.4-2 form the cornerstone for much of the further development and deserve some discussion. First note that the poles of P are always a subset of the poles of $\bar{P}$. Generically the subset is proper and therefore generically the number of unstable poles of P and $\bar{P}$ is not the same except trivially for stable systems. Therefore for all practical purposes Thm. 14.4-1 is limited to open-loop stable systems. By the same arguments Thm. 14.4-2 is limited to minimum phase systems.

Requirement (14.4–9) can be interpreted as a "robustness" condition for $\bar{H}$ to remain stable under the multiplicative perturbation L_H. Ideally one would want to select $\bar{H} = I$ — i.e., perfect control. If $P = \bar{P}$ — i.e. P itself is block diagonal, then (14.4–9) is trivially satisfied. Thus the closed loop system is stable regardless of how $\bar{H}$ is chosen. If $P \neq \bar{P}$, $\bar{H}$ has to be chosen such that (14.4–9) remains satisfied. Qualitatively, at least, it is clear that when L_H is "large" $\bar{H}$ has to be made "small" to avoid encirclements. A small $\bar{H}$ implies poor performance. IMs derived from (14.4–9) provide a quantitative indication of the constraints on $\bar{H}$ imposed by L_H.

Similarly (14.4–13) can be interpreted as a "robustness" condition for $\bar{H}$ to remain stable under the inverse multiplicative perturbation L_E. Because $\bar{E}$ should be small for good performance (14.4–13) seems easier to satisfy than (14.4–9). However, for strictly proper systems $\lim_{s \to +\infty} E = I$ and therefore encirclements can occur if L_E is too large.

14.4.2 Sufficient Stability Conditions

By invoking the Small Gain Theorem (Thm. 10.2-2) sufficient conditions for the stability of the decentralized control system can be easily derived from Thm. 14.4-1 and 14.4-2.

Theorem 14.4-3. *Assume that P and $\bar{P}$ have the same RHP poles and that $\bar{H}$ is stable. Then the closed loop system H is stable if*

$$\rho(L_H(i\omega)\bar{H}(i\omega)) < 1 \qquad \forall \omega \qquad (14.4-17)$$

or if

$$\|L_H(i\omega)\bar{H}(i\omega)\| < 1 \qquad \forall \omega \qquad (14.4-18)$$

where $\|\cdot\|$ denotes any compatible matrix norm.

Theorem 14.4-4. *Assume that P and $\bar{P}$ have the same RHP zeros and that $\bar{H}$ is stable. Then the closed loop system is stable if*

$$\rho(\bar{E}(i\omega)L_E(i\omega)) < 1 \qquad \forall \omega \qquad (14.4-19)$$

or if

$$\|\bar{E}(i\omega)L_E(i\omega)\| < 1 \qquad \forall \omega \qquad (14.4-20)$$

Several approaches can be taken to derive IMs — i.e., explicit bounds on $\bar{H}$ from (14.4–17) and (14.4–18). There is a trade-off between the assumptions made about the structure of $\bar{H}$ and the restrictiveness of the derived bounds. Less restrictive bounds on the magnitude of $\bar{H}$ are obtained as more restrictive assumptions are made about the structure of $\bar{H}$. Assuming a highly structured form for $\bar{H}$ — i.e., $\bar{H}(s) = \bar{h}(s)I$, leads to the following bound.

Corollary 14.4-1. *Assume $\bar{H}(s) = \bar{h}(s)I$. Under the assumptions of Thm. 14.4-3 the closed-loop system H is stable if*

$$|\bar{h}(i\omega)| < \rho^{-1}(L_H(i\omega)) \qquad \forall \omega \qquad (14.4-21)$$

The form of $\bar{H}(s)$ assumed for Cor. 14.4-1 is very restrictive but $\rho^{-1}(L_H(i\omega))$ is the least restrictive magnitude bound that can be derived for $\bar{H}(s)$ from (14.4–17). (A similar corollary can be derived from Thm. 14.4-4.)

If integral action is employed in all channels then $\bar{h}(0) = 1$ and the requirement (14.4–21) becomes for $\omega = 0$, $\rho(L_H(0)) < 1$. If the system is open-loop stable then it is always possible to satisfy (14.4–21) at all other frequencies ($\omega \neq 0$) by rolling off $\bar{h}(s)$ sufficiently fast. Thus we have the following corollary.

Corollary 14.4-2. *The stable system P is DIC if*

$$\rho(L_H(0)) < 1 \qquad (14.4-22)$$

The small gain condition (14.4–18) does not take into account the special structure of the matrix $\bar{H}$. Therefore it tends to be conservative though the conservativeness depends on the type of norm used. It is natural to look for an expression which takes into account the structure of $\bar{H}$ and represents an optimal bound in the following sense. Let $\bar{H}$ be block diagonal and norm bounded

$$\bar{H}(s) = \text{diag}\left\{\bar{H}_1(s), \ldots, \bar{H}_m(s)\right\} \qquad (14.4-23)$$

$$\bar{\sigma}(\bar{H}_i(i\omega)) < \delta(\omega) \qquad \forall i, \omega \qquad (14.4-24a)$$

or equivalently

$$\bar{\sigma}(\bar{H}(i\omega)) < \delta(\omega) \qquad \forall \omega \qquad (14.4-24b)$$

A real positive function $\mu(L_H(i\omega))$ is desired with the property that (14.4–9) is satisfied for *all* matrices $\bar{H}(i\omega)$ satisfying (14.4–23) and (14.4–24) if and only if

$$\bar{\sigma}(\bar{H}(i\omega)) < \mu^{-1}(L_H(i\omega)) \qquad \forall \omega \qquad (14.4-25)$$

It follows directly from Thm. 11.2-1 that μ is the structured singular value (SSV).

Theorem 14.4-5. *Assume that P and $\bar{P}$ have the same RHP poles and that $\bar{H}$ is stable. Then the closed loop system H is stable if*

$$\bar{\sigma}(\bar{H}(i\omega)) < \mu^{-1}(L_H(i\omega)) \qquad \forall \omega \qquad (14.4-26)$$

This is the tightest norm bound, in the sense that if there is a system $\bar{H}_1$ which violates (14.4–26)

$$\bar{\sigma}(\bar{H}_1(i\omega)) > \mu^{-1}(L_H(i\omega)) \qquad (14.4-27)$$

then there exists another system $\bar{H}_2$ such that

$$\bar{\sigma}(\bar{H}_1(i\omega)) = \bar{\sigma}(\bar{H}_2(i\omega)) \qquad (14.4-28)$$

for which (14.4–9) is violated and H is unstable.

A similar result can be derived from Thm. 14.4-2.

Theorem 14.4-6. *Assume that P and $\bar{P}$ have the same RHP zeros and that $\bar{H}$ is stable. Then the closed loop system H is stable if*

$$\bar{\sigma}(\bar{E}(i\omega)) < \mu^{-1}(L_E(i\omega)) \qquad \forall\omega \qquad (14.4-29)$$

Different IM's can be derived readily from Thms. 14.4-1 through 14.4-6.

14.4.3   Diagonal Dominance Interaction Measures

Assume that $\bar{P}$ and $\bar{H}$ are diagonal and that the 1-norm is used in (14.4–18). Then (14.4–18) becomes

$$\sum_{i \neq j} |(p_{ij}(i\omega)/p_{jj}(i\omega))\bar{h}_j(i\omega)| < 1 \qquad \forall j, \omega \qquad (14.4-30)$$

or

$$|\bar{h}_j(i\omega)| < \left[\sum_{i \neq j} |p_{ij}(i\omega)/p_{jj}(i\omega)|\right]^{-1} \qquad \forall j, \omega \qquad (14.4-31)$$

This is the "IMC interaction measure" for column dominance. It expresses the constraints the individual loops, $\bar{h}_j(s)$ must satisfy for the overall system to be stable. A plot of the RHS of (14.4–31) as a function of frequency is a good indicator of the bandwidth over which good control can be achieved.

Definition 14.4-1. *Let $\bar{P}(i\omega)$ be diagonal. The complex matrix $P(i\omega)$ is column dominant if*

$$\|L_H(i\omega)\|_1 < 1 \qquad (14.4-32a)$$

or

$$\sum_{i \neq j} |p_{ij}(i\omega)| < |p_{jj}(i\omega)| \qquad \forall j \qquad (14.4-32b)$$

When $P(i\omega)$ is column dominant for all ω, a fortunate and rare situation, then the constraint (14.4–31) on $|\bar{h}_j(i\omega)|$ is very mild

$$|\bar{h}_j(i\omega)| < \alpha \quad \text{where} \quad \alpha > 1 \qquad \forall j, \omega$$

and there are no limitations on the bandwidth imposed by the interactions. Note that

$$\det(I + (P - \bar{P})\bar{P}^{-1}\bar{H}) = \det(I + \bar{P}^{-1}\bar{H}(P - \bar{P})) \qquad (14.4 - 33)$$

Therefore another sufficient stability condition which is similar to (14.4–18) is

$$\|\bar{P}^{-1}(i\omega)\bar{H}(i\omega)(P(i\omega) - \bar{P}(i\omega))\| < 1 \qquad \forall\omega \qquad (14.4 - 34)$$

The IMC interaction measure for row dominance can be derived by employing the ∞-norm in (14.4–34).

14.4.4 Generalized Diagonal Dominance Interaction Measures

Assume again that $\bar{P}$ and $\bar{H}$ are diagonal. If the inputs of P are scaled by a diagonal nonsingular matrix D_2 and the outputs by a diagonal nonsingular matrix D_1 and if the controller C is scaled accordingly, the stability of the system should be unaffected. This can be seen easily from (14.4–9)

$$\det(I + (D_1PD_2 - D_1\bar{P}D_2)D_2^{-1}\bar{P}^{-1}D_1^{-1} \cdot D_1\bar{H}D_1^{-1}) \qquad (14.4 - 35a)$$

$$= \det(I + D_1(P - \bar{P})\bar{P}^{-1}\bar{H}D_1^{-1}) \qquad (14.4 - 35b)$$

$$= \det(I + D_1(P - \bar{P})\bar{P}^{-1}D_1^{-1}\bar{H}) \qquad (14.4 - 35c)$$

$$= \det(I + L_H\bar{H}) \qquad (14.4 - 35d)$$

A similar development holds for the right-hand side of (14.4–33). Though stability is independent of scaling the sufficient stability condition

$$\|\bar{H}(i\omega)\| < \|L_H(i\omega)\|^{-1} \qquad \forall\omega \qquad (14.4 - 36)$$

derived from (14.4–18) is not. Therefore it is natural to seek the scaling which makes (14.4–36) least conservative. Equation (14.4–35c) shows that for the error matrix $L_H(s)$ only one scaling matrix (D_1) is necessary, the other one cancels. Thus the minimization problems requiring solutions are

$$\min_{D_1} \|D_1 L_H(i\omega)D_1^{-1}\|_1 \qquad (14.4 - 37a)$$

and

$$\min_{D_2} \|D_2^{-1}L_H'(i\omega)D_2\|_\infty \qquad (14.4 - 37b)$$

The solutions of (14.4–37) are provided by the Perron-Frobenius Theorem.

Theorem 14.4-7.

$$\min_{D_1} \|D_1 L_H(i\omega)D_1^{-1}\|_1 = \min_{D_2} \|D_2^{-1}L_H'(i\omega)D_2\|_\infty = \rho(|L_H(i\omega)|) \qquad (14.4-38)$$

where $|A|$ denotes the matrix A with all its elements replaced by their magnitudes.

Corollary 14.4-3. *Assume that $P(s)$ and $\bar{P}(s)$ have the same RHP poles and that $\bar{H}(s)$ is stable. Then the closed loop system $H(s)$ is stable if*

$$|\bar{h}_i(i\omega)| < \rho^{-1}(|L_H(i\omega)|) \qquad \forall i, \omega \qquad (14.4-39)$$

Note the similarity between Cor. 14.4-1 and 14.4-3. Equation (14.4–21) is a tighter bound than (14.4–39) because the spectral radius bounds any norm — even when it is minimized — from below. However, for obtaining this tighter bound $\bar{H}(s)$ had to be restricted to $\bar{H}(s) = \bar{h}(s)I$.

Definition 14.4-2. *Assume $\bar{P}(i\omega)$ to be diagonal. The complex matrix $P(i\omega)$ is generalized diagonal dominant if*

$$\rho(|L_H(i\omega)|) < 1 \qquad (14.4-40)$$

When $G(i\omega)$ is generalized diagonal dominant for all ω then the constraint (14.4–39) is very mild and there are no limitations on the bandwidth imposed by the interactions.

It would be incorrect to view (14.4–39) as less conservative than (14.4–31). Inequality (14.4–31) provides individual bounds for each of the single loop transfer functions $\bar{h}_j(s)$ and thus allows trade-offs between the different loops. The optimization giving rise to (14.4–39) minimizes the worst bound and in the process makes all the bounds even. However (14.4–39) has the advantage that independent of the number of system inputs and outputs the design engineer can determine from a single curve whether or not the selected control structure leads to significant performance deterioration.

14.4.5 The μ-Interaction Measure

Theorem 14.4-5 states that for stability the magnitude of the diagonal blocks $\bar{H}_i(s)$ has to be constrained by the reciprocal of the SSV μ of the relative error matrix

$$\bar{\sigma}(\bar{H}_j(i\omega)) < \mu^{-1}(L_H(i\omega)) \qquad \forall j, \omega \qquad (14.4-41)$$

The value of μ depends on the structure assumed for $\bar{H}(s)$. In Sec. 11.2.2 we found the bounds

$$\rho(L_H(i\omega)) \leq \mu(L_H(i\omega)) \leq \bar{\sigma}(L_H(i\omega)) \qquad (14.4-42)$$

This confirms the finding of Cor. 14.4-1 that $\rho^{-1}(L_H(j\omega))$ constitutes the loosest bound but that it is only correct for the rather restricted structure $\bar{H}(s) = \bar{h}(s)I$. The upper bound on μ is consistent with the conservative result (14.4–36) when the spectral norm is used and when no structural constraints are put on $\bar{H}(s)$. Not surprisingly $\mu(L_H(j\omega))$ lies between the extremes when $\bar{H}(s)$ has no specific structure at all and when $\bar{H}(s) = \bar{h}(s)I$.

From Thm. 14.4-6 we obtain the interaction constraint

$$\bar{\sigma}(\bar{E}_i(i\omega)) < \mu^{-1}(L_E(i\omega)) \qquad \forall i, \omega \qquad (14.4-43)$$

for minimum phase systems. For strictly proper open-loop systems with integral control $H \to 0$, $\bar{E}_i \to I$ as $\omega \to \infty$ and $H_i \to I$, $\bar{E}_i \to 0$ as $\omega \to 0$. Therefore (14.4–41) can be easily satisfied at high frequencies and (14.4–43) at low frequencies. Unfortunately it is not possible to combine these two bounds over different frequency ranges.

If $\mu(L_H(0)) < 1$ then the system is DIC. If the individual loops are designed in accordance with (14.4–41) then they can be detuned or turned off without affecting the stability of the rest of the system. Thus, design constraint (14.4–41) leads to controllers which have attractive practical properties.

Note that μ treats both diagonal and block diagonal $\bar{H}(s)$ in a unified optimal manner. Just as in the case of generalized diagonal dominance all loops are given equal preference.

14.4.6 Interaction Measures for 2×2 Systems

The interaction measures for 2×2 systems which are most widely used industrially are the RGA and the Rijnsdorp interaction measure

$$\kappa_R(s) = \frac{p_{12}(s)p_{21}(s)}{p_{11}(s)p_{22}(s)} \qquad (14.4-44)$$

It is related to the RGA through

$$\lambda_{11}(s) = \frac{1}{1 - \kappa_R(s)} \qquad (14.4-45)$$

It can be easily shown that these empirical IMs are closely related to those we derived earlier on theoretical grounds.

Corollary 14.4-4. *Assume that $P(s)$ and $\bar{P}(s)$ have the same RHP poles and that $\bar{H}(s)$ is stable. Then the closed-loop system $H(s)$ is stable if*

$$|\bar{h}_i(i\omega)| < |\kappa_R(i\omega)|^{-1/2} = \rho^{-1}(L_H(i\omega)) = \mu^{-1}(L_H(i\omega)) \qquad \forall i, \omega \qquad (14.4-46)$$

Corollary 14.4-5. *The stable 2×2 system P is DIC if any one of the following equivalent conditions is satisfied:*

(a) $P(0)$ *is generalized diagonal dominant.*

(b) $\rho(L_H(0)) < 1$

(c) $\rho(|L_H(0)|) < 1$

(d) $\mu(L_H(0)) < 1$

(e) $|\kappa_R(0)| < 1$

(f) RGA: $\lambda_{11}(0) > \frac{1}{2}$

Compare these conditions with the necessary *and* sufficient conditions for DIC derived in Sec. 14.3.

$$\text{Re}\{\lambda_i(L_H(0))\} \geq -1 \qquad (14.4-47)$$

$$\text{RGA} : \lambda_{ii}(0) > 0 \qquad (14.4-48)$$

The conservativeness of (b) and (f), and therefore of all IMs derived in this section is immediately apparent.

14.4.7   Examples

Example 14.4-1. Consider the distillation column of Doukas and Luyben (1978). The transfer function matrix for a 3×3 subsystem is shown in Table 14.4-1.

   Line 1 in Fig. 14.4-2 shows a plot of $\mu^{-1}(L_H(i\omega))$ for the fully decentralized control system with (diagonal) pairings $((1,1), (2,2), (3,3))$. This curve shows that a fully decentralized controller with integral action cannot be designed on the basis of Thm. 14.4-5 since $\mu^{-1}(L_H(0)) < 1$. This constraint can be relaxed by considering a more complex controller structure. Line 2 in the same figure shows a plot of $\mu^{-1}(L_H(i\omega))$ for the block decentralized control system with pairings $((1-2,1-2), (3,3))$. In this case, a controller with integral action is possible since $\mu^{-1}(L_H(0)) > 1$. However, the interactions limit the achievable closed loop bandwidth to about 0.2 rad min^{-1}.

Table 14.4-1. Distillation Column Transfer Matrix.

	Reflux Ratio	Side Draw	Reboil Duty
Toluene in bottom	$\dfrac{0.37e^{-7.75s}}{(22.2s+1)^2}$	$\dfrac{-11.3e^{-3.70s}}{(21.74s+1)^2}$	$\dfrac{-9.811e^{-1.59s}}{(11.36s+1)}$
Toluene in tops	$\dfrac{-1.986e^{-0.71s}}{(66.67s+1)^2}$	$\dfrac{5.24e^{-60s}}{(400s+1)}$	$\dfrac{5.94e^{-2.24s}}{(14.29s+1)}$
Benzene in side draw	$\dfrac{0.204e^{-0.59s}}{(7.14s+1)^2}$	$\dfrac{0.33e^{-0.68s}}{(2.38s+1)^2}$	$\dfrac{2.38e^{-0.42s}}{(1.43s+1)^2}$

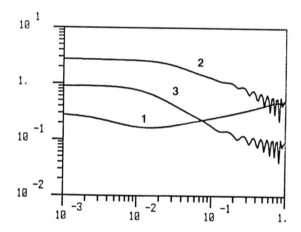

Figure 14.4-2. Interaction Measures. Line 1 = $\mu^{-1}(E(j\omega))$ for fully decentralized controller; line 2 = $\mu^{-1}(E(j\omega))$; line 3 = $\rho^{-1}(|E(j\omega)|)$ for block decentralized controller. (Reprinted with permission from *Automatica*, **22**, 316 (1986), Pergamon Press, plc.)

The generalized diagonal dominance bound $\rho^{-1}(|L_H(i\omega)|)$ for the block decentralized controller is shown as line 3 in Fig. 14.4-2. A comparison of lines 2 and 3 demonstrates the conservativeness associated with $\rho^{-1}(|L_H(j\omega)|)$ as IM.   □

Example 14.4-2. Assume now a fully decentralized control structure for the system in Table 14.4-1 implemented on the variable pairs $((1,2),(2,1),(3,3))$. The objective is to demonstrate the response of the closed-loop system when the three controllers are designed on the basis of Thm. 14.4-5. Figure 14.4-3 shows a plot of $\mu^{-1}(L_H(i\omega))$ (Line 1). The interactions limit the closed loop bandwidth to about 0.08 rad min^{-1}. Figure 14.4-3 also shows plots of $|\bar{h}_i(i\omega)|, (i = 1, 3)$, for three different sets of controllers. In the first case, the three controllers are chosen such that

$$\bar{h}_i(s) = e^{-\tau_i s}(4s + 1)^{-3} \qquad i = 1, 3 \tag{14.4 − 49}$$

while in the second and third cases the controllers are chosen such that

$$\bar{h}_i(s) = e^{-\tau_i s}(10s + 1)^{-3} \qquad i = 1, 3 \tag{14.4 − 50}$$

and

$$\bar{h}_i(s) = e^{-\tau_i s}(25s + 1)^{-3} \qquad i = 1, 3 \tag{14.4 − 51}$$

respectively. Here, τ_i, $(i = 1, 3)$, are the time delays in p_{12}, p_{21}, and p_{33}, respectively. The responses of the closed-loop systems for these three sets of controllers were tested for step changes in r_2. The responses corresponding to (14.4–49 to 51) are shown in Fig. 14.4-4. It is apparent that as the $\bar{h}_i(s)$ are moved away from the stability bound (Line 1) and thus made more conservative, the closed-loop responses become more sluggish but less oscillatory.   □

14.5   Robust Performance Conditions

In the spirit of the previous section we will derive bounds on the magnitude of the sensitivity $\bar{E}$ and the complementary sensitivity $\bar{H}$. These bounds will have the following property: If each block i of the decentralized controller is designed such that $\bar{E}_i$ or $\bar{H}_i$ satisfies the corresponding bound, then *robust performance* (in the H_∞ sense) of the overall coupled system is assured and not just nominal stability as before. Because robust stability is just a special case of robust performance, it is not discussed separately.

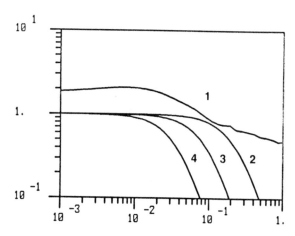

Figure 14.4-3. Line 1 = $\mu^{-1}(E(j\omega))$; lines 2-4 = amplitude ratio for (14.4–49), (14.4–50) and (14.4–51), respectively. (Reprinted with permission from *Automatica*, **22**, 317 (1986), Pergamon Press, plc.)

14.5.1 Sufficient Conditions for Robust Performance

Let us assume that the robust performance requirement is expressed as

$$\mu(M) < 1, \quad \forall\omega \tag{14.5-1}$$

We will proceed in three steps:

1. Express M as an LFT of $\tilde{H}$

2. Express $\tilde{H}$ as an LFT of $\bar{H}$

3. Express M as an LFT of $\bar{H}$. Based on this LFT state a bound on $\bar{\sigma}(\bar{H})$ which guarantees RP of the overall system — i.e., (14.5-1).

The same procedure can be defined for E instead of H. It is based on the results in Sec. 11.4 where we show how robustness conditions can be expressed in terms of bounds on specific transfer matrices related to M by LFT's.

1. *Express M as an LFT of $\tilde{H}$ or $\tilde{E}$.* In Sec. 11.4 we showed in detail how such an LFT can be constructed if it is not obvious from inspection. In general it has the form

$$M = N_{11}^T + N_{12}^T T N_{21}^T \tag{14.5-2}$$

where $T = \tilde{H}$ or $T = \tilde{E}$.

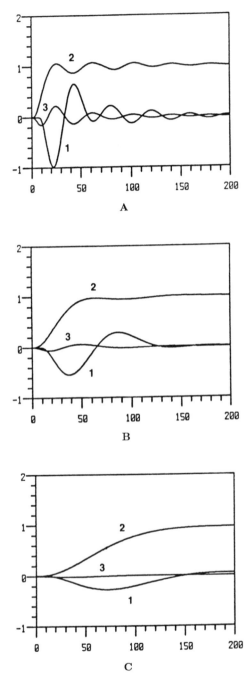

Figure 14.4-4. Closed-loop response to unit step change in $r_2(s)$ for different loop designs $\bar{h}_i$; A: (14.4–49), B: (14.4–50), C: (14.4–51). (Reprinted with permission from *Automatica*, **22**, 317 (1986), Pergamon Press, plc.)

2. *Express* $\tilde{H}(\tilde{E})$ *as an LFT of* $\bar{H}(\bar{E})$. It can be easily verified that

$$\tilde{H} = \tilde{P}\bar{P}^{-1}\bar{H}(I + L_H\bar{H})^{-1} \qquad (14.5-3)$$

$$\tilde{E} = \bar{E}(I - L_E\bar{E})^{-1}\tilde{P}\bar{P}^{-1} \qquad (14.5-4)$$

3. *Express* M *as an LFT of* $\bar{H}(\bar{E})$ *and compute a bound on* $\bar{\sigma}(\bar{H})(\bar{\sigma}(\bar{E}))$ *which guarantees robust performance.* Substitute (14.5-3) into (14.5-2) to obtain M as an LFT of $\bar{H}$:

$$M = N_{11}^H + N_{12}^H \tilde{P}\bar{P}^{-1}\bar{H}(I + L_H\bar{H})^{-1}N_{21}^H \qquad (14.5-5)$$

Similarly from (14.5-4)

$$M = N_{11}^E + N_{12}^E \bar{E}(I - L_E\bar{E})^{-1}\tilde{P}\bar{P}^{-1}N_{21}^E \qquad (14.5-6)$$

The following theorem follows directly from Thm. 11.4-1.

Theorem 14.5-1. *The overall system satisfies the robust performance condition* $\mu(M) < 1, \forall\omega$ *if the individual subsystems satisfy*

$$\sigma(\bar{H}) \le c_H \text{ or } \bar{\sigma}(\bar{E}) \le c_E \qquad \forall\omega \qquad (14.5-7)$$

where at each frequency c_H *and* c_E *solve*

$$\mu_{\hat{\Delta}}\begin{pmatrix} N_{11}^H & N_{12}^H\tilde{P}\bar{P}^{-1} \\ N_{21}^H c_H & -L_H c_H \end{pmatrix} = 1 \qquad (14.5-8)$$

$$\mu_{\hat{\Delta}}\begin{pmatrix} N_{11}^E & N_{12}^E \\ \tilde{P}\bar{P}^{-1}N_{21}^E c_E & L_E c_E \end{pmatrix} = 1 \qquad (14.5-9)$$

and μ *is computed with respect to the structure* $\hat{\Delta} = \text{diag}\{\Delta, C\}$.

We will first use an example to illustrate the construction of the LFT's. Then we will propose a design procedure based on Thm. 14.5-1.

Example 14.5-1 (Robust Performance with Input Uncertainty). In Sec. 11.3.3 we found that robust performance is achieved if and only if $\mu(M) < 1$, where

$$M = \begin{pmatrix} -\tilde{P}^{-1}\tilde{H}\tilde{P}\bar{\ell}_I & -\tilde{P}^{-1}\tilde{H}w \\ \tilde{E}\tilde{P}\bar{\ell}_I & \tilde{E}w \end{pmatrix} \qquad (14.5-10)$$

By inspection M may be written as LFT's in terms of $\tilde{H}$ and $\tilde{E}$

$$M = N_{11}^H + N_{12}^H \tilde{H} N_{21}^H = \begin{pmatrix} 0 & 0 \\ \tilde{P}\bar{\ell}_I & Iw \end{pmatrix} + \begin{pmatrix} -\tilde{P}^{-1} \\ -I \end{pmatrix} \tilde{H} (\tilde{P}\bar{\ell}_I \quad Iw) \quad (14.5-11)$$

$$M = N_{11}^E + N_{12}^E \tilde{E} N_{21}^E = \begin{pmatrix} -I\bar{\ell}_I & -\tilde{P}^{-1}w \\ 0 & 0 \end{pmatrix} + \begin{pmatrix} \tilde{P}^{-1} \\ I \end{pmatrix} \tilde{E} (\tilde{P}\bar{\ell}_I \quad Iw) \quad (14.5-12)$$

Thus c_H and c_E have to solve

$$\mu_{\hat{\Delta}} \begin{pmatrix} 0 & 0 & -\bar{P}^{-1} \\ \tilde{P}\bar{\ell}_I & Iw & -\tilde{P}\bar{P}^{-1} \\ \tilde{P}\bar{\ell}_I c_H & Iw c_H & -L_H c_H \end{pmatrix} = 1 \qquad (14.5-13)$$

and

$$\mu_{\hat{\Delta}} \begin{pmatrix} -I\bar{\ell}_I & -\tilde{P}^{-1}w & \tilde{P}^{-1} \\ 0 & 0 & I \\ \bar{P}\bar{\ell}_I c_E & \bar{P}\tilde{P}^{-1}w c_E & L_E c_E \end{pmatrix} = 1 \qquad (14.5-14)$$

$\square$

14.5.2 Design Procedure

Consistent with the other tests and techniques discussed in this section we will assume that each one of the "loops" is designed separately. We propose the following procedure: find a decentralized controller which yields individual loops ($\bar{H}$ and $\bar{E}$) which are stable and in addition satisfy

1. *Nominal Stability*: Satisfy $\bar{\sigma}(\bar{H}) < \mu^{-1}(L_H)$ (14.4–41) at *all* frequencies or satisfy $\bar{\sigma}(\bar{E}) < \mu^{-1}(L_E)$ (14.4–43) at *all* frequencies. It is *not* allowed to combine (14.4–41) and (14.4–43).

2. *Robust Performance*: At each frequency satisfy *either* $\bar{\sigma}(\bar{H}) \le c_H$ *or* $\bar{\sigma}(\bar{E}) \le c_E$ (14.5–7). Combining the two conditions over different frequency ranges is allowed.

Thus, in general, two separate conditions must be satisfied by the individual design: one for nominal stability and one for robust performance. Usually the bound $\bar{\sigma}(\bar{H}) \le c_H$ is satisfied for high and the bound $\bar{\sigma}(\bar{E}) \le c_E$ for low frequencies. In some rare cases a single bound can be satisfied over the whole frequency range. Assume $\bar{\sigma}(\bar{H}) \le c_H, \forall\omega$. Because c_H solves (14.5–8), $\mu_\Delta(N_{11}^H) \le 1, \forall\omega$. From (14.5–5) this is equivalent to $\mu_\Delta(M(\bar{H} = 0)) \le 1, \forall\omega$. Thus, in order for $\bar{\sigma}(\bar{H}) \le c_H, \forall\omega$ the performance requirements have to be such, that the robust performance condition is satisfied for $\bar{H} = 0$. This may be the case if we are interested in *robust stability* only.

Assume on the other hand $\bar{\sigma}(\bar{E}) \leq c_E, \forall \omega$. Because c_E solves (14.5–9), $\mu_\Delta(N_{11}^E) \leq 1, \forall \omega$. From (14.5–6) this is equivalent to $\mu_\Delta(M(\bar{E} = 0)) \leq 1, \forall \omega$. Thus, in order for $\bar{\sigma}(\bar{E}) \leq c_E, \forall \omega$, the uncertainty has to be so small, that the robust performance condition is satisfied for $\bar{E} = 0$. This may be the case if we are interested in *nominal performance* only.

In these extreme cases, it turns out that if the individual stable loops satisfy the robust performance bound (e.g., $\bar{\sigma}(\bar{H}) \leq c_H, \forall \omega$) for the overall system, then its nominal stability is implied automatically; Step 1 of the proposed design procedure can be omitted.

This again can be seen easily from the following argument. If $\bar{\sigma}(\bar{H}) \leq c_H, \forall \omega$, then c_H solves (14.5–8) for all values of ω and $\mu_c(L_H c_H) \leq 1$ or $c_H \leq \mu_c^{-1}(L_H), \forall \omega$. This implies that the robust performance constraint is tighter than the robust stability constraint (14.4–41). A similar derivation involving the sensitivity $\bar{E}$ is possible.

We want to stress that, in general, when neither of the *individual* bounds in (14.5–7) holds over the *whole* frequency range, then nominal stability is *not* implied by robust performance: for nominal stability either one of the two bounds (14.4–41 or 14.4–43) has to be satisfied for *all* frequencies. These bounds cannot be combined over different frequency ranges.

14.5.3   Example

We continue Example 14.5-1 on robust performance with diagonal input uncertainty. Consider the following plant

$$\hat{P} = \frac{1}{1 + 75s} \begin{pmatrix} -0.878\frac{1-0.2s}{1+0.2s} & 0.014 \\ -1.082\frac{1-0.2s}{1+0.2s} & -0.014\frac{1-0.2s}{1+0.2s} \end{pmatrix} \tag{14.5 $-$ 15}$$

Physically, this may correspond to a high-purity distillation column using distillate (D) and boilup (V) as manipulated inputs to control top and bottom composition (Appendix). We want to design a decentralized (diagonal) controller for this plant such that robust performance is guaranteed when there is 10% uncertainty on each manipulated input. The uncertainty and performance weights are

$$\bar{\ell}_I(s) = 0.1 \tag{14.5 $-$ 16}$$

$$w(s) = 0.25\frac{7s + 1}{7s} \tag{14.5 $-$ 17}$$

The weight (14.5–17) implies that we require integral action ($w(0) = \infty$) and allow an amplification of disturbances at high frequencies by a factor of four at most ($w_P(\infty) = 0.25$). A particular sensitivity function which matches the performance bound (14.5–17) exactly for low frequencies and satisfies it easily for high frequencies is $E = \frac{28s}{28s+1}I$. This corresponds to a first order response with time constant 28 min.

Nominal Stability (NS). The nominal model has $\mu(L_H(0)) = 1.11$. Consequently, it is impossible to satisfy the NS-condition (14.4–41).

The NS-condition (14.4–43) for $\bar{\sigma}(\bar{E})$ cannot be satisfied either. Firstly, $\hat{P}$ has one RHP-zero, while the diagonal plant has two. Secondly, the plant is clearly not diagonal dominant at high frequencies, and $\mu(L_E(i\omega))$ is larger than one for $\omega > 4$ min^{-1}. The simplest way to get around this problem is to treat the RHP-zeros as uncertainty. This is actually not very conservative, since RHP-zeros limit the achievable performance anyway. To this end define the following new nominal model

$$\tilde{P} = \frac{1}{1 + 75s} \begin{pmatrix} -0.878 & 0.014 \\ -1.082 & -0.014 \end{pmatrix} \qquad (14.5 - 18)$$

and include the RHP-zeros in the input uncertainty by using the following new uncertainty weight

$$\bar{\ell}_I(s) = 0.1 \frac{5s + 1}{0.25s + 1} \qquad (14.5 - 19)$$

$|\bar{\ell}_I(i\omega)|$ reaches a value at one at about $\omega = 2$ min^{-1}. This includes the neglected RHP-zeros since the multiplicative uncertainty introduced by replacing $\frac{1-0.2s}{1+0.2s}$ by 1 is $|1 - \frac{1-0.2s}{1+0.2s}|$, which reaches a value of one at about $\omega = 3$ min^{-1}.

With the new model (14.5–8) we still cannot satisfy the NS-condition (14.4–41) for $\bar{\sigma}(\bar{H})$. However, the NS-condition (14.4–43) for $\bar{\sigma}(\bar{E})$ is easily satisfied since $\tilde{P}$ and $\bar{P}$ have the same number of RHP-zeros (none), and $\mu(L_E) = 0.743$ at all frequencies. The only restriction this imposes on $\bar{E}$ is that the maximum peaks of $|\bar{\epsilon}_1|$ and $|\bar{\epsilon}_2|$ must be less than $1/0.743 = 1.35$. This is easily satisfied since both $\bar{p}_{11} = \frac{-0.878}{1+75s}$ and $\bar{p}_{22} = \frac{-0.014}{1+75s}$ are minimum phase.

In the remainder of this example the model of the plant ($\tilde{P}$) is assumed to be given by (14.5–18) and the uncertainty weight ($\bar{\ell}_I$) by (14.5–19).

Nominal Performance (NP). From the overall performance requirements

$$NP \iff \bar{\sigma}(\tilde{E}) \leq |w|^{-1} \qquad \forall \omega \qquad (14.5 - 20)$$

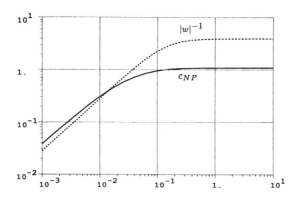

Figure 14.5-1. NP is satisfied if and only if $\bar{\sigma}(\tilde{E}) \leq |w|^{-1}$ which is satisfied if $\bar{\sigma}(\tilde{E}) \leq c_{NP}$.

we would expect that the individual loops have to satisfy *at least,* $\bar{\sigma}(\tilde{E}) \leq |w|^{-1}$. However, this is not necessarily the case, as illustrated by the example: Condition (14.5–20) is equivalent to $\mu_\Delta(M) \leq 1$ with $M = w\tilde{E}$ and $\Delta = \Delta_P$ (Δ_P is a full matrix). From (14.5–7) we find the following sufficient condition for NP in terms of $\bar{E}$:

$$NP \;\Leftarrow\; \bar{\sigma}(\bar{E}) \leq c_{NP} \qquad \forall\omega \qquad\qquad (14.5-21)$$

where c_{NP} solves at each frequency

$$\mu_{\hat{\Delta}} \begin{pmatrix} 0 & wI \\ c_{NP}\bar{P}P^{-1} & c_{NP}L_E \end{pmatrix} = I \qquad\qquad (14.5-22)$$

and $\hat{\Delta} = diag\{\Delta_P, C\}$. In our example Δ_P is a "full" 2×2 matrix, and C is a diagonal 2×2 matrix. The bound c_{NP} is shown graphically in Fig. 14.5-1 and is seen to be *larger* than $|w|^{-1}$ for low frequencies. Consequently, the performance of the overall system may be *better* than that of the individual loops — that is, the interactions may improve the performance.

Robust Performance (RP). Bound on $\bar{\sigma}(\bar{H})$. The bound c_H defined by (14.5–8) is shown graphically in Fig. 14.5-2 (μ in (14.5–8) is computed with respect to the structure $\hat{\Delta} = diag\{\Delta_I, \Delta_P, C\}$, where Δ_I is a diagonal 2×2 matrix, Δ_P is a full 2×2 matrix and C is a diagonal 2×2 matrix). It is clearly *not* possible to satisfy the bound $\bar{\sigma}(\bar{H}) < c_H$ at all frequencies. In particular, we find $c_H < 0$ for $\omega < 0.03$ min^{-1}. The reason is that the performance weight $|w| > 1$ in

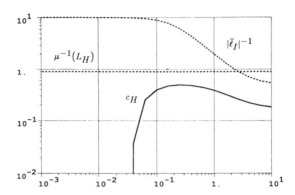

Figure 14.5-2. Bounds $\mu^{-1}(L_H)$ and c_H on $\bar{\sigma}(\bar{H})$.

this frequency range, which means that feedback is required (i.e., $\bar{H} = 0$ is not possible).

Bound on $\bar{\sigma}(\bar{E})$. The bound c_E defined by (14.5–9) is shown graphically in Fig. 14.5-3 (μ is computed with respect to the same structure as above). Again it is not possible to satisfy this bound at all frequencies. In particular, we find $c_E \leq 0$ for $\omega > 2$ min^{-1}. The reason is that the uncertainty weight $|\bar{\ell}_I| > 1$ in this frequency range, which means that perfect control ($\bar{E} = 0$) is not allowed.

Combining bounds on $\bar{\sigma}(\bar{H})$ and $\bar{\sigma}(\bar{E})$. The bound on $\bar{\sigma}(\bar{E})$ is easily satisfied at low frequencies, and the bound on $\bar{\sigma}(\bar{H})$ is easily satisfied at high frequencies. The difficulty is to find an $\bar{E} = I - \bar{H}$ which satisfies either one of the conditions in the frequency range from 0.1 to 1 min^{-1}. The following design is successful (Fig. 14.5-4).

$$\bar{\eta}_1 = \bar{\eta}_2 = \frac{1}{7.5s + 1} \quad , \quad \bar{\epsilon}_1 = \bar{\epsilon}_2 = \frac{7.5s}{7.5s + 1} \qquad (14.5 - 23)$$

The bound on $|\bar{\epsilon}_i|$ is satisfied for $\omega < 0.3$ min^{-1}, and the bound on $|\bar{\eta}_i|$ for $\omega > 0.23$ min^{-1}. The controller which yields (14.5–23) is

$$C = k\frac{(1 + 75s)}{s} \begin{pmatrix} \frac{-1}{0.878} & 0 \\ 0 & \frac{-1}{0.014} \end{pmatrix}, \quad k = 0.133 \qquad (14.5 - 24)$$

Because the bounds c_H and c_E are almost flat in the cross-over region, the result is fairly insensitive to the particular choice of controller gain: for $0.06 < k < 0.25$ the design satisfies either $\bar{\sigma}(\bar{E}) < c_E$ or $\bar{\sigma}(\bar{H}) < c_H$ at each frequency and thus has RP.

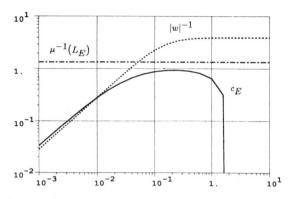

Figure 14.5-3. Bounds $\mu^{-1}(L_E)$ and c_E on $\bar{\sigma}(\bar{E})$.

Figure 14.5-5 shows that RP of the overall system is guaranteed ($\mu(M) \leq 0.63$) as expected from the design procedure. The fact that $\mu(M)$ is much smaller than one, demonstrates some of the conservativeness of conditions (14.5–7) which are only sufficient for RP.

14.6   Summary

Decentralized controllers are popular in practice because they involve fewer tuning parameters than full multivariable control systems and because it is easier to make them fault tolerant. Our approach is to design the controllers for the individual subsystems *independently* but subject to some constraint which guarantees the stability and (robust) performance of the overall system. In order for this to be possible with integral action in all channels the system has to be *Decentralized Integral Controllable* (DIC). The following conditions are *necessary* for DIC (Sec. 14.3.1) and can be used to identify attractive control structure candidates

$$\text{Re}\{\lambda_i(P^+(0)\} \geq 0 \qquad \forall i$$

$$\text{Re}\{\lambda_i(L(0))\} \geq -1 \qquad \forall i$$

$$\text{RGA}: \ \lambda_{ii}(P(0)) \geq 0 \qquad \forall i$$

where

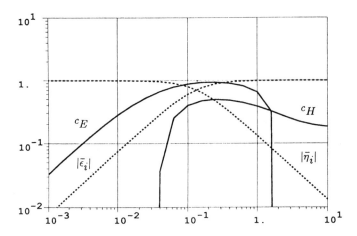

Figure 14.5-4. RP is guaranteed since $|\bar{\epsilon}_i| < c_E$ for $\omega < 0.3$ min^{-1} and $|\bar{\eta}_i| < c_H$ for $\omega > 0.23$ min^{-1}.

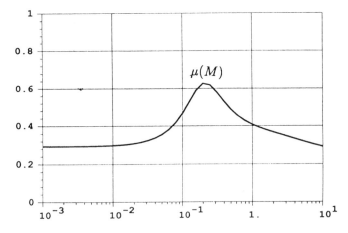

Figure 14.5-5. $\mu(M)$ as a function of frequency. RP of the overall system is guaranteed since $\mu(M) < 1$ for all frequencies.

$$L = (P - \bar{P})\bar{P}^{-1}$$

From different variations of the Small Gain Theorem we derived different bounds on the magnitude of the individual loop transfer functions which, when satisfied, guarantee the stability of the overall system. We call these bounds which constrain the individual designs *interaction measures*. For example, the IMC interaction measure for column dominance requires

$$|\bar{h}_j(i\omega)| < \left[\sum_{i \neq j} |p_{ij}(i\omega)/p_{jj}(i\omega)|\right]^{-1} \qquad \forall j, \omega \qquad (14.4-31)$$

and general dominance requires

$$|\bar{h}_j(i\omega)| < \rho^{-1}(|L_H(i\omega)|) \quad \forall j, \omega \qquad (14.4-39)$$

The least conservative norm bound is provided by the μ-interaction measure

$$\bar{\sigma}(\bar{H}_j(i\omega)) < \mu^{-1}(L_H(i\omega)) \quad \forall j, \omega \qquad (14.4-41)$$

Given an uncertainty description in terms of a block diagonal Δ and an H_∞ performance specifications, bounds c_H and c_E of a similar type can be derived on $\bar{\sigma}(\bar{H}_j)$ and $\bar{\sigma}(\bar{E}_j)$

$$\bar{\sigma}(\bar{H}) \leq c_H \text{ or } \bar{\sigma}(\bar{E}) \leq c_E \qquad (14.5-7)$$

They guarantee robust performance of the overall system if the nominal system $\tilde{P}$ with the decentralized controller is stable. Here c_H and c_E are defined implicitly through (14.5–8) and (14.5–9).

14.7 References

14.1. A good discussion of the importance of the pairing problem was presented by Nett & Spang (1987).

14.2. Davison (1976) used a concept similar to integral controllability. The definition of decentralized integral controllability was proposed by Skogestad.

14.3. This section is based on the papers by Grosdidier, Morari & Holt (1985) and Morari (1983b, 1985). The determinant condition for IS (Thm. 14.3-1) was derived independently by Lunze (1982). Eigenvalue conditions similar to that in Thm. 14.3-2 were published by Locatelli et al. (1982) and Lunze (1983, 1985). Condition (c) for DIC was obtained by Mijares et al. (1986) in a very different setting and under more restrictive assumptions. Condition (a) for DIC was stated by Niederlinksi (1971) but the claims about necessity *and* sufficiency in that paper are incorrect. Example 14.3-4 is due to Koppel as cited by McAvoy (1983).

13.3.2. Postlethwaite & MacFarlane (1979) interpret the MIMO Nyquist criterion in terms of characteristic loci, which are used for the proof of Thm. 14.3-2.

14.4. This section follows closely the paper by Grosdidier & Morari (1986).

14.4.3. The stability condition based on the IMC interaction measure was derived by Economou & Morari (1986). Rosenbrock (1974) expresses an entirely equivalent stability condition in terms of "Gershgorin bands".

14.4.4. A proof of the Perron-Frobenius theorem is available from Seneta (1973). The concept of generalized diagonal dominance for control system design was introduced by Mees (1981) and Limebeer (1982).

Both the diagonal dominance and generalized diagonal dominance concept can be extended to block diagonal system (Feingold and Varga, 1962; Limebeer 1982). The bounds on $\|\bar{H}_i(i\omega)\|$ obtained by this approach are excessively conservative and therefore not very useful in most practical applications.

14.4.5. Grosdidier & Morari (1986) discuss how one or the other loop can be given preferential treatment (i.e., tightened) by employing weights in the computation of the μ-interaction measure.

14.4.6. The interaction measure κ_R was defined by Rijnsdorp (1965). Nett & Uthgenannt (1988) determine "optimal" weights for the μ-interaction measure for the 2×2 block case.

14.5. Skogestad & Morari (1988b) developed the design procedure for robust performance.

Part IV

SAMPLED DATA MULTI-INPUT MULTI-OUTPUT SYSTEMS

Chapter 15

MIMO Sampled-Data Systems

We assume that the reader has mastered the chapters on SISO sampled-data systems as well the chapters on MIMO continuous systems. When the extension of results to MIMO sampled-data systems is straightforward the discussion will be brief and sometimes limited to simply defining the appropriate notation.

15.1 Fundamentals of MIMO Sampled-Data Systems

15.1.1 Sampled-Data Feedback

The block diagram of a typical sampled-data feedback loop is shown in Fig. 15.1-1A. Thick lines are used to represent the paths along which the signals are continuous. Equations (7.1–1) to (7.10–11) carry through to the MIMO case, with vectors instead of scalars. $C(z)$ denotes the discrete controller implemented through a digital computer. $H_0(s)$ models the D/A converter. We have

$$H_0(s) = h_0(s)I \qquad (15.1-1)$$

where $h_0(s)$ is the zero-order hold given by (7.1–12) and I is the identity matrix with dimension equal to the number of controller outputs. The block $\Gamma(s)$ represents an anti-aliasing prefilter. The problem of aliasing was discussed in Sec. 7.1. Assuming that the the same sampling time is used for all the process outputs, it is reasonable to choose

$$\Gamma(s) = \gamma(s)I \qquad (15.1-2)$$

where I has dimension equal to the number of the process outputs. $P(s)$ is the continuous system transfer matrix described in Sec. 10.1.1.

When the continuous output y is not observed directly but after the prefilter and only at the sampling points, then Fig. 15.1-1A can be simplified to Fig. 15.1-1B, where

$$d_\gamma^*(z) = \mathcal{ZL}^{-1}\{\Gamma(s)d(s)\} = \mathcal{ZL}^{-1}\{\gamma(s)d(s)\} \qquad (15.1-3)$$

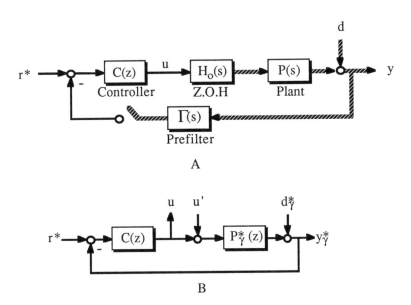

Figure 15.1-1. Block diagram of computer controlled system A: Sampled-data structure with thick lines indicating analog signals. B: Discrete structure with all signals discrete.

$$y_\gamma^*(z) = \mathcal{ZL}^{-1}\{\Gamma(s)y(s)\} = \mathcal{ZL}^{-1}\{\gamma(s)y(s)\} \qquad (15.1-4)$$

and all signals are discrete. Note that when the operator $\mathcal{ZL}^{-1}$ is applied to a vector or matrix, it is simply applied to each element separately. We define

$$P_\gamma^*(z) = \mathcal{ZL}^{-1}\{\Gamma(s)P(s)H_0(s)\} = \mathcal{ZL}^{-1}\{h_0(s)P(s)\gamma(s)\} \qquad (15.1-5)$$

$$P^*(z) = \mathcal{ZL}^{-1}\{P(s)H_0(s)\} = \mathcal{ZL}^{-1}\{h_0(s)P(s)\} \qquad (15.1-6)$$

All the elements of a vector or matrix pulse transfer function are always rational in z, although the continuous transfer functions may include time delays. In order to be physically realizable the transfer matrices (or vectors) have to be proper or causal.

Definition 15.1-1. *A vector or matrix $G^*(z)$ is proper or causal if all its elements are proper and strictly proper if all its elements are strictly proper. All systems $G^*(z)$ which are not proper are called improper or noncausal .*

15.1.2 Poles and Zeros

Let

$$G^*(z) = \mathcal{ZL}^{-1}\{h_0(s)G(s)\} \qquad (15.1-7)$$

where $G(s)$ is the transfer matrix representation of the system of differential and algebraic equations of Sec. 10.1.1. Then $G^*(z)$ is the z-transfer matrix that describes the system of difference equations

$$x(kT + T) = \Phi x(kT) + \Gamma u(kT) \qquad (15.1-8)$$

$$y(kT) = Cx(kT) + Du(kT) \qquad (15.1-9)$$

where T is the sampling time and

$$\Phi = e^{AT} \qquad (15.1-10)$$

$$\Gamma = \int_0^T e^{At}\, dt\ B \qquad (15.1-11)$$

Taking the z-transform of (15.1-8), (15.1-9) we get

$$x(z) = (zI - \Phi)^{-1}\Gamma u(z) \qquad (15.1-12)$$

$$y(z) = Cx(z) + Du(z) \qquad\qquad (15.1-13)$$

and substituting (15.1–12) into (15.1–13) yields

$$y(z) = G^*(z)u(z) \qquad\qquad (15.1-14)$$

where

$$G^*(z) \triangleq C(zI - \Phi)^{-1}\Gamma + D \qquad\qquad (15.1-15)$$

The matrix $G^*(z)$ will be assumed to be of full normal rank. The poles and zeros of $G^*(z)$ are defined in exactly the same way as those of $G(s)$.

Definition 15.1-2. *The eigenvalues $\pi_i, i = 1, \ldots, n_p$, of the matrix Φ are called the poles of the system (15.1–8), (15.1–9). The pole polynomial $\pi(z)$ is defined as*

$$\pi(z) = \prod_{i=1}^{n_p}(z - \pi_i) \qquad\qquad (15.1-16)$$

Definition 15.1-3. *ζ is a zero of $G^*(z)$ if the rank of $G^*(\zeta)$ is less than the normal rank of $G^*(z)$.*

The zero polynomial is defined as

$$\zeta(z) = \prod_{i=1}^{n_z}(z - \zeta_i) \qquad\qquad (15.1-17)$$

where n_z is the number of finite zeros of $G^*(z)$.

15.1.3   Internal Stability

Assuming that no unstable poles of the continuous process have become unobservable after sampling, the internal stability of the system in Fig. 15.1-1A can be assessed from the internal stability of the system in Fig. 15.1-1B.

Theorem 15.1-1. *The sampled-data system in Fig. 15.1-1A is internally stable if and only if the transfer matrix in (15.1-18)*

$$\begin{pmatrix} y_\gamma^* \\ u \end{pmatrix} = \begin{pmatrix} P_\gamma^* C(I + P_\gamma^* C)^{-1} & (I + P_\gamma^* C)^{-1}P_\gamma^* \\ C(I + P_\gamma^* C)^{-1} & -C(I + P_\gamma^* C)^{-1}P_\gamma^* \end{pmatrix} \begin{pmatrix} r^* \\ u' \end{pmatrix} \qquad (15.1-18)$$

is stable — i.e. if and only if all its poles are strictly inside the unit circle.

Another test for internal stability is the Nyquist criterion, which was discussed for continuous systems in Sec. 10.2.2. The derivation follows exactly the same steps. The difference is that when we are dealing with z-transfer functions instead of Laplace transfer functions, the Nyquist D-contour encircles the area outside the UC instead of the RHP.

Theorem 15.1-2 (Nyquist Stability Criterion). *Let the map of the Nyquist D-contour under* $\det(I + P_\gamma^*(z)C(z))$ *encircle the origin n_F times in the clockwise direction. Let the number of open-loop unstable poles of P_γ^*C be n_{PC}. Then the closed-loop system is stable if and only if*

$$n_F = -n_{PC} \qquad (15.1-19)$$

15.1.4 IMC Structure

The block diagram of the sampled-data MIMO IMC structure is shown in Fig. 15.1-2A, where

$$\tilde{P}_\gamma^*(z) = \mathcal{ZL}^{-1}\{\Gamma(s)\tilde{P}(s)H_0(s)\} = \mathcal{ZL}^{-1}\{\gamma(s)\tilde{P}(s)h_0(s)\} \qquad (15.1-20)$$

$$\tilde{P}^*(z) = \mathcal{ZL}^{-1}\{\tilde{P}(s)H_0(s)\} = \mathcal{ZL}^{-1}\{\tilde{P}(s)h_0(s)\} \qquad (15.1-21)$$

When the IMC controller Q and the feedback controller C are related through

$$C = Q(I - \tilde{P}_\gamma^*Q)^{-1} \qquad (15.1-22)$$

$$Q = C(I + \tilde{P}_\gamma^*C)^{-1} \qquad (15.1-23)$$

then $u(z)$ and $y(s)$ react to inputs $r^*(z)$ and $d(s)$ in exactly the same way for both the classic feedback and the IMC structure.

Figure 15.1-2B is a different representation of the sampled-data IMC structure, which is equivalent to that in Fig. 15.1-2A, but not suitable for computer implementation because of the presence of the continuous model $\tilde{P}(s)$. If only the sampled signals are of interest, then Fig. 15.1-2A is equivalent to Fig. 15.1-2C, where all signals are digital.

15.1.5 Model Uncertainty Description

In Sec. 7.3.2, we demonstrated how the modeling error in the description of the discretized plant is related to that in the continuous plant description. We pointed out that some conservativeness is introduced when the uncertainty bounds for the discrete plant are derived from those for the continuous plant. However, the conservativeness is quite small for the type of unstructured SISO system

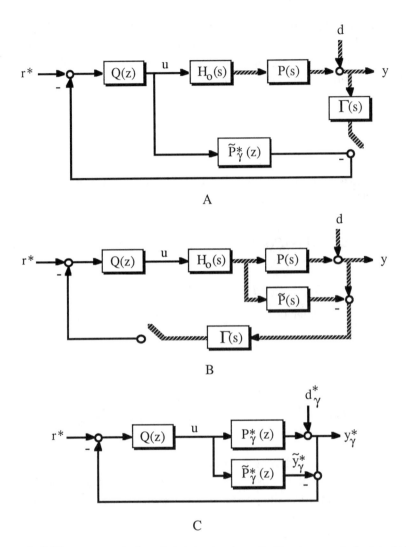

Figure 15.1-2. IMC structure: A: Sampled-data structure; B: Structure equivalent to (A) but not implementable; C: Discrete structure (all signals discrete).

uncertainty that was used in Chaps. 7 through 9. The same is true for a few types of MIMO-system uncertainty. For example, let us assume that the additive uncertainty for the continuous plant is bounded by $\bar{\ell}_A$:

$$\bar{\sigma}(P(i\omega) - \tilde{P}(i\omega)) \le \bar{\ell}_A(\omega) \qquad (15.1-24)$$

For the discretized plant we have from (15.1–6), (15.1–21)

$$P^*(z) - \tilde{P}^*(z) = \mathcal{ZL}^{-1}\{h_0(s)(P(s) - \tilde{P}(s))\} \qquad (15.1-25)$$

Then from the z-transform property (7.1–5), the singular value property $\bar{\sigma}(A + B) \le \bar{\sigma}(A) + \bar{\sigma}(B)$, and (15.1–24), (15.1–25), it follows that

$$\bar{\sigma}(P^*(e^{i\omega T}) - \tilde{P}^*(e^{i\omega T})) \le \frac{1}{T} \sum_{k=-\infty}^{\infty} |h_0(i\omega + ik2\pi/T)| \bar{\ell}_A(i\omega + ik2\pi/T) \triangleq \bar{\ell}_a^*(\omega)$$

$$(15.1-26)$$

The ZOH $h_0(s)$ is small at frequencies higher than π/T and goes to 0 as fast as $1/\omega$ as $\omega \to \infty$. Therefore only a few terms around $k = 0$ are important in the infinite sum. Also note that for a physical system, $\bar{\ell}_A(\omega) \to 0$ at least as fast as $1/\omega$ as $\omega \to \infty$, and hence the sum converges.

However, it is not always possible to obtain a mathematical description for the uncertainty in the z-domain in a non-conservative way, starting from the uncertainty in the s-domain. In the absence of first-principles models, these descriptions may be the result of experiments conducted with different sampling rates, one of which may be small enough to approximate the continuous system. A discussion of identification techniques is beyond the scope of this book. We will assume in this chapter that non-conservative uncertainty descriptions for the discrete and the continuous plant are available.

15.2 Nominal Internal Stability

15.2.1 IMC Structure

The same arguments as in Sec. 12.2 imply that the following matrix must be stable for internal stability of the IMC structure in Fig. 15.1-2C.

$$S_1 = \begin{pmatrix} P_\gamma^* Q & (I - P_\gamma^* Q)P_\gamma^* & P_\gamma^* \\ Q & -Q P_\gamma^* & 0 \\ P_\gamma^* Q & -P_\gamma^* Q P_\gamma^* & P_\gamma^* \end{pmatrix} \qquad (15.2-1)$$

Note that stability of the structure in Fig. 15.1-2C implies stability of that in Fig. 15.1-2A, provided that no open-loop unstable poles of the plant become unobservable after sampling.

Theorem 15.2-1. *For* $P = \tilde{P}$, *the IMC system in Fig. 15.1-2A is internally stable if and only if both the plant P and the controller Q are stable.*

Hence for open-loop unstable plants, the IMC structure cannot be implemented. In such cases, the IMC design procedure is used to design the controller, which is then implemented through the classic feedback structure.

15.2.2 Feedback Structure

When there is no modeling error, substitution of (15.1–22) into (15.1–18) yields for the internal stability matrix

$$S_2 = \begin{pmatrix} P_\gamma^* Q & (I - P_\gamma^* Q)P_\gamma^* \\ Q & -Q P_\gamma^* \end{pmatrix} \qquad (15.2-2)$$

All four transfer matrices in (15.2–2) have to be stable for nominal stability of the classic feedback structure in Fig. 15.1-1A.

Theorem 15.2-2 provides a parametrization of all proper stabilizing controllers in terms of a stable transfer matrix Q_1. The following assumptions are analogous to those made in Sec. 12.3.

Assumption A1. *If π is a pole of $\tilde{P}^*$ outside the UC, then (a) The order of π is equal to 1 and (b) $\tilde{P}$ has no zeros at $z = \pi$.*

Assumption A2. *Any poles of $\tilde{P}^*$ or P^* on the UC are at $z = 1$. Also $\tilde{P}^*$ has no zeros on the UC.*

Theorem 15.2-2. *Assume that Assns. A1 and A2 hold and that $Q_0(z)$ is a proper transfer matrix that stabilizes $\tilde{P}^*$ — i.e., it yields a stable S_2. Then all proper Q's that make S_2 stable are given by*

$$Q(z) = Q_0(z) + Q_1(z) \qquad (15.2-3)$$

where $Q_1(z)$ is any proper and stable transfer matrix such that $P^*(z)Q_1(z)P^*(z)$ is stable.

Proof. The fact that Q_1 has to be proper in order for Q to be proper and vice versa, follows from the properness of Q_0. For the following part of the proof we will use the fact that P^* and P_γ^* have the same unstable poles.

⇒ We shall show that any Q given by (15.2–3) makes S_2 stable. From substitution of (15.2–3) into (15.2–2) it follows that all that is required is that $(\, P^* Q_1 \quad Q_1 P^* \quad P^* Q_1 P^* \,)$ be stable. From the properties of Q_1, it follows that the third element in the above matrix is stable. Stability of the other two elements follows by pre- and post-multiplication of $P^*Q_1P^*$ by $(P^*)^{-1}$, since according to assumptions A1 and A2, P^* has no zeros at the location of its unstable poles and these are the only possible unstable poles of S_2.

$\Leftarrow$ Assume that Q makes S_2 stable. Then the difference matrix

$$\Delta S_2 = S_2(Q) - S_2(Q_0) = \begin{pmatrix} P^*(Q - Q_0) & P^*(Q - Q_0)P^* \\ (Q - Q_0) & (Q - Q_0)P^* \end{pmatrix} \quad (15.2 - 4)$$

is stable. This implies that $(Q - Q_0) = Q_1$ and $P^*Q_1P^*$ are stable. $\qquad \square$

15.3 Nominal Performance

15.3.1 Sensitivity and Complementary Sensitivity Function

The development in this section follows that in Sec. 7.5.1. Therefore we shall limit ourselves to simply setting the appropriate notation for the MIMO systems.

From Fig. 15.1-2A we get for $P = \tilde{P}$

$$y(s) = h_0(s)P(s)Q(e^{sT})(r^*(e^{sT}) - d_\gamma^*(e^{sT})) + d(s) \quad (15.3 - 1)$$

Define

$$e(s) = y(s) - r(s) \quad (15.3 - 2)$$

Then for an $r(s)$ that remains constant between sampling points, we have $r(s) = h_0(s)r^*(e^{sT})$ and we can define the sensitivity and complementary sensitivity operators that relate $r(s)$ to $-e(s)$ and $y(s)$, correspondingly as

$$\tilde{E}_r(s) \triangleq I - P(s)Q(e^{sT}) \quad (15.3 - 3)$$

$$\tilde{H}_r(s) \triangleq P(s)Q(e^{sT}) \quad (15.3 - 4)$$

An approximate sensitivity function for the relation between $y(s)$ and $d(s)$ can only be obtained when the assumption is made that the disturbance is limited to the frequency band up to π/T.

$$y(i\omega) \cong \tilde{E}_d(i\omega)d(i\omega) \quad (15.3 - 5)$$

where

$$\tilde{E}_d(s) = I - P(s)\hat{Q}(s) \quad (15.3 - 6)$$

$$\hat{Q}(s) = \frac{1}{T}h_0(s)Q(e^{sT})\gamma(s) \tag{15.3 - 7}$$

Sampling of (15.3–1) yields

$$y^*(z) = P^*(z)Q(z)(r^*(z) - d^*_\gamma(z)) + d^*(z) \tag{15.3 - 8}$$

from which we can obtain the *pulse* sensitivity and complementary sensitivity functions, relating $e^*(z)$ to $r^*(z)$ and $d^*(z)$ (for $\gamma(s) = 1$) or $d^*_\gamma(z)$ (when d^*_γ is substituted for d^* in (15.3–8)):

$$\tilde{E}^*(z) \triangleq I - P^*(z)Q(z) \tag{15.3 - 9}$$

$$\tilde{H}^*(z) \triangleq P^*(z)Q(z) \tag{15.3 - 10}$$

15.3.2 H_2^* Performance Objectives

We define as $L_2^{*,n}$ the Hilbert space of complex valued vector functions $y(z)$ with n elements, defined on the unit circle and square integrable with respect to θ — i.e., for which the following quantity is finite:

$$\|y\|_2 = \left(\frac{1}{2\pi} \int_{-\pi}^{\pi} y(e^{i\theta})^H y(e^{i\theta}) \ d\theta\right)^{1/2} \tag{15.3 - 11}$$

Note that (15.3–11) defines a norm on $L_2^{*,n}$. In the case where $y(z)$ has no poles outside the UC, Parseval's theorem yields a time domain expression for $\|y\|_2$:

$$\|y\|_2 = \left(\sum_{k=0}^{\infty} y_k^T y_k\right)^{1/2} \tag{15.3 - 12}$$

For matrix valued functions $G(z)$ of dimensions $n \times m$, the space $L_2^{*,n \times m}$ is defined similarly with norm

$$\|G\|_2 = \left(\frac{1}{2\pi} \int_{-\pi}^{\pi} \mathrm{trace}[G(e^{i\theta})^H G(e^{i\theta})] \ d\theta\right)^{1/2} \tag{15.3 - 13}$$

The spaces $H_2^{*,n}$ and $H_2^{*,n \times m}$ are defined as subspaces of the corresponding L_2 spaces as in the scalar case.

The H_2^* performance objective is to minimize over all stabilizing $\tilde{Q}$ the weighted sum of squared errors for the response to an input or a set of inputs of interest. Several H_2^*–type objective functions will be considered. For a specific external input v^* (r^* or d^* or d^*_γ) define by using (15.3-9)

$$\Phi(v^*) \triangleq \|We^*\|_2^2 = \|W\tilde{E}^*v^*\|_2^2 = \|W(I - P^*\tilde{Q})v^*\|_2^2 \qquad (15.3-14)$$

where W is a frequency dependent matrix or scalar weight. One objective could be

Objective O1:

$$\min_{\tilde{Q}} \Phi(v^*)$$

for a particular input $v^* = (\ v_1 \quad v_2 \quad \ldots \quad v_n\ )^T$.

A more meaningful objective would be to minimize $\Phi(v)$ not just for one input vector v^*, but for every input in a set $\mathcal{V}$:

$$\mathcal{V} = \{v^i(z) | i = 1, \ldots n\} \qquad (15.3-15)$$

where $v^1(z), \ldots, v^n(z)$ are vectors that describe the directions and the frequency content of the expected external system inputs and n is the dimension of P. Thus, the objective is

Objective O2:

$$\min_{\tilde{Q}} \Phi(v^*) \qquad \forall v^* \in \mathcal{V}$$

However a linear time invariant $\tilde{Q}(z)$ that solves *O2* does not always exist. The conditions necessary for its existence are expressed in Thm. 15.6-3. An alternative is

Objective O3:

$$\min_{\tilde{Q}}[\Phi(v^1) + \ldots + \Phi(v^n)]$$

In this case the objective is the sum of the squared errors caused by each of the v^i's, when applied separately.

For every external input v^* that will be considered in this chapter the following assumptions will be made. They are analogous to those discussed in Sec. 12.6.1 and their physical meaning is identical.

Assumption A3. *Every nonzero element of v^* includes all the poles of $\tilde{P}$ outside the UC, each with degree 1, and those are the only poles of v^* outside the UC.*

Assumption A4. *Let ℓ_i be the maximum number of poles at $z = 1$ of any element in the ith row of P. Then the ith element of v^*, v_i, has at least ℓ_i poles at $z = 1$. Also v^* has no other poles on the UC and its elements have no zeros on the UC.*

For the case, where a set $\mathcal{V}$ of inputs is considered, define

$$V \triangleq \begin{pmatrix} v^1 & v^2 & \dots & v^n \end{pmatrix} \tag{15.3 - 16}$$

where $v^1, \dots, v^n$ satisfy Assn. A3. An additional assumption on V is needed:

Assumption A5. *V has no zeros at the location of its unstable poles or on the UC and V^{-1} cancels the unstable poles of $\tilde{P}$ in $V^{-1}\tilde{P}$.*

15.3.3 H_∞ Performance Objective

The H_∞ objective discussed in Sec. 10.4.4 can now be extended to discrete systems with band-limited disturbances. The development is similar to that for the SISO case (Sec. 7.5.5). The objective can be written as

$$\|W\tilde{E}_v\|_\infty < 1 \tag{15.3 - 17}$$

where W is the frequency weight and $\tilde{E}_v(s)$ is either the approximate disturbance sensitivity function $\tilde{E}_d$ given by (15.3–6) or the setpoint sensitivity function $\tilde{E}_r$ given by (15.3–3). Since the disturbance is assumed to be limited to the frequency band up to π/T and $h_0(s)r^*(e^{sT})$ is also limited because of h_0, the weight should satisfy

$$\bar{\sigma}(W(\omega)) \ll 1, \qquad \omega > \pi/T \tag{15.3 - 18}$$

Hence (15.3–17) can be written as

$$\bar{\sigma}(W(\omega)\tilde{E}_v(i\omega)) < 1, \qquad 0 \leq \omega \leq \pi/T \tag{15.3 - 19}$$

If W is a scalar, (15.3–19) becomes

$$\bar{\sigma}(\tilde{E}_v(i\omega)) < |w(\omega)|^{-1}, \qquad 0 \leq \omega \leq \pi/T \tag{15.3 - 20}$$

15.4   Robust Stability

In Sec. 15.1 we explained that if no open-loop unstable poles of the plant or the model become unobservable after sampling, then stability of the structures in Figs. 15.1-1A and 15.1-2A is equivalent to stability of those in Figs. 15.1-1B and 15.1-2C correspondingly. In Sec. 15.2 we developed the nominal internal stability conditions. We shall now concentrate on the robust stability of completely discrete structures like the ones in Figs. 15.1-1B and 15.1-2C. The development of robustness conditions follows exactly the same steps as those in Chap. 11.

First the $M - \Delta$ structure (Fig. 11.1-2) which is needed in the structured singular value theory, has to be generated from the given discrete control structure. For this purpose the same type of block manipulations have to be carried out as were demonstrated in Chap. 11. Then, if M and Δ are stable, the condition for robust stability is that the map of the discrete Nyquist D-contour under $\det(I - M\Delta)$ does not encircle the origin. Recall that for discrete systems the D-contour encircles the area outside the UC. We can now use the SSV to obtain the following theorem.

Theorem 15.4-1. *Assume that the nominal systems M is stable and that the perturbation Δ is such that the perturbed closed-loop system is stable if and only if the map of the Nyquist contour under $\det(I - M\Delta)$ does not encircle the origin. Then the system in Fig. 11.1-2 is stable for all $\Delta \in X_1$ if and only if*

$$\mu(M(e^{i\omega T})) < 1 \qquad 0 \leq \omega \leq \pi/T \qquad (15.4-1)$$

Note that because of the periodicity of the z-transforms and the property described by (7.1-8), only the frequencies up to π/T need to be considered.

15.5   Robust Performance

15.5.1   Sensitivity Function Approximation

First, we shall obtain an approximate sensitivity function in a similar way as in Sec. 15.3.1. Then we shall use this function to assess robust performance. From Fig. 15.1-2A it follows that

$$e(s) \overset{\Delta}{=} y(s) - r(s)$$

$$= (d(s) - r(s)) - P(s)H_0(s)Q(e^{sT})$$

$$(I + (P_\gamma^*(e^{sT}) - \tilde{P}_\gamma^*(e^{sT}))Q(e^{sT}))^{-1}(d_\gamma^*(z) - r^*(z)) \qquad (15.5-1)$$

We shall now obtain an approximation to (15.5–1) by considering the frequencies $0 \leq \omega \leq \pi/T$. Note that because of the periodicity of $Q(z)$, these are the only frequencies which one can influence independently by using a digital controller. It follows from (7.1–5) that if $a(s)$ is small for $\omega > \pi/T$, then

$$a^*(e^{i\omega T}) \cong \frac{1}{T}a(i\omega), \qquad 0 \leq \omega \leq \pi/T \qquad (15.5-2)$$

Use of (15.5–2) for all the z-transforms in (15.5–1) except for r^* for which we assume $r(s) = h_0(s)r^*(e^{sT})$, yields the approximation

$$e(i\omega) \cong E_d(i\omega)d(i\omega) - E_r(i\omega)r(i\omega), \qquad 0 \leq \omega \leq \pi/T \qquad (15.5-3)$$

where

$$E_r(i\omega) \triangleq I - P(i\omega)Q(e^{i\omega T}) \cdot$$
$$\left[I + (P(i\omega) - \tilde{P}(i\omega))Q(e^{i\omega T})\gamma(i\omega)h_0(i\omega)/T\right]^{-1} \qquad (15.5-4)$$

$$E_d(i\omega) \triangleq I - P(i\omega)Q(e^{i\omega T})\gamma(i\omega)h_0(i\omega)/T \cdot$$
$$\left[I + (P(i\omega) - \tilde{P}(i\omega))Q(e^{i\omega T})\gamma(i\omega)h_0(i\omega)/T\right]^{-1} \qquad (15.5-5)$$

Note that the above approximation is valid when the input signals r and d are small for $\omega > \pi/T$. If we assume that $r(t)$ is a staircase function then it has the desired property. If one expects disturbances with high frequency content at $\omega > \pi/T$ then one should reduce T or use the anti-aliasing prefilter whose function is to cut off signals with frequencies higher than π/T.

15.5.2 H_∞ Performance Objective

We require that the objective defined in Sec. 15.3.3 be satisfied for all plants $P(s)$ in the uncertainty set Π (note that $E_v(\tilde{P}) = \tilde{E}_v$).

$$\max_{0 \le \omega \le \pi/T} \bar{\sigma}(W(\omega)E_v(i\omega)) < 1 \qquad \forall P \in \Pi \qquad (15.5-6)$$

From this point on the treatment of the problem is identical to that presented in Sec. 11.3.1. Note that only the continuous plant $P(s)$ appears in $E_v(s)$ and therefore all the uncertain Δ's are continuous transfer functions. Hence the need mentioned in Sec. 15.1.4 for continuous as well as discrete (used for test of robust stability) uncertainty bounds.

15.6   IMC Design: Step 1 ($\tilde{Q}$)

15.6.1 H_2^*-Optimal Control

The plant P^* can be factored into an allpass portion P_A^* and a minimum phase portion P_M^*:

$$P^* = P_A^* P_M^* \qquad (15.6-1)$$

Here P_A^* is stable and such that $P_A^*(e^{i\theta})^H P_A^*(e^{i\theta}) = I$. Also $(P_M^*)^{-1}$ is stable. P_M^* has the additional property that both P_M^* and $(P_M^*)^{-1}$ are proper. In the case where P^* is scalar, this factorization can be easily accomplished as described by (8.1–2). In the general multivariable case, this "inner-outer factorization" can be accomplished by using the bilinear transformation $z = (1+s)/(1-s)$, to reduce the problem to the one for the *s*-domain, which was discussed in Sec. 12.6.4. The steps involved in this procedure are explained in Sec. 15.6.4.

Objective O1: Specific Input

Let $v_0(z)$ be the scalar allpass with the property $v_0(1) = 1$, which includes the *common* zeros outside the UC and the *common* delays of the elements of $v^*(z)$. Write

$$v^*(z) = v_0(z)\hat{v}(z) \qquad (15.6-2)$$

where $\hat{v}(z)$ is a vector. Hence $\hat{v}$ is proper with at least one element semi-proper and there is no point z outside the UC where $\hat{v}$ becomes identically zero.

Theorem 15.6-1. *Assume that Assns. A1-A4 hold. Any stabilizing $\tilde{Q}$ that solves Obj. O1 satisfies*

$$\tilde{Q}\hat{v} = z(WP_M^*)^{-1}\{z^{-1}W(P_A^*)^{-1}\hat{v}\}_* \qquad (15.6-3)$$

where the operator $\{\cdot\}_*$ denotes that after a partial fraction expansion of the operand, only the strictly proper terms are retained except those corresponding to poles of $(P_A^*)^{-1}$. Furthermore, for $n \geq 2$ the number of stabilizing controllers that satisfy (15.6–3) is infinite. Guidelines for the construction of such a controller are given in the proof.

Note that not every $\tilde{Q}$ satisfying (15.6–3) is necessarily a stabilizing controller. Equation (15.6–3) should be compared to (9.2–4) for SISO systems. If we assume that the disturbance and the plant have the same open-loop poles outside the UC, then the two equations are identical.

Proof of Theorem 15.6-1. We shall assume $W = I$. The proof of the weighted case is left as an exercise. Let V_0 be a diagonal matrix where each column satisfies Assn. A3 and every element has ℓ_v poles at $z = 1$, where ℓ_v is the maximum number of such poles in any element of v. Assume that there exists Q_0, which stabilizes P^* in the sense of Thm. 15.2-2 and also makes $(I - P^*Q_0)V_0$ stable. Its existence will be proven by construction. Substitution of (15.2–3) into (15.3–14) and use of the fact that pre- or post-multiplication of a function with an allpass does not change its L_2-norm, yields:

$$\Phi(v^*) = \|z^{-1}(P_A^*)^{-1}(I - P^*Q_0)\hat{v} - z^{-1}P_M^*Q_1\hat{v}\|_2^2$$

$$\triangleq \|f_1 - f_2Q_1\hat{v}\|_2^2 \qquad (15.6-4)$$

The term f_1 has no poles at $z = 1$ because $(I - P^*Q_0)V_0$ has no such poles. Any rational function $f_1(z)$ with no poles on the UC, can be uniquely decomposed into a strictly proper, stable part $\{f_1\}_+$ in H_2^* and a strictly unstable part $\{f_1\}_-$ in $(H_2^*)^\perp$:

$$f_1 = \{f_1\}_+ + \{f_1\}_- \qquad (15.6-5)$$

Note that according to the definition of $H_2^*, (H_2^*)^\perp$, any improper terms as well as the constant term in a partial fraction expansion of f_1, belong in $\{f_1\}_-$. Next we want to show that $f_2 Q_1 \hat{v}$ has to be stable. The fact that $(I - P^* Q_0) V_0$ is stable implies that $(I - P^* Q_0)\hat{v}$ is stable. We require that $(I - P^* Q)v$ has no poles outside the UC and therefore that $(I - P^* Q)\hat{v} = (I - P^* Q_0)\hat{v} - P^* Q_1 \hat{v}$ have no poles outside the UC. But since $(I - P^* Q_0)\hat{v}$ is stable, this requirement reduces to $P^* Q_1 \hat{v}$ having no poles outside the UC. Also in order for $\Phi(v^*)$ to be finite, Q_1 must be such that $(I - P^* Q)\hat{v}$ has no poles on the UC. But since $(I - P^* Q_0)\hat{v}$ is stable, this is equivalent to $P^* Q_1 \hat{v}$ having no poles on the UC. Hence the optimal Q_1 must be such that $P^* Q_1 \hat{v}$ is stable. Then the only possible unstable poles of $f_2 Q_1 \hat{v} = z^{-1}(P_A^*)^{-1} P^* Q_1 \hat{v}$ are the poles of $(P_A^*)^{-1}$. But Assns. A1, A2 imply that the poles of $(P_A^*)^{-1}$ are not among those of $f_2 Q_1 \hat{v}$ and therefore $f_2 Q_1 \hat{v}$ has to be stable. To proceed we will assume that Q_1 has this property. We will verify later that the solution indeed has this property.

Hence we can write

$$\Phi(v^*) = \|\{f_1\}_-\|_2^2 + \|\{f_1\}_+ - f_2 Q_1 \hat{v}\|_2^2 \qquad (15.6-6)$$

The first term on the RHS of (15.6–6) does not depend on Q_1. Hence for solving O1 we only have to look at the second term. The obvious solution is

$$Q_1 \hat{v} = f_2^{-1}\{f_1\}_+ \qquad (15.6-7)$$

Clearly such a Q_1 produces a stable $f_2 Q_1 \hat{v}$ as was assumed. It should now be proved that Q_1's that satisfy the internal stability requirements exist among those described by (15.6–7), so that the obvious solution is a true solution. For $n = 1$, (15.6–7) yields a unique Q_1, which can be shown to satisfy the requirements by following the arguments in the proof of Thm. 15.6-2. For $n \geq 2$ write

$$\hat{v} \triangleq \begin{pmatrix} \hat{v}_1 & \hat{v}_2 & \cdots & \hat{v}_n \end{pmatrix}^T \qquad (15.6-8)$$

$$\hat{V}_2 \triangleq \begin{pmatrix} \hat{v}_2 & \cdots & \hat{v}_n \end{pmatrix}^T \qquad (15.6-9)$$

$$Q_1 \triangleq \begin{pmatrix} q_1 & q_2 \end{pmatrix} \qquad (15.6-10)$$

where without loss of generality the first element of v^*, and thus $\hat{v}_1$, is assumed to be nonzero. Also q_1 is $n \times 1$ and q_2 is $n \times (n-1)$. Then from (15.6–7) it follows that

$$Q_1 = (\, \hat{v}_1^{-1}(f_2^{-1}\{f_1\}_+ - q_2 \hat{V}_2) \quad q_2 \,) \qquad (15.6-11)$$

We now need to show that a proper, stable q_2 exists such that Q_1 is proper, stable and produces a stable $P^* Q_1 P^*$. Select a q_2 of the form:

$$q_2(z) = \hat{q}_2(z)(1 - z^{-1})^{3l_v} \prod_{i=1}^{k} (1 - \pi_i z^{-1})^3 \qquad (15.6-12)$$

where $\hat{q}_2$ is proper and stable and $\{\pi_1, \ldots, \pi_k\}$ are the poles of P^* outside the UC. Then from (15.6–11) it follows that in order for $P^* Q_1 P^*$ to be stable it is sufficient that $P^* \hat{v}_1^{-1} f_2^{-1}\{f_1\}_+\{P^*\}_{1^{st}row}$ has no poles on or outside the UC. But $P^* f_2^{-1} = z P_A^*$ is stable and the only possible poles of $\hat{v}_1^{-1}\{P^*\}_{1^{st}row}$ on or outside the UC are poles of $\hat{v}_1^{-1}$ outside the UC, because of Assns. A3 and A4. These are also the only possible unstable poles of Q_1. Let α be such a pole (zero of $\hat{v}_1$). Then for stability we need to find $\hat{q}_2$ such that

$$\hat{q}_2(\alpha)\hat{V}_2(\alpha) = (1 - \alpha^{-1})^{-3l_v} \prod_{i=1}^{k} (1 - \pi_i \alpha^{-1})^{-3} f_2^{-1}(\alpha)\{f_1\}_+ \Big|_{z=\alpha} \qquad (15.6-13)$$

The above equation always has a solution because the vector $\hat{V}_2(\alpha)$ is not identically zero since any common zeros in v^* outside the UC were factored out in v_0.

We now need to examine the properness of Q_1. Since $(P_M^*)^{-1}$ is proper and $\{f_1\}_+$ is strictly proper, $f_2^{-1}\{f_1\}_+$ is proper. Then if $\hat{v}_1^{-1}$ is improper ($\hat{v}_1$ strictly proper) there exists at least one element in $\hat{V}_2$ that is semi-proper. Hence by solving a system of linear equations we can always select a $\hat{q}_2(z)$ such that of the first impulse response coefficients of $f_2^{-1}\{f_1\}_+ - q_2 \hat{V}_2$, as many are zero as needed to make the first element of the matrix in (15.6–11) proper.

We shall now proceed to obtain an expression for $Q\hat{v}$. (15.2–3) and (15.6–11) yield

$$Q\hat{v} = z(P_M^*)^{-1} \left[z^{-1}(P_A^*)^{-1} P^* Q_0 \hat{v} - \{z^{-1}(P_A^*)^{-1} P^* Q_0 \hat{v}\}_+ + \{z^{-1}(P_A^*)^{-1}\hat{v}\}_+ \right]$$

$$= z(P_M^*)^{-1} \left[\{z^{-1}(P_A^*)^{-1} P^* Q_0 \hat{v}\}_{0-} + \{z^{-1}(P_A^*)^{-1}\hat{v}\}_+ \right] \qquad (15.6-14)$$

where $\{\cdot\}_{0-}$ indicates that in the partial fraction expansion all poles on or outside the UC are retained. For (15.6–14), these poles are the poles of $\hat{v}$ on or outside the UC; $(P_A^*)^{-1}P^*Q_0 = P_M^*Q_0$ is strictly stable and proper because of Assn. A1 and the fact that Q_0 is a stabilizing controller. The fact that $(I - P^*Q_0)V_0$ has no poles at $z = 1$ imply that $(I - P^*Q_0)$ and its derivatives up to and including the $(\ell_v - 1)^{th}$ are equal to zero at $z = 1$. Also, the fact that $(I - P^*Q_0)V_0$ is stable and that the columns of this diagonal V_0 satisfy Assn. A3, imply that $(I - P^*Q_0) = 0$ at $1, \pi_1, \ldots, \pi_k$. Thus (15.6-14) simplifies to (15.6–3).

We now need to establish that a stabilizing controller Q_0 exists with the property that $(I - P^*Q_0)V_0$ is stable. The selection of a V_0 with the properties mentioned at the beginning of this section and its use instead of V in (15.6–16) yields such a controller.   □

Objectives O2 and O3: Set of v^*'s.

Factor V similarly to P^* (see Sec. 15.6.4):

$$V = V_M V_A \qquad\qquad (15.6 - 15)$$

Theorem 15.6-2. *Assume that Assns. A1-A5 hold. The controller*

$$\tilde{Q} = z(WP_M^*)^{-1}\{z^{-1}W(P_A^*)^{-1}V_M\}_* V_M^{-1} \qquad (15.6 - 16)$$

is the unique solution to O3. Here the operator $\{\cdot\}_*$ denotes that after a partial fraction expansion of the operand, only the strictly proper terms are retained except those corresponding to poles of $(P_A^*)^{-1}$.

Proof. Again we assume $W = I$ and leave the weighted case as an exercise. From (15.3–13), (15.3–14), and (15.3–16) it follows that

$$\Phi(v^1) + \Phi(v^2) + \ldots + \Phi(v^n) = \|(I - P\tilde{Q})V\|_2^2 \triangleq \Phi(V) \qquad (15.6 - 17)$$

The minimization of $\Phi(V)$ follows the steps in the proof of Thm. 15.6-1 up to (15.6–7), with V_M used instead of $\hat{v}$. In this case ℓ_v is the maximum number of poles at $z = 1$ in any element of V. From the equivalent to (15.6–7) we obtain

$$Q_1 = f_2^{-1}\{f_1\}_+ V_M^{-1} \qquad\qquad (15.6 - 18)$$

We now have to establish that Q_1 is stable, proper and produces a stable $P^*Q_1P^*$. In $P^*Q_1P^*$ the unstable poles of the P^* on the left cancel with those of $(P_M^*)^{-1}$ in f_2^{-1}. As for the P^* on the right, cancellation follows from Assn. A5. Then in the same way that (15.6–3) follows from (15.6–14), (15.6–16) follows from (15.6–18).   □

A more meaningful objective would be to solve Obj. O2. However a $\tilde{Q}$ that solves Obj. O2 will also solve Obj. O3. Then from Thm. 15.6-2 it follows that if a solution to O2 exists, it is given by (15.6–16). Factor each of the v^i in the same way as in (15.6–2):

$$v^i(z) = v_0^i(z)\hat{v}^i(z) \qquad (15.6-19)$$

Define

$$\hat{V} \triangleq (\hat{v}^1 \quad \hat{v}^2 \quad \dots \quad \hat{v}^n) \qquad (15.6-20)$$

Theorem 15.6-3. *Assume that Assns. A1-A5 hold.*

(i) *If $\hat{V}(z)$ is non-minimum phase (i.e., $\hat{V}^{-1}$ is unstable or improper), then there exists no solution to Obj. O2.*

(ii) *If $\hat{V}(z)$ is minimum phase, then use of $\hat{V}$ instead of V_M in (15.6–16) yields exactly the same $\tilde{Q}$, which also solves Obj. O2. In addition $\tilde{Q}$ minimizes $\Phi(v^*)$ for any $v^*(z)$ that is a linear combination of v^i's that have the same v_0^i's.*

Proof. ($W = I$). A stabilizing controller that solves Obj. O2 has to solve Obj. O1 for all v^i, $i = 1, \dots, n$. Satisfying (15.6–3) for every v^i is equivalent to

$$\tilde{Q} = z(P_M^*)^{-1}\{z^{-1}(P_A^*)^{-1}\hat{V}\}_*\hat{V}^{-1} \qquad (15.6-21)$$

Hence the above $\tilde{Q}$ is the only potential solution for Obj. O2. However, it is not necessarily a stabilizing controller since not only stabilizing $\tilde{Q}$'s satisfy (15.6–3) for some v^*. Indeed, if $\hat{V}$ is non-minimum phase, $\hat{V}^{-1}$ is unstable and/or improper and this results in an unstable and/or improper $\tilde{Q}$, which is therefore unacceptable. Hence in such a case, there exists no solution for Obj. O2, which completes the proof of part (i) of the theorem.

In the case where $\hat{V}^{-1}$ is stable and proper ($\hat{V}$ minimum phase), the controller given by (15.6–21) is stable and proper and therefore it is the same as the one given by (15.6–16). This fact can be explained as follows. We have

$$V = \hat{V}V_0 \qquad (15.6-22)$$

where

$$V_0 = \text{diag}\{v_0^1, v_0^2, \dots, v_0^n\} \qquad (15.6-23)$$

Since $\hat{V}^{-1}$ is stable and proper, (15.6–22) represents a factorization of V similar to that in (15.6–15). From spectral factorization theory it follows that

$$\hat{V}(z) = V_M(z)A \qquad\qquad (15.6-24)$$

where A is a constant matrix such that $AA^H = I$. Then (15.6–16) is not altered when $\hat{V}$ is used instead of V_M because A cancels.

Let us now assume without loss of generality that the first j v^i's have the same v_0^i's. Consider a v^* that is a linear combination of $v^1, \ldots, v^j$:

$$v^*(z) = \alpha_1 v^1(z) + \ldots + \alpha_j v^j(z) \qquad\qquad (15.6-25)$$

Then it follows that

$$v_0(z) = v_0^1(z) = \ldots = v_0^j(z) \qquad\qquad (15.6-26)$$

$$\hat{v}(z) = \alpha_1 \hat{v}^1(z) + \ldots + \alpha_j \hat{v}^j(z) \qquad\qquad (15.6-27)$$

One can easily check that a $\tilde{Q}$ that satisfies (15.6–3) for $\hat{v}^1, \ldots, \hat{v}^j$, will also satisfy (15.6–3) for the $\hat{v}$ given by (15.6–27) because of the property

$$\{\alpha_1 f_1(z) + \ldots + \alpha_j f_j(z)\}_* = \alpha_1 \{f_1(z)\}_* + \ldots + \alpha_j \{f_j(z)\}_* \qquad (15.6-28)$$

But then from Thm. 15.6-1 it follows that if a stabilizing controller $\tilde{Q}$ satisfies (15.6–3) for $\hat{v}$, then it minimizes the L_2 error $\Phi(v^*)$.   □

The following corollary to Thm. 15.6-3 holds for a specific choice of V.

Corollary 15.6-1. *Let*

$$V = \mathrm{diag}\{v_1, v_2, \ldots, v_n\} \qquad\qquad (15.6-29)$$

where $v_1(z), \ldots, v_n(z)$ *are scalars. Then use of* $\hat{V}$ *instead of* V_M *in (15.6–16) yields exactly the same* $\tilde{Q}$, *which minimizes* $\Phi(v^*)$ *for the following n vectors:*

$$v^* = \begin{pmatrix} v_1 \\ 0 \\ \vdots \\ 0 \end{pmatrix}, \begin{pmatrix} 0 \\ v_2 \\ \vdots \\ 0 \end{pmatrix}, \ldots, \begin{pmatrix} 0 \\ 0 \\ \vdots \\ v_n \end{pmatrix} \qquad\qquad (15.6-30)$$

and their multiples, as well as for the linear combinations of those directions that correspond to v_i's *with the same zeros outside the UC with the same degree and the same time delays.*

Example 15.6-1 (Minimum phase P). $P^*(z)$ cannot be truly MP for a physical system. Even if the Laplace transfer matrix representing the continuous plant is MP but strictly proper, the discretized plant $P^*(z)$ will still have a delay of one unit because of sampling. Hence $P_A^* = z^{-1}I$, $P_M^* = zP^*$ and (15.6-16) yields for $W = $ constant

$$\tilde{Q} = (P^*)^{-1}(I - KV_M^{-1}) \qquad (15.6 - 31)$$

where K is the constant term in a partial fraction expansion of V_M. This is equal to the first non-zero matrix in the impulse response description of $V(z)$, which can be obtained by long division.   □

15.6.2   Setpoint Prediction

In the case of setpoint tracking, future values of r^* are often known and supplied to the computer ahead of time. If at time t the setpoint value that is provided to the control algorithm as $\mathcal{Z}^{-1}\{r^*(z)\}$ is the one we wish the plant output to reach at time $t + NT$, then the objective function has to be modified to:

$$\Phi_N(r^*) = \|W(z^{-N}I - P^*\tilde{Q})r^*\|_2^2 \qquad (15.6 - 32)$$

If the above objective function is used for Objs. O1, O2, O3, then the resulting expressions for the H_2^*-optimal controller are the same as in Thms. 15.6-1, 15.6-2, and 15.6-3, but with the term z^{-N-1} instead of z^{-1} inside $\{\cdot\}_*$. All the steps in the proofs remain the same when (15.6–32) is used rather than (15.3–14).

15.6.3   Intersample Rippling

The H_2^*-optimal controller minimizes the sum of squared errors and completely disregards the plant's output behavior between the sample points. Therefore the performance of the H_2^*-optimal controller may be excellent at the sample points but may suffer from severe intersample rippling. This problem was demonstrated in Sec. 7.5.3. A modification was introduced in Secs. 8.1.2 and 9.2.2 to substitute poles in $\tilde{q}$ close to (-1,0) with poles at $z = 0$. The new $\tilde{q}$ was shown to be free of the problem of intersample rippling and to combine desirable deadbeat type characteristics with those of the H_2^*-optimal controller. This section extends the modification to MIMO systems and general open-loop stable and unstable plants. It should be pointed out that this modification is sufficient only if there are no open-loop oscillatory poles in the continuous plant transfer function, which have become unobservable after sampling.

Let $\tilde{Q}_H(z)$ be the H_2^*-optimal $\tilde{Q}$ obtained according to the previous sections. Also let $\delta(z)$ be the least common denominator of the elements of $P^*(z)$, and κ_i, $i = 1, \ldots, \rho$ be the roots of $\delta(z)$ close to (-1,0) (or in general with negative real part). Define

$$\tilde{q}_-(z) = z^{-\rho} \prod_{j=1}^{\rho} \frac{z - \kappa_j}{1 - \kappa_j} \qquad\qquad (15.6-33)$$

Then $\tilde{Q}_H$ is modified as follows:

$$\tilde{Q}(z) = \tilde{Q}_H(z)\tilde{q}_-(z)B(z) \qquad\qquad (15.6-34)$$

where the scalar $B(z)$ is selected so that the matrix S_2 (15.2–2) and $(I - P^*\tilde{Q})V$ remain stable. Let $\pi_i, i = 1, \ldots, \xi$ be the unstable roots (including $\pi_1 = 1$) of the least common denominator of $P^*(z), V(z)$. Let the multiplicity of each of them be m_i. Note that the poles outside the UC have multiplicity one, according to Assns. A1 and A3. Remember also that according to Assns. A3 and A4, V has at least as many poles at $z = 1$ as P^* and that each pole of V outside the UC is also a pole of P^*. Then, since $\tilde{Q}_H$ makes S_2 and $(I - P^*\tilde{Q}_H)V$ stable, it follows that the requirements on $B(z)$ are:

$$\frac{d^k}{dz^k}(1 - \tilde{q}_-(z)B(z))\Big|_{z=\pi_i} = 0, \qquad k = 0, \ldots, m_i - 1; \quad i = 1, \ldots, \xi \quad (15.6-35)$$

We can write

$$B(z) = \sum_{j=0}^{M-1} b_j z^{-j} \qquad\qquad (15.6-36)$$

where

$$M = \sum_{i=0}^{\xi} m_i \qquad\qquad (15.6-37)$$

and then compute the coefficients $b_j, j = 0, \ldots, M - 1$ from (15.6–35). Note that since none of the π_i's is 0 or ∞, (15.6–35) is equivalent to

$$\frac{d^k}{d\lambda^k}(1 - \tilde{q}_-(\lambda^{-1})B(\lambda^{-1}))\Big|_{\lambda=\pi_i^{-1}} = 0, \qquad k = 0, \ldots, m_i - 1; \quad i = 1, \ldots, \xi$$
$$(15.6-38)$$

Both $\tilde{q}_-(\lambda^{-1})$ and $B(\lambda^{-1})$ are polynomials in λ and therefore their derivatives with respect to λ can be computed easily. Then (15.6–38) yields a system of M linear equations with M unknowns $(b_0, b_1, \ldots, b_{M-1})$. The resulting controller $\tilde{Q}$ combines the desirable properties of the H_2^*-optimal controller and deadbeat type controllers.

Example 15.6-2. This example is presented to demonstrate the problem of intersample rippling in the H_2^*-optimal controller and the modification discussed above. Consider the continuous system

$$P(s) = \begin{pmatrix} \frac{0.50}{s+1} & \frac{1.42}{6s+1} \\ \frac{1.00}{2s+1} & \frac{1.00}{4s+1} \end{pmatrix} \tag{15.6 − 39}$$

The discretized system (zero order hold included) for a sampling time of $T = 1$, is

$$P^*(z) = \begin{pmatrix} \frac{0.316}{z-0.368} & \frac{0.218}{z-0.846} \\ \frac{0.393}{z-0.607} & \frac{0.221}{z-0.779} \end{pmatrix} \tag{15.6 − 40}$$

Computation of the roots of $\det P(z) = 0$ shows that the system in (15.6–40) has two finite zeros, at $a_1 = -0.95$ and $a_2 = 0.75$. The first zero is close to (-1,0) and is expected to cause intersample rippling when the H_2-optimal controller is used.

We find from (15.6-40) that $P_A^* = z^{-1}I$, $P_M^* = zP$. We shall consider step setpoint changes as external inputs – i.e.,

$$V(z) = \frac{z}{z-1}I \tag{15.6 − 41}$$

Then (15.6-16) yields

$$\tilde{Q}_H(z) = z^{-1}P^{-1} \tag{15.6 − 42}$$

Figure 15.6-1A shows the time response of this control system for a unit step change in the setpoint of output 1:

$$v^*(z) = r^*(z) = \begin{pmatrix} z/(z-1) \\ 0 \end{pmatrix} \tag{15.6 − 43}$$

The prediction of intersample rippling is verified. Note that at the sample points the outputs are indeed exactly at the setpoints yielding the minimum sum of squared errors.

The IMC controller is now obtained from (15.6–34) with $B(z) = 1$ and

$$\tilde{q}_-(z) = \frac{z + 0.95}{1.95z} \tag{15.6 − 44}$$

The response for this control system is shown in Fig. 15.6-1B. Clearly the rippling problem has disappeared. Note the inverse responses caused by the RHP zero of the continuous system $P(s)$. □

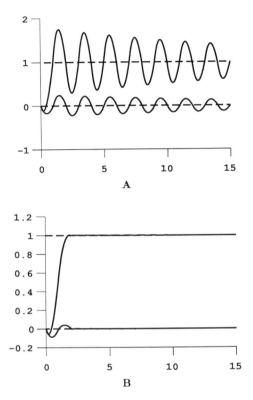

Figure 15.6-1. Response to unit setpoint change $r = (1,0)^T$. A: H_2^*-optimal controller, B: IMC.

15.6.4   Inner-Outer Factorization

As mentioned in Sec. 15.6.1, the factorization (15.6–1) is accomplished by employing the bilinear transformation

$$z = \frac{1+s}{1-s} \qquad (15.6-45)$$

to reduce the problem into the one discussed in detail in Sec. 12.6.4. The theorems below provide the formulas for the transformation of state space descriptions implied by (15.6–45) or its inverse

$$s = \frac{-1+z}{1+z} \qquad (15.6-46)$$

The following lemma is used in the proofs of the theorems.

Lemma 15.6-1. *Let* $G(x) = C(xI - A)^{-1}B + xD$. *Then*

$$xG(x) = C(xI - A)^{-1}AB + CB + xD \qquad (15.6-47)$$

Proof. We have

$$xG(x) = C(xI - A)^{-1}(A + xI - A)B + xD$$
$$= C(xI - A)^{-1}AB + CB + xD$$

$\square$

Theorem 15.6-4. *Let* $G^*(z) = C(zI - \Phi)^{-1}\Gamma + D$ *have no poles at* $z = -1$. *Then*

$$\hat{G}(s) \triangleq G^*\left(\frac{1+s}{1-s}\right) = \hat{C}(sI - \hat{A})^{-1}\hat{B} + \hat{D} \qquad (15.6-48)$$

where

$$\hat{A} = (\Phi + I)^{-1}(\Phi - I) \qquad (15.6-49)$$

$$\hat{B} = 2(\Phi + I)^{-2}\Gamma \qquad (15.6-50)$$

$$\hat{C} = C \qquad (15.6-51)$$

$$\hat{D} = D - C(\Phi + I)^{-1}\Gamma \qquad (15.6-52)$$

Proof. Since $P^*(z)$ is assumed to have no poles at $z = -1$, $\Phi + I$ is nonsingular. We can then write

$$\hat{G}(s) = C\left(\frac{1+s}{1-s}I - \Phi\right)^{-1}\Gamma + D$$

$$= (1-s)C(s(\Phi + I) + I - \Phi)^{-1}\Gamma + D$$

$$= (1-s)C(sI - (\Phi + I)^{-1}(\Phi - I))^{-1}(\Phi + I)^{-1}\Gamma + D \qquad (15.6-53)$$

Use of Lem. 15.6-1 in (15.6-53) yields

$$\hat{G}(s) = C(sI - (\Phi + I)^{-1}(\Phi - I))^{-1}(\Phi + I)^{-1}\Gamma$$

$$- \left[C(sI - (\Phi + I)^{-1}(\Phi - I))^{-1}(\Phi + I)^{-1}(\Phi - I)(\Phi + I)^{-1}\Gamma + C(\Phi + I)^{-1}\Gamma\right] + D$$

$$= C(sI - (\Phi+I)^{-1}(\Phi - I))^{-1}(I - (\Phi+I)^{-1}(\Phi - I))(\Phi+I)^{-1}\Gamma + D - C(\Phi+I)^{-1}\Gamma$$

$$= C(sI - (\Phi + I)^{-1}(\Phi - I))^{-1}2(\Phi + I)^{-2}\Gamma + D - C(\Phi + I)^{-1}\Gamma$$

□

Theorem 15.6-5. *Let* $\hat{G}(s) = \hat{C}(sI - \hat{A})^{-1}\hat{B} + \hat{D}$ *have no poles at* $s = 1$. *Then*

$$G^*(z) \triangleq \hat{G}\left(\frac{-1+z}{1+z}\right) = C(zI - \Phi)^{-1}\Gamma + D \qquad (15.6-54)$$

where

$$\Phi = (I - \hat{A})^{-1}(I + \hat{A}) \qquad (15.6-55)$$

$$\Gamma = 2(I - \hat{A})^{-2}\hat{B} \qquad (15.6-56)$$

$$C = \hat{C} \qquad (15.6-57)$$

$$D = \hat{D} + \hat{C}(I - \hat{A})^{-1}\hat{B} \qquad (15.6-58)$$

Proof. $I - \hat{A}$ is nonsingular because $\hat{G}(s)$ is assumed to have no poles at $s = 1$.
We have

$$G^*(z) = \hat{C}(\frac{-1+z}{1+z}I - \hat{A})^{-1}\hat{B} + \hat{D}$$
$$= (1+z)\hat{C}(z(I - \hat{A}) - (I + \hat{A}))^{-1}\hat{B} + \hat{D}$$

$$= (1+z)\hat{C}(zI - (I - \hat{A})^{-1}(I + \hat{A}))^{-1}(I - \hat{A})^{-1}\hat{B} + \hat{D} \qquad (15.6-59)$$

Application of Lem. 15.6-1 to (15.6–59) yields

$$G^*(z) = \hat{C}(zI - (I - \hat{A})^{-1}(I + \hat{A}))^{-1}(I - \hat{A})^{-1}\hat{B}$$

$$+ \left[\hat{C}(zI - (I - \hat{A})^{-1}(I + \hat{A}))^{-1}(I - \hat{A})^{-1}(I + \hat{A})(I - \hat{A})^{-1}\hat{B} + \hat{C}(I - \hat{A})^{-1}\hat{B}\right] + \hat{D}$$

$$= \hat{C}(zI - (I - \hat{A})^{-1}(I + \hat{A}))^{-1}(I + (I - \hat{A})^{-1}(I + \hat{A}))\hat{B} + \hat{D} + \hat{C}(I - \hat{A})^{-1}\hat{B}$$

$$= \hat{C}(zI - (I - \hat{A})^{-1}(I + \hat{A}))^{-1}2(I - \hat{A})^{-2}\hat{B} + \hat{D} + \hat{C}(I - \hat{A})^{-1}\hat{B}$$

$\square$

The factorization

$$P^*(z) = P_A^*(z)P_M^*(z) \qquad (15.6-1)$$

involves the following steps:

Step 1: Use the variable transformation (15.6–45) on $P^*(z)$ to obtain $\hat{P}(s)$. Note that the assumption of Thm. 15.6-4 that $P^*(z)$ has no poles at $z = -1$ holds for the P^*'s under consideration in this chapter because of Assn. A2.

Step 2: Apply Thm. 12.6-4 on $\hat{P}(s)$ to obtain the factorization

$$\hat{P}(s) = \hat{P}_A(s)\hat{P}_M(s) \qquad (15.6-60)$$

Note that for a strictly proper system $D = 0$ and therefore from (15.6–52) we have $\hat{D} = -C(\Phi + I)^{-1}\Gamma = P^*(-1)$. According to Assn. A2, $P^*(z)$ has no zeros on the UC and therefore $P^*(-1)$ has full rank. Hence, the assumption of no zeros on the imaginary axis including infinity in Thm. 12.6-4, holds for $\hat{P}(s)$.

Step 3: Use the variable transformation (15.6–46) on $\hat{P}_A(s)$ and $\hat{P}_M(s)$ to obtain $P_A^*(z)$ and $P_M^*(z)$ correspondingly. Note that $\hat{P}_A(s)$ satisfies the assumption of no poles at $s = 1$, since by construction all its poles are in the LHP. Also, $\hat{P}_M(s)$ has the poles of $\hat{P}(s)$, which do not include a pole at $s = 1$, since $P^*(z)$ has no poles at $z = \infty$.

The result of the above steps is a stable, all-pass P_A^* and a minimum phase P_M^*. Both P_A^* and P_M^* are proper because $\hat{P}_A$ and $\hat{P}_M$ have no poles at $s = 1$. Also note that since $\hat{P}_M(s)$ is minimum phase, it does not have a zero at $s = 1$ and therefore $P_M^*(z)$ has no zero at $z = \infty$, which means that $(P_M^*(z))^{-1}$ is proper.

To obtain the factorization

$$V = V_M V_A \qquad (15.6-15)$$

one should follow the same steps as above with the difference that in Step 2, Thm. 12.6-4 is applied on $\hat{V}(s)^T$ as described in Sec. 12.6-4.

15.7 IMC Design: Step 2 (F)

The controller $\tilde{Q}$ obtained in Step 1 of the IMC design procedure is detuned in Step 2 to satisfy the robustness conditions by augmenting it with the IMC filter $F(z)$:

$$Q = Q(\tilde{Q}, F) \qquad (15.7-1)$$

First we will postulate reasonable filter structures. Then we will define appropriate minimization problems to be solved over the filter parameters and discuss the computational issues involved.

15.7.1   Filter Structure

Some structure has to be assumed for F, which can be as general as the designer wishes. However, in order to keep the number of variables in the optimization problems small, a rather simple structure like a diagonal F with first- or second-order terms would be recommended. In most cases this is not restrictive because the controller $\tilde{Q}$ that was designed in the first step of the IMC procedure is in general a full high-order transfer matrix. More complex filter structures may be necessary in cases of ill-conditioned systems ($\bar{\sigma}(\tilde{P}^*)/\underline{\sigma}(\tilde{P}^*)$ very large). For such systems a two-filter structure was discussed in detail in Sec. 12.7.1. The elements of each of the two filters in that structure can be designed as described below.

The filter $F(z)$ is chosen to be a diagonal rational function that satisfies the following requirements.

a. Internal Stability. S_1 in (15.2–1) must be stable.

b. Asymptotic setpoint tracking and/or disturbance rejection. $(I - \tilde{P}^*\tilde{Q}F)v^*$ must be stable.

Write

$$F(z) = \mathrm{diag}\{f_1(z), \ldots, f_n(z)\} \qquad (15.7-2)$$

Then, Assns. A1-A5 and the fact that by construction $\tilde{Q}(z)$ makes S_1 and $(I - \tilde{P}^*\tilde{Q})V$ stable, imply that the requirements on an element f_ℓ of F are:

$$\frac{d^j}{dz^j}(1 - f_\ell(z))\Big|_{z=\pi_1} = 1, \qquad j = 0, \ldots, m_{1\ell} - 1 \qquad (15.7-3)$$

$$f_\ell(\pi_i) = 1, \qquad i = 2, \ldots, \xi \qquad (15.7-4)$$

where $\pi_1 = 1$ and $m_{1\ell}$ is the highest multiplicity of such a pole in any element of the ℓth row of V and π_i, $i = 2, \ldots, \xi$ are the poles of P^* outside the UC, each with multiplicity 1, according to Assn. A1.

One can now select the filter elements to be of the form discussed in Sec. 9.3

$$f(z) = \phi(z)f_1(z) \qquad (9.3-3)$$

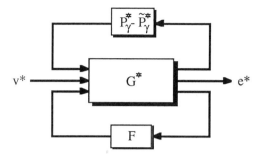

Figure 15.7-1. Discrete interconnection structure.

where

$$f_1(z) = \frac{(1-\alpha)z}{z-\alpha} \qquad (8.2-1)$$

$$\phi(z) = \sum_{j=0}^{w} \beta_j z^{-j} \qquad (9.3-4)$$

and the coefficients $\beta_0, \ldots, \beta_w$ are computed so that (15.7–3), (15.7–4) are satisfied for some specified α. The parameter α can be used as a tuning parameter.

Note that for $\xi = 1, \pi_1 = 1, m_{1\ell} = 1$, we only need $\phi(z) = 1$. For the general case, (15.7–3) and (15.7–4) a system of M_ℓ linear equations with $\beta_0, \ldots \beta_w$ as unknowns where M_ℓ is given by

$$M_\ell = m_{1\ell} + \xi \qquad (15.7-5)$$

The procedure for solving these equations is identical to the one described in Sec. 9.3, with $m_{1\ell}$ and M_ℓ replacing m_1 and M. Also, when the two-filter structure of Sec. 12.7.1 is used in the case of ill-conditioned plants, $\hat{m}_1 = \max_\ell m_{1\ell}$ should be used for all ℓ in the place of $m_{1\ell}$ in $F_1(z)$. This is required for internal stability and no steady-state offset.

15.7.2 Robust Stability Interconnection Structure

Consider the block diagram in Fig. 15.1-2C. The same block manipulations that were used to obtain Fig. 12.7-1A from Fig. 12.1-1B, can be used on Fig. 15.1-2C to obtain Fig. 15.7-1. The only difference is that P_γ^* and $\tilde{P}_\gamma^*$ take the place of P and $\tilde{P}$. All the transfer matrices in Fig. 15.7-1 are discrete.

The development described in Sec. 12.7.2 can now be applied to the block structure of Fig. 15.7-1 to put it in the form shown in Figs. 12.7-1B, C. The difference is that the uncertainty block Δ represents a discrete transfer function. If simple uncertainty descriptions of the form discussed in Sec. 12.7.2 are available for the discrete plant P_γ^*, then the corresponding formulas carry over to the discrete case, where $\bar{\ell}_A$, $\bar{\ell}_O$, $\bar{\ell}_I$, W_{e1}, W_{e2} are now discrete. Therefore only their values up to π/T need be considered.

Theorem 15.4-1 provides the robust stability condition. The matrix M in (15.4–1) is G_{11}^{*F} and therefore for robust stability the filter has to be designed such that

$$\mu_\Delta(G_{11}^{*F}(i\omega)) < 1, \qquad 0 \le \omega \le \pi/T \qquad (15.7-6)$$

The superscript $*$ is used to indicate that in this case G^F is a discrete transfer matrix.

15.7.3   Robust Performance Interconnection Structure

If one only cared about the performance at the sample points, then one could use Fig. 15.7-1 to state the robust performance conditions. However, because of the intersample rippling problem, one has to consider the continuous output of the plant and express the robust performance requirements in terms of the approximate sensitivity functions $E_r(s)$ or $E_d(s)$ given by (15.5–4) and (15.5–5).

One can obtain the appropriate interconnection structures of Fig. 12.7-1 by starting from Fig. 15.1-2B. The use of (15.5–2) in the derivation of (15.5–4) and (15.5–5), is equivalent to approximating the function of the sampling operator by $1/T$ for $0 \le \omega \le \pi/T$. This approximation is reasonable for signals with small power for $\omega > \pi/T$. Use of $1/T$ in the place of the sampling switch in Fig. 15.1-2B, allows us to derive the matrix G in the block diagram in Fig. 12.7-1A, which is slightly different from the one given by (12.7–13) for the continuous controller.

For $v = d$, $e = y$, we have

$$G_d = \begin{pmatrix} 0 & 0 & h_0\tilde{Q} \\ I & I & h_0\tilde{P}\tilde{Q} \\ -\frac{\gamma}{T}I & -\frac{\gamma}{T}I & 0 \end{pmatrix} \qquad (15.7-7)$$

For $v = -r$, $e = y - r$, with $r(s) = h_0(s)r^*(e^{sT})$:

$$G_r = \begin{pmatrix} 0 & 0 & h_0\tilde{Q} \\ I & I & h_0\tilde{P}\tilde{Q} \\ -\frac{\gamma}{T}I & h_0^{-1}I & 0 \end{pmatrix} \qquad (15.7-8)$$

Note that in this case the uncertainty block Δ represents continuous transfer functions as in Sec. 12.7.2.

15.7.4   Robust H_∞ Performance Objective

The next step is to transform (15.5–6) into an equivalent SSV condition. By using the equations in Sec. 12.7.2, but with the appropriate G_v instead of G and with $W_2 = W$, $W_1 = I$, we can obtain the corresponding G_v^F. Then (15.5–6) is satisfied if and only if

$$\mu_\Delta(G_v^F(i\omega)) < 1, \qquad 0 \le \omega \le \pi/T \qquad (15.7-9)$$

where $\Delta = \text{diag}\{\Delta_u, \Delta_p\}$, with Δ representing the uncertainty block Δ of Fig. 12.7-1 and Δ_p the additional block introduced for performance.

We can now write

$$F \triangleq F(z; \Lambda) \qquad (15.7-10)$$

where Λ is an array with the adjustable filter parameters. The filter design problem can be formulated as an minimization problem over the elements of Λ. In the filter structure proposed in Sec. 15.7.1, there is one adjustable parameter α for each element of the diagonal filter, or of each of the two diagonal filters, if two are used. Each one of these real parameters, say α_j, has to be inside the UC for F to be stable. The stability constraints can be removed from the minimization problem by setting

$$\alpha_j = e^{-T/\lambda_j^2} \qquad (15.7-11)$$

where λ_j is an element of Λ. Then any λ_j in $(-\infty, \infty)$ produces an α_j in $[0, 1)$. Note that if the parametrization (15.7–11) is used, then it is λ_j^2 and not λ_j that corresponds to a time constant. If one wishes to use a higher order $f_1(z)$ with more parameters in (8.2–1), one can write the denominator of each element of F as the product of polynomials of degree 2 and one of degree 1 if the order is odd. A polynomial of degree 2 with roots inside the UC can be written as $z^2 - (e^{Tp_1} + e^{Tp_2})z + e^{Tp_1 + Tp_2}$, where p_1, p_2 are the roots of $\lambda_2^2 x^2 + \lambda_1^2 x + 1 = 0$ for some value of λ_1, λ_2. In this way, the optimization problem is unconstrained in the optimization variables λ_1, λ_2, which can take any value in $(-\infty, \infty)$.

Our goal is to satisfy (15.7–9). The filter parameters can be obtained by solving

Objective O4:

$$\min_\Lambda \max_{0 \le \omega \le \pi/T} \mu_\Delta(G_v^F) \qquad (15.7-12)$$

It should be noted however that the optimal solution for Obj. O4, may still not satisfy (15.7–9). The reason is usually that the performance requirements set by the selection of W in (15.5–6) are too tight be to satisfied in the presence of the model-plant mismatch. In this case one should choose a less tight W and resolve Obj. O4.

Another important point is that satisfaction of the robust performance condition (15.7–9) does not necessarily imply satisfaction of the robust stability condition (15.7–6), which was the case in the continuous controller design. This is so even if the uncertainty descriptions for the continuous plant [used in (15.7–9)] and the discretized plant [used in (15.7–6)] correspond to exactly the same sets of possible plants. The reason is that (15.7–9) was obtained by using the approximations discussed in detail in Sec. 15.5.1, while there are no approximations involved in the derivation of (15.7–6). Note however, that if the uncertainty descriptions for the continuous and the discrete plant are equivalent in the sense discussed in Sec. 15.1.5, then satisfaction of (15.7–9) is usually sufficient for satisfaction of (15.7–6), although this is not guaranteed. As a result of the above discussed possibility, when a solution to Obj. O4 is found, one should check if (15.7–6) holds. If this does not happen, then one can always substitute the robust stability μ (15.7–6) in Obj. O4 and proceed with the minimization until (15.7–6) becomes less than one.

The type of problem defined by (15.7–12) is nearly identical to that defined by (12.7–25). The only difference is that the search over ω is limited to $0 \leq \omega \leq \pi/T$ in (15.7–12). This difference disappears, when only a finite number of frequencies is considered, as described in Obj. O4′ in Sec. 12.7.3. Hence the entire procedure and equations of Sec. 12.7.3 carry over to this case.

15.8   Illustration of the Design Procedure

The purpose of this section is to demonstrate the IMC design procedure by applying it to a 2×2 open-loop unstable system.

15.8.1   System Description

Let the continuous system be modeled by

$$\dot{x} = Ax + Bu \qquad (10.1-1)$$

$$y = Cx + Du \qquad (10.1-2)$$

where

$$A = \begin{pmatrix} 2.375 & 0.857 & 1.000 \\ -17.719 & -5.500 & -5.250 \\ -14.766 & -6.750 & -7.375 \end{pmatrix} \qquad (15.8-1)$$

$$B = \begin{pmatrix} 0 & 0 \\ 1 & 0 \\ -1 & 1 \end{pmatrix} \qquad (15.8-2)$$

$$C = \begin{pmatrix} 0 & 0.3 & 1.8 \\ 0 & 0 & -4.0 \end{pmatrix} \qquad (15.8-3)$$

$$D = \begin{pmatrix} 0 & 0 \\ 0 & 0 \end{pmatrix} \qquad (15.8-4)$$

The eigenvalues of A, which are the poles of the system (see Def. 10.1-3), are located at -1, -10, +0.5. Hence the open-loop system has an unstable pole of multiplicity 1 at 0.5.

From (10.1–7) we obtain the transfer matrix of the system:

$$\tilde{P}(s) = \begin{pmatrix} \frac{-1.5(s-0.2)}{(s-0.5)(s+1)} & \frac{0.3(6s+7.5)}{(s-0.5)(s+10)} \\ \frac{4s-0.5}{(s-0.5)(s+1)} & \frac{-(4s+8.5)}{(s-0.5)(s+10)} \end{pmatrix} \qquad (15.8-5)$$

Note that the unstable pole ($s = 0.5$) appears in *all* elements of $\tilde{P}(s)$, though it has only multiplicity 1. This is not an artifact of the example but rather the generic case for systems described by equations of the type (10.1–1), (10.1–2).

Let us now compute the zero-order hold discrete equivalent of $\tilde{P}(s)$ for a sampling time of $T = 0.1$. This is a reasonable choice, equal to $1/10$ of the dominant stable time constant and $1/20$ of the unstable time constant of the system. From (15.1–6) we find $\tilde{P}^*(z)$, which can be written in the form (15.1–15):

$$\tilde{P}^*(z) = C(zI - \Phi)^{-1}\Gamma + D \qquad (15.8-6)$$

where C and D are given by (15.8–3) and (15.8–4) and

$$\Phi = \begin{pmatrix} 1.2757 & 1.1138 & 1.0 \\ -0.15462 & 0.44053 & -0.41687 \\ -0.079536 & -0.44598 & 0.60772 \end{pmatrix} \qquad (15.8-7)$$

$$\Gamma = \begin{pmatrix} 0 & 0 \\ 0.071429 & 0 \\ -0.094864 & 0.071429 \end{pmatrix} \qquad (15.8-8)$$

For the design we need some information on the potential model error. We will assume a diagonal input multiplicative uncertainty

$$P^*(z) = \tilde{P}^*(z)(I + L_I^*(z)) \qquad (15.8-9)$$

where

$$L_I^*(z) = \text{diag}\{\ell_1^*(z), \ell_2^*(z)\} \qquad (15.8-10)$$

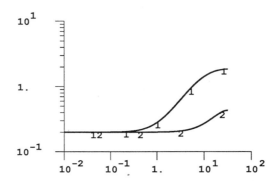

Figure 15.8-1. Multiplicative uncertainties $\bar{\ell}_1^*$ and $\bar{\ell}_2^*$.

and ℓ_1^*, ℓ_2^* are bounded by

$$|\ell_i^*(z)| \le \bar{\ell}_i^*(z) = \left| 0.2 \frac{z - p_i}{z - p_i^{10}} \cdot \frac{1 - p_i^{10}}{1 - p_i} \right| \qquad (15.8-11)$$

with $p_i = e^{-T/\tau_i}$. We will also assume that *all* plants $P^*(z)$ have exactly one unstable pole. The bound (15.8–11) implies that the uncertainty starts to increase around $\omega = 1/\tau_i$ with slope 1 and flattens out after one decade. Also, the low frequency uncertainty can be as much as 20%. The τ_i's are selected here to correspond to the dominant stable time constants of $\tilde{P}(s)$ associated with the respective inputs, i.e., $\tau_1 = 1$ and $\tau_2 = 0.1$. Bode plots of $\bar{\ell}_1^*, \bar{\ell}_2^*$ are shown in Fig. 15.8-1, for $0 \le \omega \le \pi/T$.

15.8.2   Design of $\tilde{Q}$

First one has to decide on the type of external input v for which $\tilde{Q}$ will be designed. Here we will consider *step-like* disturbances entering at the plant inputs. The diagonal $V(z)$ is of the form described in Cor. 15.6-1 with

$$v_1(z) = v_2(z) = v^*(z) = \mathcal{Z}\mathcal{L}^{-1}\{v(s)\} \qquad (15.8-12)$$

where $v(s)$ is an appropriate transfer function. Since the v_i's represent the effect of step-like inputs on the plant outputs, $v(s)$ should include both an integrator and a pole at $s = 0.5$. The simplest choice would be $v(s) = s^{-1}(-s+0.5)^{-1}$. However, such an input is "sluggish" and will result in poor robustness (see observation 3 in Sec. 4.1.2). To avoid this problem we select

$$v(s) = \frac{s + 0.5}{s(-s + 0.5)}$$

The next task is the factorization of $\tilde{P}^*$ into P_A^* and P_M^* (15.6–1). We follow the steps described in Sec. 15.6.4. This procedure yields the matrices $\Phi_A, \Gamma_A, C_A, D_A$ and $\Phi_M, \Gamma_M, C_M, D_M$ that define $P_A^*(z)$ and $P_M^*(z)$ respectively through (15.1–15):

$$\Phi_A = \begin{pmatrix} 1.54714 & 1.50513 & 1.41162 \\ -0.69253 & -0.67372 & -0.63186 \\ -0.098133 & -0.095468 & -0.089537 \end{pmatrix} \quad (15.8-13)$$

$$\Gamma_A = \begin{pmatrix} -8.27667 & -3.06852 \\ -15.61316 & -4.02293 \\ -3.78625 & -3.05041 \end{pmatrix} \quad (15.8-14)$$

$$C_A = \begin{pmatrix} -7.0645 \times 10^{-4} & -0.012260 & 0.20435 \\ 0.027830 & 0.13477 & -0.46555 \end{pmatrix} \quad (15.8-15)$$

$$D_A = \begin{pmatrix} 0 & 0 \\ 0 & 0 \end{pmatrix} \quad (15.8-16)$$

$$\Phi_M = \begin{pmatrix} 1.27570 & 1.11380 & 1.0 \\ -0.15462 & 0.44053 & -0.41687 \\ -0.079536 & -0.44598 & 0.60772 \end{pmatrix} \quad (15.8-17)$$

$$\Gamma_M = \begin{pmatrix} 0 & 0 \\ 1 & 0 \\ -1.32810 & 1 \end{pmatrix} \quad (15.8-18)$$

$$C_M = \begin{pmatrix} -0.017300 & -0.060253 & 0.050708 \\ 0.021810 & 0.12528 & -0.18355 \end{pmatrix} \quad (15.8-19)$$

$$D_M = \begin{pmatrix} -0.13652 & 0.086180 \\ 0.39111 & -0.30647 \end{pmatrix} \quad (15.8-20)$$

We also need to factor $V(z)$ according to (15.6–15), but this is trivial since $V(z)$ is diagonal and $v^*(z)$ can be factored as described by (8.1–3).

The final task is to determine $\tilde{Q}(z)$ from (15.6–16). For $W = I$ a state space description (15.1–15) of $\tilde{Q}(z)$ is given by

$$\Phi_Q = \begin{pmatrix} 1.27570 & 1.11380 & 1 & 0 & 0 \\ -0.57541 & -0.50239 & -0.45106 & -2.25465 & -0.80172 \\ 0.013486 & 0.011775 & 0.010572 & 0.058490 & 3.2232 \times 10^{-3} \\ 0 & 0 & 0 & 0.94873 & 0 \\ 0 & 0 & 0 & 0 & 0.94873 \end{pmatrix} \quad (15.8-21)$$

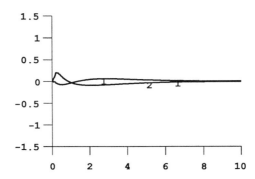

Figure 15.8-2. Nominal response to a step change at the plant input. (No filter).

$$\Gamma_Q = \begin{pmatrix} 0 & 0 \\ 81.6566 & 26.2327 \\ -3.09800 & 2.64970 \\ 1 & 0 \\ 0 & 1 \end{pmatrix} \qquad (15.8-22)$$

$$C_Q = \begin{pmatrix} 0.42079 & 0.94292 & 0.034188 & 2.25465 & 0.80172 \\ 0.46584 & 0.79454 & 0.64255 & 2.93590 & 1.06154 \end{pmatrix} \qquad (15.8-23)$$

$$D_Q = \begin{pmatrix} -81.6556 & -26.2327 \\ -105.3488 & -37.4893 \end{pmatrix} \qquad (15.8-24)$$

Figure 15.8-2 shows the response with this controller for the disturbance

$$u'(s) = \begin{pmatrix} s^{-1} \\ s^{-1} \end{pmatrix} \qquad (15.8-25)$$

entering at the plant inputs, when $P = \tilde{P}$. (The same disturbance will be used in all subsequent simulations).

15.8.3 Design of F

In this section we will design a filter $F(z)$ that guarantees robust stability in the presence of the model-plant mismatch described by (15.8-9). The condition for robust stability is given by (15.7-6). Here Δ consists of two scalar blocks and

$$G_{11}^{*F} = -\tilde{Q}F\tilde{P}^*\bar{L}_I^* \qquad (15.8-26)$$

where

$$\bar{L}_I^* = \text{diag}\{\bar{\ell}_1^*, \bar{\ell}_2^*\} \qquad (15.8-27)$$

The selection of the filter structure follows Sec. 15.7.1. A simple scalar filter will be used:

$$F(z) = f(z)I \qquad (15.8-28)$$

where $f(z)$ is given by (9.3–3) with $w = 29$. The tuning parameter α in $f_1(z)$ must be in $[0,1)$ and can be parametrized as

$$\alpha = e^{-T/\lambda} \qquad (15.8-29)$$

where λ is a positive time constant which becomes the new tuning parameter. We prefer λ over α because λ has a clear physical meaning and effect as was illustrated in Sec. 9.3.2. Note that the coefficients of $\phi(z)$ are functions of λ and are obtained from (9.3–6) and (9.3–9). If one wishes to remove the positivity constraint from the design parameter λ, then one should use (15.7–11) instead of (15.8–29). In this example however, as in the SISO case, we only have a single design variable to search over, which is a simple optimization problem. Hence (15.8–29) is used here to maintain a clear physical meaning for the optimization variable λ.

For $F = I$ ($\lambda = \alpha = 0$) we find $\mu(G_{11}^{*F}) = 3.75$, which implies that there exist plants among those described by (15.8–9) for which the closed loop system is unstable. A plot of μ is shown in Fig. 15.8-3. A search over the parameter λ shows that one has to increase λ to at least 0.5 to get $\mu = 1.0$ so that robust stability is guaranteed. Further increase of λ can reduce $\mu(G_{11}^{*F})$ even further. Plots of μ for $\lambda = 0.5$ and $\lambda = 1$ can be seen in Fig. 15.8-3.

Note, however, that the lower μ for $\lambda = 1$ does not necessarily mean that the performance of the system is superior because $\mu(G_{11}^{*F})$ is not the robust performance index. For determining robust performance, one has to select an appropriate performance weight W and compute $\mu(G^F)$ (Sec. 15.7.4). For our particular example, $\tilde{P}(s)$ has an unstable pole at $s = 0.5$ and the uncertainty becomes significant for $\omega > 1$. Therefore there is not much room for performance improvement. The question of robust performance will not be examined any further in this section. The reader is referred to Sec. 12.8 for a detailed example on the design of a filter for robust performance.

Let us now look at some simulations to examine the behavior of the control system when there is model-plant mismatch. The following transfer function was chosen for the "real" continuous plant $P(s)$:

$$P(s) = \tilde{P}(s)(I + L_I(s)) \qquad (15.8-30)$$

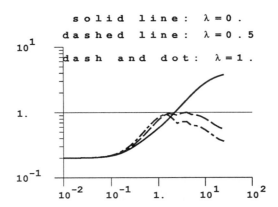

Figure 15.8-3. μ (Robust Stability) for different filter parameters λ.

where

$$L_I(s) = \begin{pmatrix} -0.2\frac{s+1}{0.1s+1} & 0 \\ 0 & -0.2\frac{0.1s+1}{0.01s+1} \end{pmatrix} \qquad (15.8-31)$$

Note that this $L_I(s)$ does not generate a plant that falls exactly in the class described by (15.8–9, 10, 11), although the steady-state gains and time constants of $L_I(s)$ match those used in (15.8–11) exactly. The reason is that no simple and non-conservative method is available for translating a type of uncertainty description (input multiplicative in this case) from the s-domain to exactly the same type in the z-domain. As explained in Sec. 15.1.5, such descriptions may be obtained either from first – principles models or via experiments conducted with different sampling rates. For the purposes of this example, (15.8–31) yields a plant sufficiently close to the class described by (15.8–9) to serve our illustration goals.

The responses to the input disturbance (15.8–22) are shown in Fig. 15.8-4 for $\lambda = 0.5$ and in Fig. 15.8-5 for $\lambda = 1$, for both the nominal case $(P = \tilde{P})$ and the case of model-plant mismatch with P given by (15.8–30). Without the IMC filter, the system is unstable for the "real" plant P in (15.8–30) as expected from the large value of $\mu(G_{11}^{*F})$. The nominal response is shown in Fig. 15.8-1. The responses for $\lambda = 1$ are not significantly better than that for $\lambda = 0.5$, although the robust stability μ is smaller for $\lambda = 1$. This is not surprising because $\mu(G_{11}^{*F})$ is an indicator of stability only.

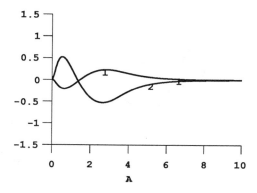

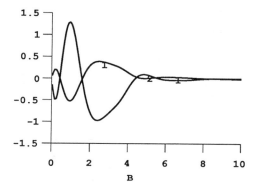

Figure 15.8-4. Responses (A) for nominal system and (B) the plant given by (15.8–30) for IMC filter time constant $\lambda = 0.5$.

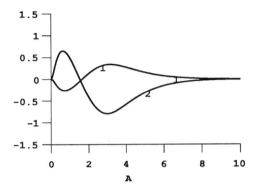

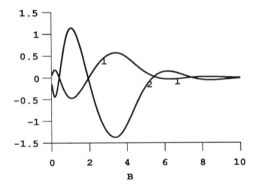

Figure 15.8-5. Responses (A) for nominal system and (B) the plant given by (15.8–30) for IMC filter time constant $\lambda = 1.0$.

15.9 Summary

For internal stability of the IMC structure, both the plant P and the IMC controller Q have to be stable. For open-loop unstable plants, it is convenient to use the IMC design procedure to design Q and then obtain the classic feedback controller C from (15.1–22) for implementation. Under some mild assumptions about pole-zero cancellations (Sec. 15.2.2), all stabilizing controllers Q for the plant P^* are parametrized by

$$Q(z) = Q_0(z) + Q_1(z) \qquad (15.2-3)$$

where Q_0 is an arbitrary proper controller that stabilizes P^* and Q_1 is any stable, proper transfer matrix such that $P^* Q_1 P^*$ is stable.

Design Procedure

Step 1: Nominal Performance

First, the stabilizing H_2^*-optimal controller $\tilde{Q}_H(z)$ is determined which minimizes the sum of the Sums of Squared Errors that each of the inputs v^i in a set $\mathcal{V} = \{v^i(z) : i = 1, \ldots, n\}$ would cause, when applied to the system separately.

Objective O3:

$$\min_{\tilde{Q}}[\Phi(v^1) + \ldots + \Phi(v^n)]$$

where

$$\Phi(v^i) \triangleq \|W e^i\|_2^2 = \|W \tilde{E}^* v^i\|_2^2 = \|W(I - P^* \tilde{Q}) v^i\|_2^2 \qquad (15.3-14)$$

The unique controller which meets Obj. O3 is given by

$$\tilde{Q}_H = z(P_M^*)^{-1}\{z^{-1}(P_A^*)^{-1} V_M\}_* V_M^{-1} \qquad (15.6-16)$$

where the operator $\{\cdot\}_*$ denotes that after a partial fraction expansion of the operand, only the strictly proper terms are retained, except for those that correspond to poles of $(P_A^*)^{-1}$. The factorization of the plant

$$P^* = P_A^* P_M^* \qquad (15.6-1)$$

into an allpass portion P_A^* and a minimum phase portion portion P_M^* can be accomplished through "inner-outer" factorization (Sec. 15.6.4). The input matrix $V = (\, v^1 \quad v^2 \quad \ldots \quad v^n \,)$ can be factored similarly

$$V = V_M V_A \qquad (15.6-15)$$

In some special cases (Thm. 15.6-3), $\tilde{Q}_H$ is also H_2^*-optimal for each of the inputs v^i separately, as well as their linear combinations.

Next, $\tilde{Q}_H$ is modified as described in Sec. 15.6.3 to eliminate the potential problem of intersample rippling:

$$\tilde{Q}(z) = \tilde{Q}_H(z)\tilde{q}_-(z)B(z) \qquad (15.6-33)$$

Step 2: Robust Stability and Robust Performance

In this step, the controller $\tilde{Q}$ is augmented by a low-pass filter F such that for the detuned controller $Q = Q(\tilde{Q}, F)$ both the robust stability

$$\mu_\Delta(G_{11}^{*F}(\tilde{P}^*, Q)) < 1, \qquad 0 \le \omega \le \pi/T, \quad \Delta = \Delta_u \qquad (15.7-6)$$

and the robust performance

$$\mu_\Delta(G_v^F(\tilde{P}, Q)) < 1, \qquad 0 \le \omega \le \pi/T, \quad \Delta = \text{diag}\{\Delta_u, \Delta_p\} \qquad (15.7-9)$$

conditions are satisfied. A nonlinear program was formulated (Sec. 15.7-4) to minimize $\mu_\Delta(G^F(\tilde{P}, Q))$ as a function of the filter parameters for a filter with a fixed diagonal structure. For unstable plants the filter has to be identity at the unstable poles of $\tilde{P}^*$. For ill-conditioned plants, two diagonal filters may be necessary to meet the requirement (15.7-9). It should be noted that satisfaction of (15.7-9) does not guarantee satisfaction of (15.7-6), although this is usually so. Hence it should be verified that the optimal solution to (15.7-9) also satisfies (15.7-6).

15.10   References

15.1.2. For a discussion on the computation of the matrices Φ and Γ see Åström & Wittenmark (1984).

15.1.3. See the same reference for more details on the Nyquist D-contour for discrete systems.

15.2. For modeling and identification methods for discrete systems see Åström & Wittenmark (1984). Jenkins and Watts (1969) is an excellent reference for identification techniques that result in norm uncertainty bounds for each element or a whole row of the system transfer matrix.

Part V

CASE STUDY

Chapter 16

LV-CONTROL OF A HIGH-PURITY DISTILLATION COLUMN

In this chapter the high-purity distillation column described in the Appendix will be studied, when reflux (L) and boilup (V) are manipulated to control the top (y_D) and bottom (x_B) compositions. This column was used as an example on several occasions earlier in this book. The LV-configuration is selected because this choice of manipulated inputs is most common in industrial practice. This does not mean that this is necessarily the best configuration; for example, the $\frac{L}{D}\frac{V}{B}$-configuration may be preferrable.

The distillation column investigated here was chosen to be representative of a large class of moderately high-purity distillation columns. The goal of this chapter is to provide a realistic control design and simulation study for the column. To be realistic at least the issues of uncertainty and nonlinearity must be addressed.

The reader is assumed to be thoroughly familiar with the material in Part III of this book.

16.1 Features

16.1.1 Uncertainty

We showed in Sec. 13.3.4 that the closed-loop system may be extremely sensitive to input uncertainty when the LV-configuration is used. In particular, inverse-based controllers were found to display severe robustness problems. In a similar manner as in Secs. 11.3.5 and 12.8 we will take uncertainty explicitly into account here when designing and analyzing the controllers via the structured singular value (μ). We will demonstrate that μ provides a much more efficient tool for comparing and analyzing the effect of various combinations of controllers, uncertainty and disturbances than the traditional simulation approach.

16.1.2   Nonlinearity

High-purity distillation columns are known to be strongly nonlinear (see Appendix), and any realistic study should take this into account. Our approach is to base the controller design on a linear model. The effect of nonlinearity is taken care of by analyzing this controller for linearized models at *different operating points*. Furthermore, all simulations will be based on the full nonlinear model.

16.1.3   Logarithmic Compositions

Several authors found that the high-frequency behavior of distillation columns is only weakly affected by operating conditions when the *scaled* transfer matrix is considered

$$\begin{pmatrix} dy_D^S \\ dx_B^S \end{pmatrix} = G^S \begin{pmatrix} dL \\ dV \end{pmatrix} \quad , G^S = \begin{pmatrix} \frac{1}{1-y_D^o} & 0 \\ 0 & \frac{1}{1-x_B^o} \end{pmatrix} G \qquad (16.1-1)$$

All plant models and controllers in this chapter are scaled in this manner. G^S is obtained by scaling the outputs with respect to the amount of impurity in each product

$$y_D^S = \frac{y_D}{1 - y_D^o}, \quad x_B^S = \frac{x_B}{x_B^o} \qquad (16.1-2)$$

Here x_B^o and y_D^o are the compositions at the nominal operating point. This relative scaling is obtained automatically by using logarithmic compositions

$$Y_D = \ln(1 - y_D) \qquad (16.1-3)$$

$$X_B = \ln x_B$$

because

$$dY_D = -\frac{dy_D}{1 - y_D}, \quad dX_B = \frac{dx_B}{x_B} \qquad (16.1-4)$$

Furthermore, the use of logarithmic compositions (Y_D and X_B) effectively eliminates the effect of nonlinearity at high frequency and also reduces its effect at steady-state. For control purposes the high-frequency behavior is of principal importance. Consequently, if logarithmic compositions are used we expect a linear controller to perform satisfactorily even when the operating conditions are far removed from the nominal operating point for which the controller was designed. Another objective of this chapter is to confirm that this is indeed true.

   In most cases the column is operated close to its nominal operating point and there is hardly any advantage in using logarithmic compositions which merely corresponds to a rescaling of the outputs in this case. However, if, for some reason, the column is taken far away from this nominal operating point, for example, during startup or due to a temporary loss of control, the use of logarithmic compositions may bring the column safely back to its nominal operating point, whereas a controller based on unscaled compositions (y_D and x_B) may easily yield an unstable response.

16.1.4 Choice of Nominal Operating Point

The design approach suggested by the above discussion is to design a linear controller based on a linearized model for some nominal operating point. What operating point should be used? If an operating point corresponding to both products of high and equal purities is chosen (i.e., $1 - y_D = x_B$ is small), it is easily shown that the steady-state gains and the linearized time constant will change drastically for small perturbations from this operating point. We may therefore question if acceptable closed-loop control can be obtained by basing the controller design on a linearized model at such an operating point. Some authors indicate that this is not advisable, and that a model based on a perturbed operating point should be used. However, as we just discussed, the high-frequency behavior, which is of primary importance for feedback control, shows much less variation with operating conditions. Therefore, *provided* the model gives a good description of the high-frequency behavior, we expect to be able to design an acceptable controller also when the nominal operating point has both products of high purity. This is also confirmed by the results in this chapter.

Therefore, a main conclusion is that acceptable closed-loop performance may be obtained by designing a linear controller based on a linear model at *any* nominal operating point. If large perturbations from steady state are expected then logarithmic compositions should be used to reduce the effect of nonlinearity.

16.2 The Distillation Column

The column model is derived in the Appendix. The following simplifying assumptions are made: binary separation; constant relative volatility; constant molar flows; constant holdups on all trays; perfect pressure and level control. The last assumption results in immediate flow response, that is, flow dynamics are neglected. This is somewhat unrealistic, and in order to avoid unrealistic controllers, we will add "uncertainty" at high frequencies to include the effect of neglected flow dynamics when designing and analyzing the controllers.

We will investigate the column at two different operating points. At the nominal operating point, A, both products are high-purity and $1 - y_D^o = x_B^o = 0.01$.

Operating point C is obtained by increasing D/F from 0.500 to 0.555 which yields a less pure top product and a purer bottom product; $1 - y^o_{DC} = 0.10$ and $x^o_{BC} = 0.002$ (subscript C denotes operating point C while no subscript denotes operating point A). We will study the column for the following three assumptions regarding reboiler and condenser holdup

Case 1: Almost negligible condenser and reboiler holdup ($M_D/F = M_B/F =$ 0.5 min).

Case 2: Large condenser and reboiler holdup ($M_D/F = 32.1$ min, $M_B/F = 11$ min).

Case 3: Same holdup as in Case 2, but the composition of the overhead vapor (y_T) is used as a controlled output instead of the composition in the condenser (y_D).

These three cases will be denoted by subscripts 1, 2 and 3, respectively. The holdup on each tray inside the column is $M_i/F = 0.5$ min in all three cases.

16.2.1   Modelling

Nominal operating point (A). A 41st order linear model for the columns is easily derived

$$\begin{pmatrix} dy_D \\ dx_B \end{pmatrix} = G(s) \begin{pmatrix} dL \\ dV \end{pmatrix} \qquad (16.2-1)$$

The *scaled* steady-state gain matrix is

$$G(0) = \begin{bmatrix} 87.8 & -86.4 \\ 108.2 & -109.6 \end{bmatrix} \qquad (16.2-2)$$

which yields the following values for the condition number and the 1,1-element in the RGA

$$\gamma(G(0)) = \bar{\sigma}(G(0))/\underline{\sigma}(G(0)) = 141.7 \qquad \lambda_{11}(G(0)) = 35.1$$

However, $\gamma(G)$ and $\lambda_{11}(G)$ are much smaller at high frequencies as seen from Fig. 16.2-1. A very crude model used throughout the earlier chapters is

$$\text{Model 0}: \quad G_0(s) = \frac{1}{1+75s}G(0) \qquad (16.2-3)$$

This model has the same $\gamma(G)$ and $\lambda_{11}(G)$ for all frequencies and is therefore a poor description of the actual plant at high frequency.

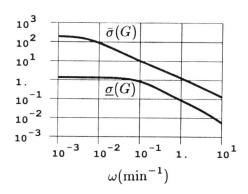
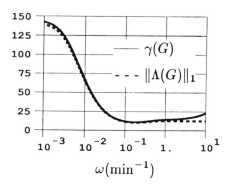

Figure 16.2-1. Column A, Case 1 ($G = G_1$). The condition number of the plant is about 10 times lower at high frequencies than at steady state. (Reprinted with permission from *Chem. Eng. Sci.* **43**, 35 (1988), Pergamon Press, plc.)

Case 1. For the case of negligible reboiler and condenser holdup the following simple two time-constant model yields an *excellent* approximation of the 41st order linear model.

$$\text{Model 1}: \quad G_1(s) = \begin{pmatrix} \frac{87.8}{1+\tau_1 s} & -\frac{87.8}{1+\tau_1 s} + \frac{1.4}{1+\tau_2 s} \\ \frac{108.2}{1+\tau_1 s} & \frac{-108.2}{1+\tau_1 s} - \frac{1.4}{1+\tau_2 s} \end{pmatrix} \quad \begin{array}{l} \tau_1 = 194 \text{ min} \\ \tau_2 = 15 \text{ min} \end{array} \quad (16.2-4)$$

This two state model uses two time constants: τ_1 is the time constant for changes in the external flows and is dominant. τ_2 is the time constant for changes in internal flows (simultaneous change in L and V with constant product rates, D and B). The simple model (16.2–4) matches the observed variation of the condition number with frequency (Fig. 16.2-1).

The effect of the reboiler and condenser holdups (Case 2) can be partially accounted for with Model 1 by multiplying $G_1(s)$ by diag$\{(1+\tau_D s)^{-1}, (1+\tau_B s)^{-1}\}$, where in our case $\tau_D = M_D/V_T = 10$ min and $\tau_B = M_B/L_B = 3$ min. However, in practice the top composition is often measured in the overhead vapor line (Case 3), rather than in the condenser. $G_1(s)$ provides a good approximation of the plant in such cases.

Cases 2 and 3. In order to obtain a low-order model for Cases 2 and 3, we performed a model reduction on the full 41st order model. These reduced order models are denoted by $G_2(s)$ and $G_3(s)$ respectively. A good approximation was obtained with a 5th-order model as illustrated in Fig. 16.2-2.

Operating point C. We will return with a discussion of the model for this case in Sec. 16.5 when we also discuss the control of the plant.

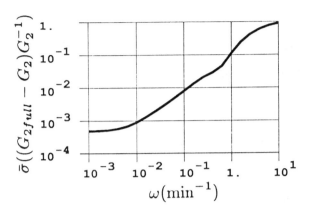

Figure 16.2-2. Column A, Case 2. Relative difference between the 5th order model $G_2(s)$ and the 41st order plant $G_{2\text{full}}(s)$. The 5th-order model provides an excellent approximation within the frequency range of interest ($\omega < 1\text{min}^{-1}$). (Reprinted with permission from *Chem. Eng. Sci.*, **43**, 36 (1988), Pergamon Press, plc.)

16.2.2   Simulations

The design and analysis of the controllers are based on the linear models $G_1(s), G_2(s)$, and $G_3(s)$. However, except for the five simplifying assumptions stated above, all simulations are carried out with the full *nonlinear* model. (In some cases the changes are so small, however, that the results are equivalent to linear simulations.) To get a realistic evaluation of the controllers *input uncertainty* must be included. Simulations are therefore shown both with and without 20% uncertainty with respect to the change of the two inputs. The following uncertainties are used:

$$\Delta L = (1 + \Delta_1)\Delta L_c, \quad \Delta_1 = 0.2 \qquad (16.2-5a)$$

$$\Delta V = (1 + \Delta_2)\Delta V_c, \quad \Delta_2 = -0.2 \qquad (16.2-5b)$$

Here ΔL and ΔV are the actual changes in manipulated flow rates, while ΔL_c and ΔV_c are the desired values as computed by the controller. $\Delta_1 = -\Delta_2$ was chosen to represent the worst combination of the uncertainties (Sec. 13.3.4).

16.3   Formulation of the Control Problem

16.3.1   Performance and Uncertainty Specifications

The uncertainty and performance specifications are the same as those used elsewhere in this book.

Uncertainty. The only source of uncertainty which is considered here is uncertainty on the manipulated inputs (L and V) with a magnitude bound:

$$w_I(s) = 0.2\frac{5s+1}{0.5s+1} \qquad (16.3-1)$$

The bound (16.3-1) allows for an input error of up to 20% at low frequency as was assumed for the simulations (16.2-5). The uncertainty bound (16.3-1) increases with frequency. This allows, for example, for a time delay of about 1 min in the response between the inputs, L and V, and the outputs, y_D and x_B. In practice, such delays may be caused by the flow dynamics. Therefore, although flow dynamics are not included in the models or in the simulations, they are partially accounted for in the μ-analysis and in the controller design.

Performance. Robust performance is satisfied if

$$\bar{\sigma}(E) = \bar{\sigma}((I+GC)^{-1}) \leq \frac{1}{|w_p|} \qquad (16.3-2)$$

is satisfied for all possible plants G. We use the performance weight

$$w_p(s) = 0.5\frac{10s+1}{10s} \qquad (16.3-3)$$

A particular E which exactly matches the bound (16.3-2) at low frequencies and satisfies it easily at high frequencies is $E = 20s/20s+1$. This corresponds to a first-order response with closed-loop time constant 20 min.

16.3.2   Analysis of Controllers

Comparison of controllers is based mainly on μ for robust performance (μ_{RP}). Simulations are used only to support conclusions found using the μ-analysis. The main advantage of the μ-analysis is that it provides a well-defined basis for comparison. On the other hand, simulations are strongly dependent on the choice of setpoints, uncertainty, etc.

The value of μ_{RP} is indicative of the worst-case response. If $\mu_{RP} > 1$ then the "worst case" does not satisfy our performance objective, and if $\mu_{RP} < 1$ then the "worst case" is better than required by our performance objective. Similarly, if $\mu_{NP} < 1$ then the performance objective is satisfied for the *nominal* case. However, this may not mean very much if the system is sensitive to uncertainty and μ_{RP} is significantly larger than one. We will show below that this is the case, for example, if an inverse-based controller is used for our distillation column.

16.3.3   Controllers

We will study the distillation column with the following six controllers:

1. *Diagonal PI-controller.*

$$C_{PI}(s) = \frac{0.01}{s}(1 + 75s)\begin{pmatrix} 2.4 & 0 \\ 0 & -2.4 \end{pmatrix} \qquad (16.3-4)$$

This controller was studied in Sec. 11.3.5 and was tuned to achieve as good a performance as possible while maintaining robust stability (see also Fig. 16.4-1).

2. *Steady-state decoupler plus two PI-controllers.*

$$C_{oinv}(s) = 0.7\frac{(1 + 75s)}{s}G(0)^{-1} = \frac{0.01(1 + 75s)}{s}\begin{pmatrix} 27.96 & -22.04 \\ 27.60 & -22.40 \end{pmatrix} \qquad (16.3-5)$$

This controller was tuned to achieve good *nominal* performance. However, the *controller* has large RGA-elements ($\lambda_{11}(C) = 35.1$) at *all* frequencies and we expect the controller to be extremely sensitive to input uncertainty (see Sec. 13.3).

3. *Inverse-based controller based on the linear model $G_1(s)$ for Case 1.*

$$C_{1inv}(s) = \frac{0.7}{s}G_1(s)^{-1} \qquad (16.3-6)$$

At low frequency this controller is equal to $C_{oinv}(s)$. Note that $C_{1inv}(s)$ and $G_1(s)^T$ have the same RGA-elements. Therefore from Fig. 16.2-1 we expect $C_{1inv}(s)$ to be sensitive to input uncertainty at low frequency, but not at high frequency.

4, 5, and 6. *μ-optimal controllers* based on the models $G_0(s), G_1(s)$ and $G_2(s)$. The controllers are denoted $C_{0\mu}(s), C_{1\mu}(s)$ and $C_{2\mu}(s)$, respectively.

These controllers were obtained by minimizing $\sup_\omega \mu(N_{RP})$ for each model using the input uncertainty and performance weights given above. The numerical procedure used for the minimization is the same as mentioned in Sec. 11.3.5. The μ-plots for robust performance for the μ-optimal controllers are of particular interest since they indicate the best *achievable* performance for the plant. Bode plots of the transfer matrix elements of $C_{1\mu}(s)$ and $C_{2\mu}(s)$ are shown in Fig. 16.3-1. Note the similarities between these controllers and the simple diagonal PI controller (16.3-4).

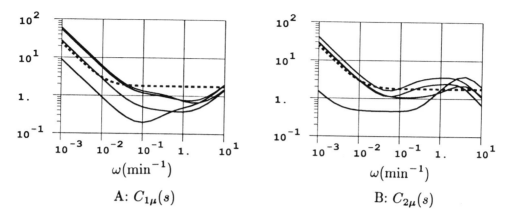

Figure 16.3-1. Magnitude plots of elements in μ−optimal controllers $C_{1\mu}(s)$ and $C_{2\mu}(s)$. Dotted line: $C_{PI}(s)$. (Reprinted with permission from *Chem. Eng. Sci.*, **43**, 39 (1988), Pergamon Press, plc.)

At low frequency ($s \to 0$) the six controllers are approximately

$$C_{PI} = \frac{0.01}{s} \begin{pmatrix} 2.4 & 0 \\ 0 & -2.4 \end{pmatrix}$$

$$C_{0inv} = C_{1inv} = \frac{0.01}{s} \begin{pmatrix} 27.96 & -22.04 \\ 27.80 & -22.40 \end{pmatrix}$$

$$C_{0\mu} = \frac{0.01}{s} \begin{pmatrix} 3.82 & -0.92 \\ 1.73 & -3.52 \end{pmatrix}; C_{1\mu} = \frac{0.01}{s} \begin{pmatrix} 6.07 & -0.90 \\ 2.80 & -2.93 \end{pmatrix}; C_{2\mu} = \frac{0.01}{s} \begin{pmatrix} 4.06 & +0.15 \\ 2.85 & -2.93 \end{pmatrix}$$

$\|\Lambda(C)\|_1$ is shown as a function of frequency for the six controllers in Fig. 16.3-2. As expected, the μ-optimal controllers have small RGA-elements, which make them insensitive to input uncertainty. For example, $C_{2\mu}$ is nearly triangular at low frequency and consequently has $\Lambda \cong I$.

16.4   Results for Operating Point A

In this section we will study how the six controllers perform at the nominal operating point A for the three assumptions regarding condenser and reboiler holdup (corresponding to the models $G_1(s), G_2(s)$, and $G_3(s)$). The μ-plots for the 18 possible combinations are given in Fig. 16.4-1. A number of interesting observations can be derived from these plots. These are presented below. In some cases the simulations in Figs. 16.4-2 to 16.4-4 are used to support the claims.

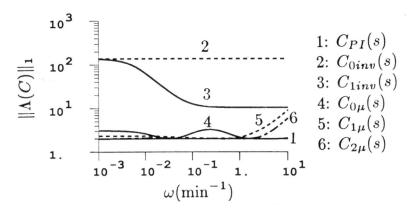

Figure 16.3-2. Magnitude of RGA elements of controllers. $\|\Lambda(C)\|_1 = \sum_{i,j} |\lambda_{ij}(C)|$. (Reprinted with permission from *Chem. Eng. Sci.*, **43**, 39 (1988), Pergamon Press, plc.)

16.4.1   Discussion of Controllers

$C_{PI}(s)$. The simple diagonal PI-controller performs reasonably well in all cases. μ_{NP} is higher than one at low frequency, which indicates a slow return to steady state. This is confirmed by the simulations in Fig. 16.4-3 for a feed rate disturbance; after 200 min the column has still not settled. Operators are usually unhappy about this kind of response. The controller is insensitive to input uncertainty and to changes in reboiler and condenser holdup.

$C_{0inv}(s)$. This controller uses a steady-state decoupler. The nominal response is very good for Case 1 (Fig. 16.4-2), but the controller is extremely sensitive to input uncertainty. In practice, this controller will yield an unstable system.

$C_{1inv}(s)$. This controller gives an excellent nominal response for Case 1 (Fig. 16.4-1). This is also confirmed by the simulations in Fig. 16.4-2; the response is almost perfectly decoupled with a time constant of about 1.4 min. Since the simulations are performed with the full-order model, while the controller was designed based on the simple two time-constant model, $G_1(s)$ (16.2-4), this confirms that $G_1(s)$ yields a very good approximation of the linearized plant when the reboiler and condenser holdups are small. The controller is sensitive to the input uncertainty as expected from the RGA analysis. Also note that the controller performs very poorly when the condenser and reboiler holdups are increased. This shows that the controller is also very sensitive to other sources of model-plant mismatch.

$C_{0\mu}(s)$. This is the μ-optimal controller from our previous study which was designed based on the very simplified model $G_0(s)$. The controller performs

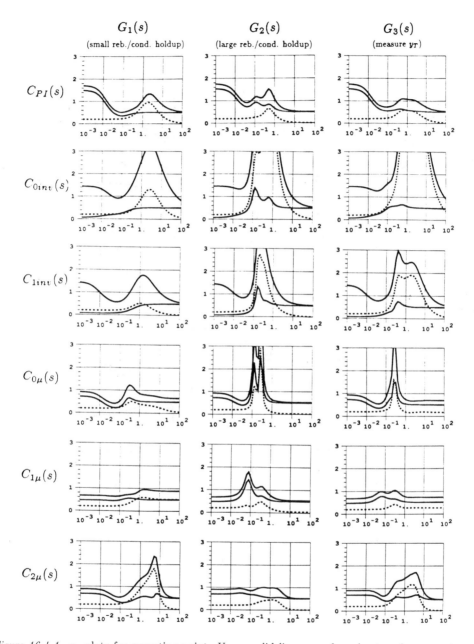

Figure 16.4-1. μ−plots for operating point. Upper solid line: μ_{RP} for robust performance; lower solid line:μ_{NP} for nominal performance; dotted line: μ_{RS} for robust stability. (Reprinted with permission from *Chem. Eng. Sci.*, **43**, 40 (1988), Pergamon Press, plc.)

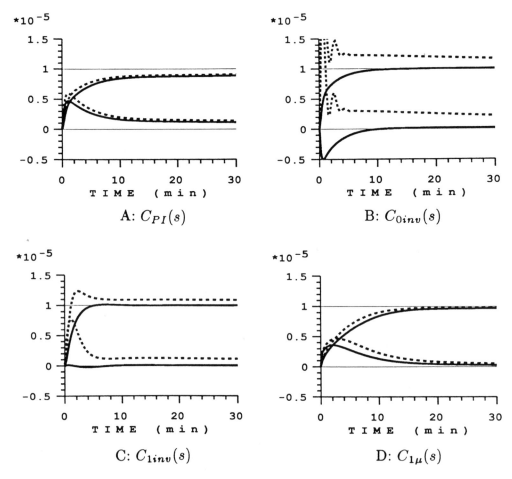

Figure 16.4-2. Operating point A, Case 1. Closed-loop response to small setpoint change in y_D. Solid lines: no uncertainty; dotted lines: 20% uncertainty on inputs L and V (16.2–5). (Reprinted with permission from *Chem. Eng. Sci.*, **43**, 41 (1988, Pergamon Press, plc.)

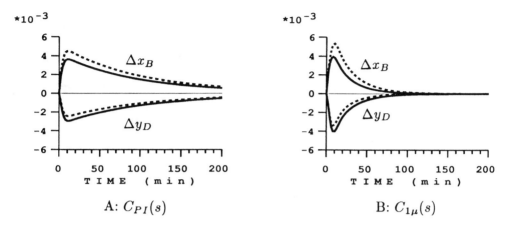

A: $C_{PI}(s)$ B: $C_{1\mu}(s)$

Figure 16.4-3. Operating point A, Case 1. Closed-loop response to a 30% increase in feed rate. Solid lines: no uncertainty; dotted lines: 20% uncertainty on inputs L and V (16.2-5). (Reprinted with permission from *Chem. Eng. Sci.*, **43**, 41 (1988), Pergamon Press, plc.)

surprisingly well on the actual plant $(G_1(s))$ when the holdups are negligible. However, the controller is seen to perform very poorly when the holdups in the reboiler and condenser are increased, which shows that the controller is very sensitive to other sources of model inaccuracies (for which it was not designed).

$C_{1\mu}(s)$. This is the μ-optimal controller when there is negligible holdup $(G_1(s))$. The robust performance condition is satisfied for this case since $\mu_{RP} \cong 0.95$. The nominal performance is not as good as for the inverse-based controller $C_{1inv}(s)$; we have to sacrifice nominal performance to make the system robust with respect to uncertainty. The controller shows some performance deterioration when the reboiler and condenser holdups are increased (Case 2). This is not surprising since the added holdup makes the response of y_D and x_B more sluggish; the open-loop response for y_D changes from approximately $(1+194s)^{-1}$ to $((1+194s)(1+10s))^{-1}$ [recall discussion following (16.2-4)]. As expected, the controller is much less sensitive to changes in condenser holdup if the overhead composition is measured in the vapor line (Case 3). Overall, this is the best of the six controllers.

$C_{2\mu}(s)$. This is the μ-optimal controller for the case with considerable reboiler and condenser holdup, and with y_D measured in the condenser $(G_2(s))$. $\mu_{RP} \cong$ 1.00 for this case. The nominal response is good in all cases (Fig. 16.4-1), but the controller is very sensitive to uncertainty when the plant is $G_1(s)$ or $G_3(s)$ rather than $G_2(s)$. This is clearly not desirable since changes in condenser and reboiler holdup are likely to occur during normal operation. The observed behavior is not

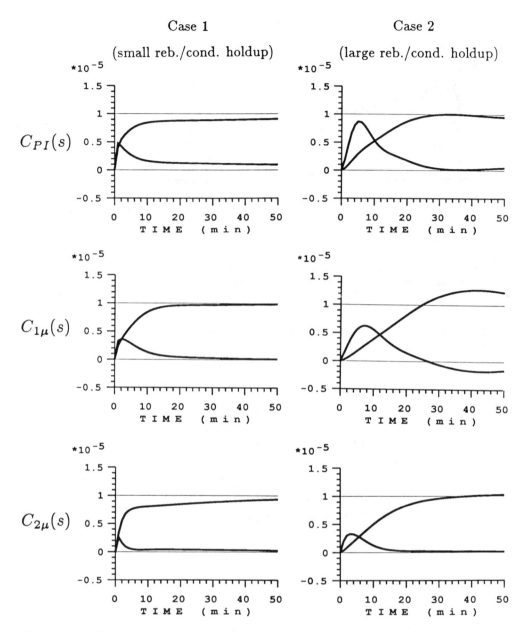

Figure 16.4-4. Operating point A. Effect of reboiler and condenser holdup on closed-loop response. No uncertainty. (Reprinted with permission from *Chem. Eng. Sci.*, **43**, 42 (1988), Pergamon Press, plc.)

surprising since the controller includes lead elements at $\omega \cong 0.1$ (Fig. 16.3-1B) to counteract the lags caused by the reboiler and condenser holdups. If these lags are not present in the plant ($G_1(s)$ or $G_3(s)$), the "derivative" action caused by the lead elements results in a system which is very sensitive to uncertainty.

16.4.2 Conclusions

- The μ-optimal controller $C_{0\mu}(s)$ for the plant $G_0(s)$ has $\mu_{RP} \cong 1.06$ while the μ-optimal controller $C_{1\mu}(s)$ for the plant $G_1(s)$ has $\mu_{RP} \cong 0.95$. Thus, somewhat surprisingly, the achievable performance is not much better for $G_1(s)$ than for $G_0(s)$, even though $G_0(s)$ is ill-conditioned and has large RGA elements at *all* frequencies, while $G_1(s)$ has large RGA elements only at low frequencies (Fig. 16.2-1). This seems to indicate that large RGA-elements at low frequency imply limitations on the achievable control performance and partially justifies the use of steady-state values of the RGA for selecting the best control configuration.

- However, the use of the more detailed model $G_1(s)$, rather than $G_0(s)$, is still justified since the resulting μ-optimal controller is much less sensitive to changes in reboiler and condenser holdup (which will occur during operation).

- $G_1(s)$ approximates the full-order model very closely as seen from Fig. 16.4-2C; the response is almost perfectly decoupled when there is no uncertainty.

- To avoid sensitivity to the amount of condenser and reboiler holdup, the overhead composition should be measured in the overhead vapor, rather than in the condenser. In practice, temperature measurements *inside* the column are often used to infer compositions, and the dynamic response of these measurements is similar to that when the condenser and reboiler holdup is neglected.

- The simple model $G_2(s)$ is useful for controller design even when the reboiler and condenser holdup is large.

- The main advantage of the μ-optimal controllers over the simple diagonal PI controller is a faster return to steady-state. This can be seen very clearly in Fig. 16.4-3 which shows the closed-loop response to a 30% increase in feed rate.

16.5   Effect of Nonlinearity (Results for Operating Point C)

We will not treat nonlinearity as uncertainty because this approach is not rigorous and is also very conservative due to the strong correlation between all the parameters in the model which is difficult to account for. Furthermore, we know from the data in the Appendix that the column is actually *not* as nonlinear as one might expect. Though the steady-state gains may change dramatically, the high frequency behavior, which is of principal importance for feedback control, is much less affected. In particular, this is the case if relative (logarithmic) compositions are used. To demonstrate this fact we compute μ and show simulations for some of the controllers when the "plant" is $G_C(s)$ rather than $G(s)$.

16.5.1   Modelling

The model $G_C(s)$ describes the same column as $G(s)$, but the distillate flow rate $(\frac{D}{F})$ has been increased from 0.5 to 0.555 such that $y_D = 0.9$ and $x_B = 0.002$. For Case 1 ($M_D/F = M_B/F = 0.5$ min), the following approximate model is derived when scaled compositions ($dy_D/0.1, dx_B/0.002$) are used:

$$G_{C1}(s) = \begin{pmatrix} \frac{16.0}{1+\tau_1 s} & \frac{16.0}{1+\tau_1 s} + \frac{0.023}{1+\tau_2 s} \\ \frac{9.3}{1+\tau_1 s} & \frac{-9.3}{1+\tau_1 s} - \frac{1.41}{1+\tau_2 s} \end{pmatrix} \qquad \begin{array}{l} \tau_1 = 24.5 \text{ min} \\ \tau_2 = 10 \text{ min} \end{array} \qquad (16.5-1)$$

The steady-state gains and time constants are entirely different from those at operating point A (16.2–4). Also note that at steady state $\lambda_{11}(G(0)) = 35.1$ for operating point A, but only 7.5 for operating point C. However, at high-frequency the scaled plants at operating points A and C are very similar. Equations (16.2–4) and (16.5–1) yield:

$$G_1(\infty) = \frac{1}{s}\begin{pmatrix} 0.45 & -0.36 \\ 0.56 & -0.65 \end{pmatrix} \qquad \lambda_{11}(\infty) = 3.2 \qquad (16.5-2a)$$

$$G_{C1}(\infty) = \frac{1}{s}\begin{pmatrix} 0.65 & -0.65 \\ 0.38 & -0.52 \end{pmatrix} \qquad \lambda_{11}(\infty) = 3.7 \qquad (16.5-2b)$$

Therefore, as we will show, controllers which were designed based on the model $G(s)$ (operating point A) remain satisfactory when the plant is $G_C(s)$ rather than $G(s)$. Recall that the use of a scaled plant is equivalent to using logarithmic compositions (Y_D and X_B). The variation in gain with operating conditions is much larger if unscaled compositions are used – both at steady-state and at high frequencies:

$$G_1^{us}(\infty) = \frac{0.01}{s} \begin{pmatrix} 0.45 & -0.36 \\ 0.56 & -0.65 \end{pmatrix} \qquad (16.5 - 3a)$$

$$G_{C1}^{us}(\infty) = \frac{0.01}{s} \begin{pmatrix} 6.5 & -6.5 \\ 0.08 & -0.10 \end{pmatrix} \qquad (16.5 - 3b)$$

16.5.2 μ-Analysis

The μ-plots with the model $G_C(s)$ and four of the controllers are shown in Fig. 16.5-1 (all four controllers yield *nominally* stable closed-loop systems). For high frequencies the μ-values are almost the same as those found at operating point A. The only exception is the inverse based controller $C_{1inv}(s)$ which is robustly stable at operating point A, but not at operating point C. Again, this confirms the sensitivity of this controller to model inaccuracies. Performance is clearly worse for low frequencies at operating point C (Fig. 16.5-1) than at operating point A (Fig. 16.4-1). This is expected; the controllers were designed based on model A, and the plants are quite different in the low frequency range.

The μ-optimal controller $C_{1\mu}(s)$ satisfies the robust performance requirements also at operating point C when the reboiler and condenser holdups are small. Consequently, with scaled (logarithmic) compositions, a single linear controller is able to give acceptable performance at these two operating points although the linear models are quite different. The main difference between $C_{1\mu}(s)$ and the diagonal PI controller is again that the μ-optimal controller gives a much faster return to steady-state. This is clearly seem from Fig. 16.5-2A.

16.5.3 Logarithmic Versus Unscaled Compositions

Figure 16.5-1 shows how controllers, designed based on the *scaled* plant $G(s)$ at operating point A, perform for the *scaled* plant (different scaling factors!) at operating point C; this is equivalent to using logarithmic compositions (Y_D and X_B). We know from (16.5–3) that the plant model based on absolute compositions changes much more. Therefore we expect the closed-loop performance to be entirely different at operating ponts A and C when unscaled (absolute) compositions are used. This is indeed confirmed by Fig. 16.5-2B which shows the closed-loop response to a small setpoint change in x_B at operating point C. Fig. 16.5-2B should be compared to Fig. 16.5-2A which shows the same response, but with logarithmic compositions as controlled outputs. In Fig. 16.5-2B (absolute compositions) the response for x_B is significantly more sluggish, but the response for y_D is much faster than in Fig. 16.5-2A (logarithmic compositions). This is exactly what we would expect from a comparison of (16.5–3a) and (16.5–3b). The high-frequency gain for changes in y_D is increased by an order of magnitude and

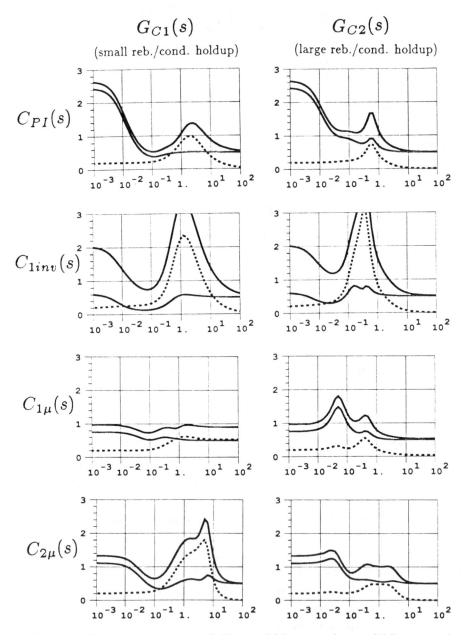

Figure 16.5-1. μ−plots for operating point C. Upper solid line: μ_{RP}; lower solid line: μ_{NP}; dotted line: μ_{RS}. (Reprinted with permission from *Chem. Eng. Sci.*, **43**, 44 (1988), Pergamon Press, plc.)

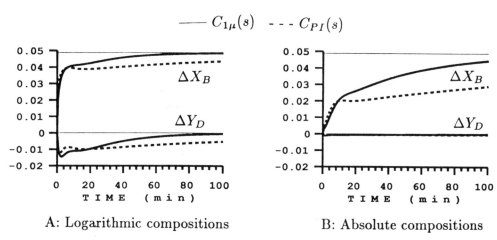

—— $C_{1\mu}(s)$ - - - $C_{PI}(s)$

A: Logarithmic compositions   B: Absolute compositions

Figure 16.5-2. Operating point C, Case 1. Closed-loop response to small setpoint change in x_B (x_B increases from 0.002 to 0.0021) using diagonal PI controller (dotted line) and the μ-optimal controller for operating point A (solid line). Left: logarithmic compositions as controlled outputs (equivalent to using scaled compositions); right: absolute (unscaled) compositions as controlled outputs. No uncertainty. (Reprinted with permission from *Chem. Eng. Sci.*, **43**, 45 (1988), Pergamon Press, plc.)

the gain for changes in x_B is reduced by an order of magnitude. However, recall from (16.5–2) that the gain changes very little when logarithmic compositions are used.

16.5.4   Transition from Operating Point A to C

Figure 16.5-3 shows a transition from operating point A ($Y_D = X_B = 4.605$) to operating point C ($Y_D = 2.303, X_B = 6.215$) using logarithmic compositions as controlled ouputs. The desired setpoint change is a first order response with time constant 10 min:

$$\Delta Y_{Ds} = \frac{2.303}{1 + 10s}\frac{1}{s} \ , \quad \Delta X_{Bs} = \frac{-1.609}{1 + 10s}\frac{1}{s}$$

The closed-loop response is seen to be very good. The diagonal controller $C_{PI}(s)$ and the μ-optimal controller $C_{1\mu}(s)$ give very similar responses in this particular case. (However, the μ-optimal controller generally performs better at operating point C as is evident from Figs. 16.5-1 and 16.5-2.) This illustrates that a linear controller, based on the nominal operating point A, can be satisfactory for a large

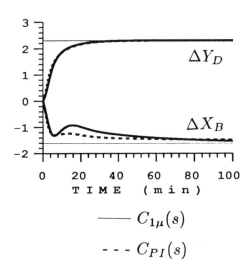

Figure 16.5-3. Transition from operating point A to C (Case 1) with controllers $C_{1\mu}$ (solid line) and C_{PI} (dotted line). Logarithmic compositions are used as controlled outputs to reduce the effect of nonlinearity. Desired trajectory is a first-order response with time constant 10 min. No uncertainty. (Reprinted with permission from *Chem. Eng. Sci.*, **43**, 45 (1988), Pergamon Press, plc.)

deviation from this operating point when logarithmic compositions are used.

16.6 Conclusions

A single linear controller is able to provide satisfactory control for this high-purity column at widely different operating conditions. To compensate for the plant nonlinearity it is advantageous to use "logarithmic compositions." For small deviations from steady state linear controllers using "absolute compositions" work also well.

The performance with a simple diagonal controller is robust with respect to model-plant mismatch but after an upset the return to steady state can be very sluggish. This particular deficiency can be removed by a μ-optimal controller. Inverse-based controllers, and specifically those involving steady-state decouplers, were shown to be very sensitive to model-plant mismatch.

16.7 References

This chapter is abstracted from a paper by Skogestad & Morari (1988a) where the state-space description of the μ-optimal controllers and the reduced order models for Cases 2 and 3 are also provided. For a general discussion of distillation control the reader is referred to the book by Shinskey (1984) or the thesis by Skogestad (1987). Skogestad and Morari (1987a) have reviewed the current industrial understanding of distillation control from the viewpoint of modern robust control.

16.1.2. The nonlinear behavior was observed specifically by Moczek, Otto & Williams (1963) and Fuentes and Luyben (1983).

16.1.3. The use of "logarithmic compositions" seems to have been first suggested by Ryskamp (1981).

16.1.4. The change of gain and time constant with operating condition was analyzed by Kapoor, McAvoy & Marlin (1986) and Skogestad & Morari (1988d).

16.2.1. The model reduction was performed via "Balanced Realization" (Moore, 1981). The dominant time constant τ_1 can be estimated, for example, from the inventory time constant introduced by Moczek, Otto & Williams (1963). The time constant τ_2 can be obtained by matching the high-frequency behavior as shown by Skogestad & Morari (1988d).

Appendix

Dynamic Model of Distillation Column

On many occasions in this book a high purity distillation column is used as an example. In this Appendix all the necessary information is summarized to enable the reader to verify any of the results reported in this book and to use the distillation model as a test case for other analysis and design procedures.

A.1 Nomenclature and Assumptions

The column is shown in Fig. A.1-1 where most symbols are also defined.

Symbols:

M	hold up
N	number of equilibrium (theoretical) stages
$N+1$	total number of stages including total condenser
N_F	feed stage location
F	feed rate
z_F	mole fraction of light component in feed
q_F	fraction liquid in feed
D	distillate flow
V	boilup
V_T	top vapor flow
L	reflux flow
B	bottom flow
p	pressure
q	mole fraction of feed which is liquid
x	mole fraction of light component in liquid
y	mole fraction of light component in vapor
α	relative volatility
κ	linearized VLE-constant

The unit of mass is kmol and the unit of time is minute.

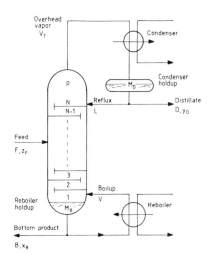

Figure A.1-1. Two product distillation column with single feed and total condenser.

Subscripts:

i	tray i (trays are numbered from bottom with the reboiler as tray 1)
F	feed
D	distillate
B	bottom
T	top

Assumptions:

- binary mixture

- constant pressure

- constant relative volatility α

- constant molar flows

- no vapor holdup (immediate vapor response, $dV_T = dV_B$)

- constant liquid holdup M_i on all trays (immediate liquid response, $dL_T = dL_B$)

- Vapor-Liquid Equilibrium (VLE) and perfect mixing on all stages

A.2   Nonlinear Model

Material balances for change in holdup of light component on each tray $i = 2, \ldots, N (i \neq N_F, i \neq N_F + 1)$:

$$M_i \dot{x}_i = L_{i+1} x_{i+1} + V_{i-1} y_{i-1} - L_i x_i - V_i y_i$$

Above feed location $i = N_F + 1$:

$$M_i \dot{x}_i = L_{i+1} x_{i+1} + V_{i-1} y_{i-1} - L_i x_i - V_i y_i + F_V y_F$$

Below feed location, $i = N_F$:

$$M_i \dot{x}_i = L_{i+1} x_{i+1} + V_{i-1} y_{i-1} - L_i x_i - V_i y_i + F_L x_F$$

Reboiler, $i = 1$:

$$M_B \dot{x}_i = L_{i+1} x_{i+1} - V_i y_i - B x_i, \quad x_B = x_1$$

Total condenser, $i = N + 1$:

$$M_D \dot{x}_i = V_{i-1} y_{i-1} - L_i x_i - D x_i, \quad y_D = x_{N+1}$$

VLE on each tray $(i = 1, \ldots, N)$, constant relative volatility:

$$y_i = \frac{\alpha x_i}{1 + (\alpha - 1) x_i}$$

Flow rates assuming constant molar flows:

$$i > N_F \text{ (above feed)}: \quad L_i = L, \quad V_i = V + F_V$$

$$i \leq N_F \text{ (below feed)}: \quad L_i = L + F_L, \quad V_i = V$$

$$F_L = q_F F, \quad F_V = F - F_L$$

$$D = V_N - L = V + F_V - L \quad \text{(constant condenser holdup)}$$

$$B = L_2 - V_1 = L + F_L - V \quad \text{(constant reboiler holdup)}$$

Compositions x_F and y_F in the liquid and vapor phase of the feed are obtained by solving the flash equations:

$$F z_F = F_L x_F + F_V y_F$$

Table A.2-1. Column Data

Relative volatility	$\alpha = 1.5$
No. of theoretical trays	$N = 40$
Feed tray (1=reboiler)	$N_F = 21$
Feed composition	$z_F = 0.5$
Condenser time constant	M_D/F: see Ch. 16
Reboiler time constant	M_B/F: see Ch. 16
Tray time constant	$M_i/F = 0.5$min

$$y_F = \frac{\alpha x_F}{1 + (\alpha - 1)x_F}$$

The column data and operating conditions used in the book are shown in Tables A.2-1 and A.2-2. The tray compositions are listed in Table A.2-3. The column behavior is highly nonlinear as Fig. A.2-1 illustrates.

Table A.2-2. Operating Variables:

Operating Point	A	C
y_D	0.99	0.90
x_B	0.01	0.002
$(L/D)_{\min}$	3.900	3.000
L/D	5.413	4.935
D/F	0.500	0.555
B/F	0.500	0.445
V/F	3.206	3.291
L/F	2.706	2.737

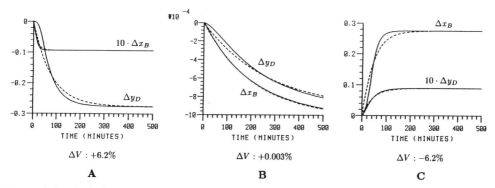

Figure A.2-1. (—) Nonlinear open loop responses Δy_D and Δx_B for changes in boilup V (reflux L constant). (- - -) Approximation with linear first order response. (Different time constants for A, B and C). A: $V + 6.2\%$, B: $V + 0.003\%$, C: $V - 6.2\%$.

Table A.2-3. Tray compositions for operating conditions A and C.

Tray	A x	y	C x	y
41	0.99000001	0.00000000	0.89999998	0.00000000
40	0.98507464	0.99000001	0.85714287	0.89999998
39	0.97891331	0.98584270	0.80946749	0.86436397
38	0.97124165	0.98064220	0.75826764	0.82472157
37	0.96174449	0.97416693	0.70532095	0.78214854
36	0.95007116	0.96615076	0.65266412	0.73812300
35	0.93584847	0.95629781	0.60229003	0.69433850
34	0.91870391	0.94429302	0.55585837	0.65245205
33	0.89830160	0.92982209	0.51450431	0.61384380
32	0.87439108	0.91260135	0.47878328	0.57945764
31	0.84686637	0.89241952	0.44873506	0.54975533
30	0.81582677	0.86918712	0.42401719	0.52477002
29	0.78162557	0.84298790	0.40405512	0.50421697
28	0.74489015	0.81412017	0.38817218	0.48761836
27	0.70649838	0.78311342	0.37568399	0.47441158
26	0.66750675	0.75070858	0.36595631	0.46402755
25	0.62903917	0.71779740	0.35843393	0.45593894
24	0.59216058	0.69532860	0.35264966	0.44968402
23	0.55776387	0.65420103	0.34822109	0.44487435
22	0.52649474	0.62516826	0.34484178	0.44119197
21	0.49872494	0.59877533	9.34226966	0.43838206
20	0.47416389	0.57493830	0.29737207	0.38832030
19	0.44553304	0.54654711	0.25339118	0.33734646
18	0.41298330	0.51345152	0.21191093	0.28741339
17	0.37701508	0.47582585	0.17416464	0.24031939
16	0.33849627	0.43424863	0.14091858	0.19746466
15	0.29860669	0.38972306	0.11246721	0.15971924
14	0.25870702	0.34361297	8.87127668E-02	0.12741737
13	0.22015929	0.29749119	6.92852587E-02	0.10044810
12	0.18414706	0.25293222	5.36631495E-02	7.83913583E-02
11	0.15154333	0.21130413	4.12711129E-02	6.06550165E-02
10	0.12285385	0.17361607	3.15471478E-02	4.65858951E-02
9	9.82341841E-02	0.14045265	2.39814352E-02	3.55459340E-02
8	7.75578097E-02	0.11199372	1.81338135E-02	2.69563195E-02
7	6.05053641E-02	8.80929977E-02	1.36372205E-02	2.03172937E-02
6	4.66509089E-02	6.83813393E-02	1.01931207E-02	1.52121522E-02
5	3.55311297E-02	5.23663759E-02	7.56312115E-03	1.13019431E-02
4	2.66932649E-02	3.95125374E-02	5.55941742E-03	8.31601024E-03
2	1.42609123E-02	2.12399177E-02	2.87815696E-03	4.31103166E-03
1	9.99999978E-03	1.49253728E-02	2.00000009E-03	2.99700303E-03

A.3 Linearized Model

We linearize the material balance on each tray ($dL_i = dL$, $dV_i = dV$):

$$M_i \dot{x}_i = L_{i+1} dx_{i+1} - (L_i + K_i V_i) dx_i + K_{i-1} V_{i-1} dx_{i-1} + (x_{i+1} - x_i) dL - (y_i - y_{i-1}) dV$$

Here K_i is the linearized VLE-constant:

$$K_i = \frac{dy_i}{dx_i} = \frac{\alpha}{(1 + (\alpha - 1)x_i)^2}$$

and y_i, x_i, L_i and V_i are the steady-state values at the nominal operating point. Written in the standard state variable form in terms of deviation variables the model becomes

$$\dot{x} = Ax + Bu, \qquad y = Cx$$

where $x = (dx_1, \ldots, dx_{N+1})^T$ are the tray compositions, $u = (dL, dV)^T$ are the manipulated inputs and $y = (dy_D, dx_B)^T$ are the controlled outputs. The state matrix $A = \{a_{i,j}\}$ is tri-diagonal:

$$
\begin{aligned}
i \neq N+1: \quad & a_{i,i+1} = L_{i+1}/M_i \\
& a_{i,i} = -(L_i + K_i V_i)/M_i \\
i \neq 1: \quad & a_{i,i-1} = K_{i-1} V_{i-1}/M_i
\end{aligned}
$$

Input matrix $B = \{b_{i,j}\}$:

$$
\begin{aligned}
& i \neq N+1: \quad b_{i,1} = (x_{i+1} - x_i)/M_i, \quad b_{N+1,1} = 0 \\
& i \neq 1, i \neq n+1: \ b_{i,2} = -(y_i - y_{i-1})/M_i, \quad b_{N+1,2} = 0, \quad b_{1,2} = (y_1 - x_1)/M_1
\end{aligned}
$$

Output matrix C:

$$C = \begin{pmatrix} 0 & 0 & 0 & \ldots & 0 & 1 \\ 1 & 0 & \ldots & & 0 & 0 \end{pmatrix}$$

The condenser drum is assumed to be under perfect level control

$$V = L + D$$

Thus, if a different set of manipulated variables, $U = (dD, dV)^T$ is employed the new model is obtained via the linear transformation

$$\begin{pmatrix} dD \\ dV \end{pmatrix} = \begin{pmatrix} -1 & 1 \\ 0 & 1 \end{pmatrix} \begin{pmatrix} dL \\ dV \end{pmatrix}$$

Table A.4-1. Gain Information

Operating Point	A	C
$G_{LV}(0)$	$\begin{pmatrix} 0.878 & -0.864 \\ 1.082 & -1.096 \end{pmatrix}$	$\begin{pmatrix} 1.604 & -1.602 \\ 0.01865 & -0.02148 \end{pmatrix}$
$G_{LV}^s(0)$ (scaled compositions)	$\begin{pmatrix} 87.8 & -86.4 \\ 108.2 & -109.6 \end{pmatrix}$	$\begin{pmatrix} 16.0 & 16.023 \\ 9.3 & -10.71 \end{pmatrix}$
RGA $\lambda_{11}(0)$	35.1	7.5
$G_{LV}(\infty)$	$\frac{0.01}{s} \begin{pmatrix} 0.45 & -0.36 \\ 0.56 & -0.65 \end{pmatrix}$	$\frac{0.01}{s} \begin{pmatrix} 6.5 & -6.5 \\ 0.08 & -0.10 \end{pmatrix}$
$G_{LV}^s(\infty)$	$\frac{1}{s} \begin{pmatrix} 0.45 & -0.36 \\ 0.56 & -0.65 \end{pmatrix}$	$\frac{1}{s} \begin{pmatrix} 0.65 & -0.65 \\ 0.38 & -0.52 \end{pmatrix}$
$\lambda_{11}(\infty)$	3.2	3.7

A.4   Gain Information

From the gain information in Tables A.4-1 through A.4-3 one can see that the non-linearity appears mostly in the low-frequency range and is much less pronounced for high frequencies. The time constant ($\tau = 75\,\text{min}$) used in the simulations represents an average value based on the simulations shown in Fig. A.2-1.

Table A.4-2. Singular Value Decomposition of Gain Matrices

Configuration	LV	DV
$G(0)$	$\begin{pmatrix} 0.878 & -0.864 \\ 1.082 & -1.096 \end{pmatrix}$	$\begin{pmatrix} -0.878 & 0.014 \\ -1.082 & -0.014 \end{pmatrix}$
RGA λ_{11}	35.1	0.45
Condition number κ	141.7	70.8
$SVD;\ G = U\Sigma V^H$		
U	$\begin{pmatrix} -0.625 & 0.781 \\ -0.781 & -0.625 \end{pmatrix}$	$\begin{pmatrix} -0.630 & 0.777 \\ -0.777 & -0.630 \end{pmatrix}$
Σ	$\begin{pmatrix} 1.972 & 0 \\ 0 & 0.0139 \end{pmatrix}$	$\begin{pmatrix} 1.393 & 0 \\ 0 & 0.0197 \end{pmatrix}$
V	$\begin{pmatrix} -0.707 & 0.708 \\ 0.708 & 0.707 \end{pmatrix}$	$\begin{pmatrix} 1.000 & -0.001 \\ 0.001 & 1.000 \end{pmatrix}$

Table A.4-3. Disturbance gains for LV-configuration.

	x_F	F	q_F	V_d
dy_D	0.881	0.394	0.868	0.864
dx_B	1.119	0.586	1.092	1.096

References

Arkun, Y., V. Manousiouthakis and A. Palazoglu, "A Case Study of Decoupling Control in Distillation." *Ind. Eng. Chem. Process Des. Dev.*, **23**, 93-101 (1984).

Åström, K. J., "Frequency Domain Properties of Otto Smith Regulators." *Int. J. Control,* **26**, 307-314 (1977).

Åström, K. J., P. Hagander and J. Sternby, "Zeros of Sampled Systems." *Automatica,* **20**, 31-38 (1984).

Åström, K. J. and B. Wittenmark, *Computer Controlled Systems,* Prentice-Hall, Englewood Cliffs, NJ (1984).

Bellman, R., *Introduction to Matrix Analysis,* McGraw-Hill Book Company, New York (1970).

Bjork, A. and T. Elfring, "Algorithms for Confluent Vandermonde Systems." *Numer. Math.*, **18**, 44-60 (1973).

Black, H. S. "Inventing the Negative Feedback Amplifier." *IEEE Spectrum,* 55-60 (1977).

Bode, H. W., "Feedback – The History of an Idea." in *Selected Papers on Mathematical Trends in Control Theory."* Dover, NY, 106-123 (1964).

Bristol, E. H., "On a New Measure of Interaction for Multivariable Process Control." *IEEE Trans. Autom. Control,* **AC-11**, 133-134 (1966).

Bristol, E. H., "A New Process Interaction Concept, Pinned Zeros." *Internal Report of the Foxboro Co.*, Boston, MA (1980).

Brosilow, C. B., "The Structure and Design of Smith Predictors from the Viewpoint of Inferential Control." *Proc. of Joint Automatic Control Conf.*, Denver, CO (1979).

Bruns, D. D. and C. R. Smith, "Singular Value Analysis: A Geometrical Structure For Multivariable Processes." *AIChE Winter Meeting*, Orlando, Fla. (1982).

Buckley, P. S., W. L. Luyben and J. P. Shunta, *Design of Distillation Column Control Systems*. Instrument Society of America, Research Triangle Park, NC (1985).

Chen, S., *Control System Design for Multivariable Uncertain Processes.*, Ph.D. Thesis, Chem. Eng. Dept., Case Western Reserve University, Cleveland, OH (1984).

Chu, C. C., H_∞-*Optimization and Robust Multivariable Control*, Ph.D. Thesis, University of Minnesota, Minneapolis, MN (1985).

Churchill, R. V. and J. W. Brown, *Complex Variables and Applications*, McGraw-Hill, New York (1984).

Cohen, G. H. and G. A. Coon, "Theoretical Considerations of Retarded Control." *Trans. ASME*, **75**, 827 (1953).

Cutler, C. R. and B. L. Ramaker, "Dynamic Matrix Control - A Computer Control Algorithm." *AIChE National Mtg.*, Houston, TX (1979); also *Proc. Joint Autom. Control Conf.*, San Francisco, CA (1980).

Dahlquist, G., A. Björck (translated by N. Anderson), *Numerical Methods*, Prentice-Hall, Englewood Cliffs, NJ (1974).

Dailey, R. L., *Conic Sector Analysis for Digital Control Systems with Structured Uncertainties*, Ph.D. Thesis, California Institute of Technology, Pasadena, CA (1987).

Davison, E. J., "Multivariable Tuning Regulators: The Feedforward and Robust Control of a General Servomechanism Problem." *IEEE Trans. Autom. Control*, **AC-21**, 35-47 (1976).

Desoer, C. A. and A. N. Gündes, "Decoupling Linear Multi-Input Multi-Output Plants by Dynamic Output Feedback." *IEEE Trans. Autom. Control*, **AC-31**, 744-750 (1986).

Desoer, C. A. and C. A. Lin, "A Comparative Study of Linear and Nonlinear MIMO Feedback Configurations". *Intl. J. Syst. Sci.*, **16**, 789-813 (1985).

Desoer, C. A. and J. D. Schulman, "Zeros and Poles of Matrix Transfer Functions and Their Dynamical Interpretation." *IEEE Trans. Circuits and Systems*, **CAS-21**, 3 (1974).

Desoer, C. A. and M. Vidyasagar, *Feedback Syst ns: Input-Output Properties*, Academic Press, New York (1975).

Doukas, N. and W. Luyben, "Control of ᴂestream Columns Separating Ternary Mixtures." *Anal. Instrum.*, **16**, 51-58 (1978).

Doyle, J. C., "Analysis of Control Systems With Structured Uncertainty." *IEE Proc., Part D*, **129**, 242 (1982).

Doyle, J. C., *Lecture Notes – ONR/Honeywell Workshop on Advances on Multivariable Control* (1984).

Doyle, J. C., "Structured Uncertainty in Control System Design." *Proc. IEEE Conf. on Decision and Control*, Ft. Lauderdale, FL (1985).

Doyle, J. C., R. S. Smith and D. F. Enns, "Control of Plants with Saturation Nonlinearities." *Proc. American Control Conf.*, Minneapolis, MN, 1034-1039 (1987).

Doyle, J. C. and G. Stein, "Multivariable Feedback Design: Concepts for a Classical/Modern Synthesis." *IEEE Trans. Autom. Control*, **AC-26**, 4 (1981).

Doyle, J. C. and J. E. Wall, "Performance and Robustness Analysis for Structured Uncertainty." *Proc. IEEE Conf. on Decision and Control*, Orlando, FL (1982).

Economou, C. G. and M. Morari, "Internal Model Control – 6. Multiloop Design." *Ind. Eng. Chem. Proc. Des. & Dev.*, **25**, 411-419 (1986).

Fan, M. K. H. and A. L. Tits, "Characterization and Efficient Computation of the Structured Singular Value." *IEEE Trans. Autom. Control*, **AC-31**, 734-743 (1986).

Feingold, D. G. and R. S. Varga, "Block Diagonally Dominant Matrices and Generalizations of the Gerschgorin Circle Theorem." *Pacific J. Maths.*, **12**, 1241-1250 (1962).

Foss, A. S., "Critique of Chemical Process Control Theory." *AIChE J.*, **19**, 209 (1973).

Francis, B. A., *A Course in H_∞ Control Theory. Lecture Notes in Control and Information Sciences,* Springer-Verlag, Berlin (1987).

Frank, P. M., *Entwurf von Regelkreisen mit vorgeschriebenem Verhalten*, G. Braun, Karlsruhe (1974).

Franklin, G. F., J. D. Powell and A. Emami-Naeini, *Feedback Control of Dynamic Systems*, Addison-Wesley, Reading, MA (1986).

Fuentes, C. and W. L. Luyben, "Control of High – Purity Distillation Columns." *Ind. & Eng. Chem. Process Des. Dev.*, **22**, 361-366 (1983).

Gantmacher, F. R., *The Theory of Matrices,* Chelsea Publishing Co., New York (1959).

Garcia, C. E. and M. Morari, "Internal Model Control – 1. A Unifying Review and Some New Results. " *Ind. Eng. Chem. Process Des. & Dev.*, **21**, 308-323 (1982).

Garcia, C. E. and M. Morari, "Internal Model Control – 2. Design Procedure for Multivariable Systems." *Ind. Eng. Chem. Process Des. & Dev.*, **24**, 472-484 (1985a).

Garcia, C. E. and M. Morari, "Internal Model Control – 3. Multivariable Control Law Computation and Tuning Guidelines." *Ind. Eng. Chem. Process Des. & Dev.*, **24**, 484-494 (1985b).

Giloi, W., *Zur Theorie und Verwirklichung einer Regelung für Laufzeitstrecken nach dem Prinzip der ergänzenden Rückführung*, Ph.D. Thesis, TH Stuttgart (1959).

Golub, G. H. and C. F. Van Loan, *Matrix Computations.* Johns Hopkins University Press, Baltimore, MD (1983).

Grosdidier, P. and M. Morari, "Interaction Measures for Systems Under Decentralized Control." *Automatica*, **22**, 309-319 (1986).

Grosdidier, P., M. Morari and B. R. Holt, "Closed Loop Properties from Steady State Gain Information." *Ind. Eng. Chem. Fundam.*, **24**, 221-235 (1985).

Hanus, R., M. Kinnaert and J. L. Henrotte, "Conditioning Technique, a General Anti-Windup and Bumpless Transfer Method." *Automatica*, **23**, 729-739 (1987).

Holt, B. R. and M. Morari, "Design of Resilient Processing Plants – V. The Effect of Deadtime on Dynamic Resilience." *Chem. Eng. Sci.*, **40**, 1229-1237 (1985a).

Holt, B. R. and M. Morari, "Design of Resilient Processing Plants – VI. The Effect of Right-Half-Plane Zeros on Dynamic Resilience." *Chem. Eng. Sci.*, **40**, 59-74 (1985b).

Horowitz, I. M., *Synthesis of Feedback Systems,* Academic Press, London (1963).

Horowitz, I. "Some Properties of Delayed Controls (Smith Predictors)." *Int. J. Control*, **38**, 977-990 (1983).

Isermann, R., *Digital Control Systems,* Springer-Verlag, Berlin (1981).

Jenkins, G. M. and D. G. Watts, *Spectral Analysis and its Applications,* Holden-Day, San Francisco, CA (1969).

Kantor, J. C. and R. N. Andres, "Characterization of 'Allowable Perturbations' for Robust Stability." *IEEE Trans. Autom. Control*, **AC-28**, 107-109 (1983).

Kapoor, N., T. J. McAvoy and T. E. Marlin, "Effect of Recycle Structure on Distillation Tower Time Constant." *AIChE J.*, **32**, 411-418 (1986).

Kestenbaum, A., R. Shinnar and F. E. Thau, "Design Concepts for Process Control." *Ind. & Eng. Chem. Proc. Des. & Dev.*, **15**, 2 (1976).

Kouvaritakis, B. and M. Latchman, "Necessary and Sufficient Stability Criterion for Systems with Structured Uncertainties: The Major Principal Direction Alignment Principle." *Int. J. Control*, **42**, 575-598 (1985).

Kucera, V., "State Space Approach to Discrete Linear Control." *Kybernetika*, **8**, 233 (1972).

Kuo, B. C., *Digital Control Systems,* Holt, Rinehart and Winston, New York (1980).

Kwakernaak, H. and R. Sivan, *Linear Optimal Control Systems,* Wiley-Interscience, New York (1972).

Laub, A. J., "A Schur Method for Solving Riccati Equations." *IEEE Trans. Automatic Control*, **AC-24**, 913-921 (1979).

Laub, A. J. and B. C. Moore, "Calculation of Transmission Zeros Using QZ Techniques." *Automatica*, **14**, 557 (1978).

Laughlin, D. L., K. G. Jordan and M. Morari, "Internal Model Control and Process Uncertainty: Mapping Uncertainty Regions for SISO Controller Design." *Int. J. Control*, **44**, 1675-1698 (1986).

Lehtomaki, N. A., *Practical Robustness Measures in Multivariable Control System Analysis.*, Ph.D. Thesis, Dept. of Electrical Eng. and Computer Sci., Massachusetts Institute of Technology, Cambridge, MA (1981).

Lewin, D. R., R. E. Heersink, A. Skjellum, D. L. Laughlin, D. E. Rivera and M. Morari, "ROBEX: Robust Control Synthesis via Expert System." *Proc. 10th IFAC World Congress, Munich*, **6**, 369-374 (1987).

Limebeer, D. J. N., "The Application of Generalized-Diagonal Dominance to Linear System Stability Theory." *Int. J. Control*, **36**, 185-212 (1982).

Locatelli, A., F. Romeo, R. Scattolini and N. Schiavoni, "A Parameter Optimization Approach to the Design of Reliable Robust Decentralized Regulators." *Laboratorio di Controlli Automatici*, Politecnico di Milano, Relazione Interna, **82-7** (1982).

Lunze, J., "Notwendige Modellkenntnisse zum Entwurf robuster Mehrgrössenregler mit I-Charakter." *Messen, Steuern, Regeln*, **25**, 608-612 (1982).

Lunze, J., "Untersuchungen zur Autonomie der Teilregler einer Dezentralen Regelung mit I-Charakter." *Messen, Steuern, Regeln*, **26**, 451-455 (1983).

Lunze, J., "Determination of Robust Multivariable I-Controllers by Means of Experiment and Simulation." *Syst. Anal. Model. Simul.*, **2**, 227-249 (1985).

Luyben, W. L., "Distillation Decoupling." *AIChE J.*, **16**, 198-203 (1970).

Maxwell, J. C., "On Governors," *Proc. of the Royal Society of London*, **16**, 270-283 (1867/68); in *Mathematical Trends in Control Theory*, R. Bellman and R. Kalaba, eds., Dover, NY, 3-17 (1964).

Mayr, O., *The Origins of Feedback Control*, MIT Press, Cambridge, MA (1970).

McAvoy, T. J., *Interaction Analysis, ISA Monograph*, ISA, Research Triangle Pk (1983).

Mees, A. I., "Achieving Diagonal Dominance." *Systems and Control Letters*, **1**, 155-158 (1982).

Mijares, G., J. D. Cole, N. W. Naugle, H. A. Preisig and C. D. Holland, "A New Criterion for the Pairing of Control and Manipulated Variables." *AIChE J.*, **32**, 1439-1449 (1986).

Moczek, J. S., R. E. Otto and T. J. Williams, "Approximation Model for the Dynamic Responses of large Distillation Columns." *Proc. 2nd IFAC Congress*, Basel (1963); also published in *Chem. Eng. Progress Symp. Series*, **61**, 136-146 (1965).

Molinari, B. P., "The Stabilizing Solution of the Algebraic Riccati Equation." *SIAM J. Control*, **11**, 262-271 (1973).

Moore, B. C., "Principal Component Analysis in Linear Systems: Controllability, Observatibility, and Model Reduction," *IEEE Trans. Autom. Control*, **AC-26**, 17-32 (1981).

Morari, M., "Design of Resilient Processing Plants-III. A General Framework for the Assessment of Dynamic Resilience." *Chem. Eng. Sci.*, **38**, 1881-1891 (1983a).

Morari, M., "Robust Stability of Systems with Integral Control." *Proc. IEEE Conf. on Decision and Control*, San Antonio, TX, 865-869 (1983b).

Morari, M., "Robust Stability of Systems with Integral Control." *IEEE Trans. Autom. Control*, **AC-30**, 574-577 (1985).

Morari, M., W. Grimm, M. J. Oglesby and I. D. Prosser, "Design of Resilient Processing Plants – VII. Design of Energy Management System for Unstable Reactors – New Insights." *Chem. Eng. Sci.*, **40**, 187-198 (1985).

Morari, M., S. Skogestad and D. E. Rivera, "Implications of Internal Model Control for PID Controllers. " *Proc. of American Control Conf.*, San Diego, CA, 661-666 (1984).

Morari, M., E. Zafiriou and B. R. Holt, "Design of Resilient Processing Plants – X. New Characterization of the Effect of RHP Zeros." *Chem. Eng. Sci.*, **42**, 2425-2428 (1987).

Nett, C. N. and V. Manousiouthakis, "Euclidean Condition and Block Relative Gain: Connections, Conjectures and Clarifications." *IEEE Trans. Autom. Control*, **5**, 405-407 (1987).

Nett, C. N. and H. Spang, "Control Structure Design: A Missing Link in the Evolution of Modern Control Theories." *Proc. American Control Conf.*, Minneapolis, MN (1987).

Nett, C. N. and J. A. Uthgenannt, "An Explicit Formula and an Optimal Weight for the 2-Block Structured Singular Value Interaction Measure." *Automatica*, **24**, 261-265 (1988).

Newton, G. C., L. A. Gould and J. F. Kaiser, *Analytical Design of Feedback Controls,* Wiley, New York (1957).

Niederlinski, A., "A Heuristic Approach to the Design of Linear Multivariable Interacting Control Systems." *Automatica*, **7**, 691-701 (1971).

Nyquist, H., "Regeneration Theory." *Bell Syst. Tech. J.*, **11**, 126-147 (1932).

Osborne, E. E., "On Pre-Conditioning of Matrices." *I. Assoc. Comput. Mach.*, **7**, 338-345 (1960).

Palmor, Z. J. and R. Shinnar, "Design of Advanced Process Controllers." *AIChE J.*, **27**, 793-805 (1981).

Postlethwaite, I. and Y. K. Foo, "Robustness with Simultaneous Pole and Zero Movement across the $j\omega$-Axis." *Automatica*, **21**, 433-443 (1985).

Postlethwaite, I. and A. G. J. MacFarlane, *A Complex Variable Approach to the Analysis of Linear Multivariable Feedback Systems,* Springer Verlag, Berlin (1979).

Prett, D. M. and R. D. Gillette, "Optimization and Constrained Multivariable Control of a Catalytic Cracking Unit." *AIChE National Mtg.*, Houston, TX (1979); also *Proc. Joint Autom. Control Conf.*, San Francisco, CA (1980).

Ramkrishna, D. and N. R. Amundson, *Linear Operator Methods in Chemical Engineering with Appication to Transport and Chemical Reaction Systems,* Prentice-Hall, Englewood Cliffs, NJ (1985).

Richalet, J. A., A. Rault, J. L. Testud and J. Papon, "Model Predictive Heuristic Control: Applications to an Industrial Process." *Automatica*, **14**, 413-428 (1978).

Rijnsdorp, J. E., "Interaction in Two-Variable Control Systems for Distillation Columns – I." *Automatica*, **1**, 15-28 (1965).

Rivera, D. E., S. Skogestad and M. Morari, "Internal Model Control. 4. PID Controller Design." *Ind. Eng. Chem. Proc. Des. & Dev.*, **25**, 252-265 (1986).

Rivera, D. E., *Modeling Requirements for Process Control.* Ph.D. Thesis, California Institute of Technology (1987).

Rosenbrock, H. H., *Computer Aided Control System Design*, Academic Press, London (1974).

Ryskamp, C. J., "Explicit Versus Implicit Decoupling in Distillation Control." *Chemical Process Control 2 Conf.*, Sea Island, GA, Jan. 18-23 (1981), (T. F. Edgar and D. E. Seborg, eds.), United Engineering Trustees (1982), available from AIChE.

Sandell Jr., N. and M. Athans, "On 'Type L' Multivariable Linear Systems." *Automatica*, **9**, 171-176 (1973).

Seneta, E., *Non-Negative Matrices*, John Wiley, New York (1973).

Shinskey, F. G., *Distillation Control, Second Edition*, McGraw Hill, New York (1984).

Skogestad, S., *Studies on Robust Control of Distillation Columns.*, Ph.D. Thesis, California Institute of Technology, Pasadena, CA (1987).

Skogestad, S. and M. Morari, "Control Configurations for Distillation Columns." *AIChE J.*, **33**, 1620-1635 (1987a).

Skogestad, S. and M. Morari, "Design of Resilient Processing Plants-IX. Effect of Model Uncertainty on Dynamic Resilience." *Chem. Eng. Sci.*, **42**, 1765-1780 (1987b).

Skogestad, S. and M. Morari, "Implications of Large RGA-Elements on Control Performance." *Ind. & Eng. Chem. Research*, **26**, 2323-2330 (1987c).

Skogestad, S. and M. Morari, "Effect of Disturbance Directions on Closed Loop Performance." *Ind. Eng. Chem. Research*, **26**, 2029-2035 (1987d).

Skogestad, S. and M. Morari, "LV-Control of a High-Purity Distillation Column." *Chem. Eng. Sci.* , **43**, 33-48 (1988a).

Skogestad, S. and M. Morari, "Robust Performance of Decentralized Control Systems by Independent Designs." *Proc. American Control Conf.*, Minneapolis, MN, 1325-1330 (1987); *Automatica*, submitted (1988b).

Skogestad, S. and M. Morari, "Some New Properties of the Structured Singular Value." *IEEE Trans. Autom. Control*, in press (1988c).

Skogestad, S. and M. Morari, "Understanding the Dynamic Behavior of Distillation Columns." *Ind. Eng. Chem. Research*, in press (1988d).

Skogestad, S., M. Morari and J. C. Doyle, "Robust Control of Ill-Conditioned Plants: High Purity Distillation." *IEEE Trans. Autom. Control*, in press (1988).

Smith, C. A. and A. B. Corripio, *Principles and Practice of Automatic Process Control*, John Wiley & Sons, New York, NY (1985).

Smith, O. J. M., "Closer Control of Loops with Dead Time." *Chem. Eng. Progress*, **53(5)**, 217-219 (1957).

Stanley, G. M., M. Marino-Gallaraga and T. J. McAvoy, "Short Cut operability Analysis. 1. The Relative Disturbance Gain." *Ind. Eng. Chem. Process Des. Dev.*, **24**, 1181-1188 (1985).

Stein, G., "Beyond Singular Values and Loop Shapes." *unpublished manuscript* (1985).

Stephanopoulos, G., *Chemical Process Control. An Introduction to Theory and Practice*, Prentice-Hall, Englewood Cliffs, NJ (1984).

Stewart, G. C., *Introduction to Matrix Computations*, Academic Press, New York (1973).

Strang, G., *Linear Algebra and Its Applications. Second Edition*, Academic Press, New York (1980).

Thompson, P. M., *Conic Sector Analysis of Hybrid Control Systems.*, Ph.D. Thesis, Massachusetts Institute of Technology, Cambridge, MA (1982).

Toijala (Waller), K. and K. Fagervik, "A Digital Simulation Study of Two-Point Feedback Control of Distillation Columns." *Kemian Teollisuus*, **29**, 1-12 (1982).

Vidyasagar, M., *Control System Synthesis - A Factorization Approach*, MIT Press, Cambridge, MA (1985).

Vidyasagar, M., H. Schneider and B. A. Francis, "Algebraic and Topological Aspects of Feedback Stabilization." *IEEE Trans. Automatic Control*, **AC-27**, 880 (1982).

Vyshnegradskii, I. A., "On Controllers of Direct Action." *Izv. SPB Tekhnolog. Inst.* (1877).

Wolfe, C. A. and J. S. Meditch, "Theory of System Type for Linear Multivariable Servomechanisms." *IEEE Trans. Automatic Control*, **AC-22**, 36-46 (1977).

Youla, D. C., J. J. Bongiorno and H. A. Jabr, "Modern Wiener-Hopf Design of Optimal Controllers – Part I. The Single Input-Output Case." *IEEE Trans. Autom. Control*, **AC-21**, 3-13 (1976).

Youla, D. C., H. A. Jabr and J. J. Bongiorno, "Modern Wiener-Hopf Design of Optimal Controllers – Part II. The Multivariable Case." *IEEE Trans. Autom. Control*, **AC21**, 319-338 (1976).

Zafiriou, E., *A Methodology for the Synthesis of Robust Control Systems for Multivariable Sampled-Data Processes*, Ph.D. Thesis, California Institute of Technology, Pasadena, CA (1987).

Zafiriou, E. and M. Morari, "Digital Controllers for SISO Systems. A Review and a New Algorithm." *Int. J. Control*, **42**, 855-876 (1985).

Zafiriou, E. and M. Morari, "Design of Robust Controllers and Sampling Time Selection for SISO Systems." *Int. J. Control*, **44**, 711-735 (1986a).

Zafiriou, E. and M. Morari, "Design of the IMC Filter by Using the Structured Singular Value Approach." *Proc. American Control Conf.*, Seattle, WA, 1-6 (1986b).

Zafiriou, E. and M. Morari, "Digital Controller Design for Multivariable Systems with Structural Closed–Loop Performance Specifications." *Int. J. Control*, **46**, 2087-2111 (1987).

Zames, G., "Feedback and Optimal Sensitivity: Model Reference Transformations, Multiplicative Seminorms and Approximate Inverses." *IEEE Trans. Autom. Control*, **AC-26**, 301-320 (1981).

Zames, G. and B. A. Francis, "Feedback, Minimax Sensitivity and Optimal Robustness." *IEEE Trans. Autom. Control*, **AC-28**, 585 (1983).

Ziegler, J. G. and N. B. Nichols, "Optimum Settings for Automatic Controllers." *Trans. ASME*, **64**, 759-768 (1942).

Zirwas, H. C., *Die ergänzende Rückführung als Mittel zur schnellen Regelung von Regelstrecken mit Laufzeit,* Ph.D. Thesis, TH Stuttgart (1958).

Index